D1768783

Urn Models and Their Application

An Approach to Modern Discrete Probability Theory

QA
273
J623

NORMAN L. JOHNSON

University of North Carolina, Chapel Hill

SAMUEL KOTZ

Temple University, Philadelphia

6/29/78

John Wiley & Sons

New York · London · Sydney · Toronto

WITHDRAWN
OHIO UNIVERSITY
LIBRARY

Copyright © 1977 by John Wiley & Sons, Inc.

All rights reserved. Published simultaneously in Canada.

No part of this book may be reproduced by any means, nor transmitted, nor translated into a machine language without the written permission of the publisher.

Library of Congress Cataloging in Publication Data
Johnson, Norman Lloyd.
Urn models and their application.
(Wiley series in probability and mathematical statistics)
Includes bibliographies and indexes.
1. Probabilities. 2. Distribution (Probability theory) I. Kotz, Samuel, joint author. II. Title.
QA273.J623 519.2 76–58846
ISBN 0–471–44630–0

Printed in the United States of America

10 9 8 7 6 5 4 3 2 1

Probability and Mathematical Statistics (Continued)

SEBER · Linear Regression Analysis
WILKS · Mathematical Statistics
ZACKS · The Theory of Statistical Inference

Applied Probability and Statistics

BAILEY · The Elements of Stochastic Processes with Applications to the Natural Sciences
BAILEY · Mathematics, Statistics and Systems for Health
BARTHOLOMEW · Stochastic Models for Social Processes, *Second Edition*
BECK and ARNOLD · Parameter Estimation in Engineering and Science
BENNETT and FRANKLIN · Statistical Analysis in Chemistry and the Chemical Industry
BHAT · Elements of Applied Stochastic Processes
BLOOMFIELD · Fourier Analysis of Time Series: An Introduction
BOX and DRAPER · Evolutionary Operation: A Statistical Method for Process Improvement
BROWN and HOLLANDER · Statistics: A Biomedical Introduction
BROWNLEE · Statistical Theory and Methodology in Science and Engineering, *Second Edition*
BURY · Statistical Models in Applied Science
CHERNOFF and MOSES · Elementary Decision Theory
CHOW · Analysis and Control of Dynamic Economic Systems
CLELLAND, deCANI, BROWN, BURSK, and MURRAY · Basic Statistics with Business Applications, *Second Edition*
COCHRAN · Sampling Techniques, *Third Edition*
COCHRAN and COX · Experimental Designs, *Second Edition*
COX · Planning of Experiments
COX and MILLER · The Theory of Stochastic Processes, *Second Edition*
DANIEL · Application of Statistics to Industrial Experimentation
DANIEL and WOOD · Fitting Equations to Data
DAVID · Order Statistics
DEMING · Sample Design in Business Research
DODGE and ROMIG · Sampling Inspection Tables. *Second Edition*
DRAPER and SMITH · Applied Regression Analysis
DUNN and CLARK · Applied Statistics: Analysis of Variance and Regression
ELANDT-JOHNSON · Probability Models and Statistical Methods in Genetics
FLEISS · Statistical Methods for Rates and Proportions
GIBBONS, OLKIN and SOBEL · Selecting and Ordering Populations. A New Statistical Methodology
GNANADESIKAN · Methods for Statistical Data Analysis of Multivariate Observations
GOLDBERGER · Econometric Theory
GROSS and CLARK · Survival Distributions
GROSS and HARRIS · Fundamentals of Queueing Theory
GUTTMAN, WILKS and HUNTER · Introductory Engineering Statistics, *Second Edition*

continued on back

Preface

It is a main purpose of this book to demonstrate how a very considerable number of results in probability theory can be derived from simple urn models. We have deliberately restricted ourselves to situations in which the use of urn models is reasonably natural and thereby have omitted some important aspects of elementary probability theory. In particular, we have not included any work on geometric probability.

We have even been, perhaps, unnecessarily restrictive in not using equivalent concepts such as "boxes," "envelopes," "cells," and so on, from time to time in place of urns. This also has been done deliberately, to make clearer both the logical comparison of different problems and the logical unity of the theory presented. When we give applications (especially in Chapter 5) we do of course describe problems in terms of the actual measurements or concepts concerned, as well as applying our (urn-based) theory.

It is not our intention to claim that urn models are the *only* way to develop probability theory—Hogben's (1950, 1955) classic book indicates in its title at least one other possible set of concepts. However, urns do have certain advantages over things like dice, card packs, and chessboards in that they are not commonly associated with certain numbers such as 6, 52, and 64. One can of course think of a 23-sided die, a 47-card pack, or a 5×3 board, but they are certainly unusual relative to our present experience. To quote from an address by Freudenthal (1960):

> No statistician present at this moment will have been in doubt about the meaning of my words when I mentioned the common statistical model. It must be a stochastic device producing random results. Tossing coins or a dice or playing at cards are not flexible enough. The most general chance instrument is the urn filled with balls of different colours or with tickets bearing some ciphers or letters. This model is continuously used in our courses as a didactic tool, and in our statistical analyses as a means of translating realistic problems into mathematical ones. In statistical language "urn model" is a standard expression.

A further quotation from the same address is also relevant:

> The urn model is to be the expression of three postulates: (1) the constancy of a probability distribution, ensured by the solidity of the vessel, (2) the random character of the choice, ensured by the narrowness of the mouth, which is to prevent visibility of the contents and any consciously selective choice, (3) the independence of successive choices, whenever the drawn balls are put back into the urn. Of course in abstract probability and statistics the word "choice" can be avoided and all can be done without any reference to such a model. But as soon as the abstract theory is to be applied, random choice plays an essential role.

Heitele (1975) supports this point of view, remarking that "In principle it is possible to assign urn models to the greater part of chance experiments, at least those with a countable sample space," and Pólya (1954) asserts that "any problem of probability appears comparable to a suitable problem about bags containing balls, and any random mass phenomenon appears as similar in certain essential respects to successive drawings of balls from a system of suitably combined bags."

The mathematical level of the book generally is that of college algebra. Except in Chapter 6, we make relatively little use of differential and integral calculus, though we have not striven to avoid it. Matrices are employed even less frequently, but when there are situations where their use greatly simplifies notation and presentation, we have not hesitated to use them. We believe, however, that even for those who are not conversant with these tools (calculus and matrices) a sufficiently large part of the book will be understandable and will provide a sound basic training in *discrete probability theory*.

We have (especially in Chapter 6, but also in other places in the book) introduced a number of continuous probability distributions. It is hoped these will enhance the value and interest of the book, but they are not our primary field of study. We regard continuous distributions as approximations to discrete ones, rather than the other way around. [See also Bartlett (1937) on this point.]

For classroom use, Chapters 1 through 4 might constitute a one-semester course. Chapters 5 and 6 are less suited to course work, but each might form the topic for a series of seminars or technical study groups.

The financial support of the U.S. Air Force Office of Scientific Research and the assistance of Dr. I. N. Shimi of the Mathematical Sciences Division are sincerely appreciated.

The authors gratefully acknowledge the assistance of Dr. W. J. Ewens (Section 5.2) and Dr. J. Galambos (Sections 1.6 and 6.4) and thank them for their contributions to the book. We are indebted to Dr. R. Srinivasan, for his help in the preparation of the tables of randomized occupancy and Pólya-Eggenberger distributions in Chapter 3 and the Appendix of Chapter 4, respectively.

We also thank J. Seeger and C. S. Soong for their able help in searching the literature and abstracting, and Ms. June Maxwell for her skill and patience in typing and retyping successive versions of the text. Last (but not least) we are much indebted to Ms. Beatrice Shube for her experienced guidance throughout the writing and production of this volume.

References

Bartlett, M. S. (1937) "Natural" mathematics, *Math. Gaz.*, **21**, 44–45.

Freudenthal, H. (1960) Models in applied probability, *Synthèse*, **12**, 204–210.

Heitele, D. (1975) An epistemological view on fundamental stochastic ideas, *Educ. Stud. Math.*, **6**, 187–205.

Hogben, L. (1950, 1955) *Chance and Choice by Cardpack and Chessboard*, Vols. 1 and 2, London: Max Parrish.

Pólya, G. (1954) *Patterns of Plausible Inference*, Princeton, N.J.: Princeton University Press.

NORMAN L. JOHNSON; SAMUEL KOTZ

Chapel Hill, North Carolina; Philadelphia, Pennsylvania
February 1977

Contents

A Reader's Guide to this Book

Chapter 1 contains information on necessary mathematical and probabilistic tools. This includes a substantial amount of combinatorics and finite difference methodology. This chapter should provide sufficient technical bases for Chapters 2 through 5. More advanced readers may skip this chapter and refer to it only as the need arises. We note, however, that a number of illuminating examples are in the chapter, and these may be of interest, even to readers with a strong technical background. The brief historical remarks in Section 1.2 may also interest such readers.

Similar remarks apply to the relatively short Chapter 2, in which are presented the basic distributions of probability theory and their characteristics *via urn models*. This chapter also includes a section on some relations between the concept of exchangeability and urn models written by Dr. J. Galambos. This topic requires more advanced mathematical techniques.

Material more specific to the book appears in Chapter 3 where various occupancy distributions are discussed. These include the following broad categories: (1) classical occupancy distributions [equations (3.6) through (3.15)]; (2) conditional classical occupancy and urns of two classes [(3.25) through (3.31)]; (3) Bose-Einstein occupancy [(3.37)]; (4) randomized occupancy [(3.62) through (3.65)]; (5) multivariate occupancy [(3.71) through (3.91)]; (6) sequential occupancy [(3.93) and (3.94)]; (7) coupon collector and waiting-time problems [(3.101)] and a rather comprehensive section on committee problems subdivided into three subsections: (a) individual membership, (b) group membership, and (c) randomized committee problems.

Chapter 4 deals with urn models involving stochastic replacements of balls (to be distinguished from the "nonreplaceable" occupancy models of Chapter 3). This can also be subdivided into the following basic categories: (1) standard Pólya-Eggenberger ("contagious") models [(4.1) through (4.6)]; (2) generalized Pólya-Eggenberger models [(4.11) through (4.18)]; (3) inverse Pólya-Eggenberger models [(4.28) through (4.31)]; (4) multivariate Pólya-Eggenberger models [(4.32) through (4.39)]; (5) multivariate inverse Pólya-Eggenberger models [(4.40) through (4.42)]; (6) urn transfer models of which the Ehrenfest urn model, its modifications, and its mixtures, constitute the major part [(4.54) through (4.75)]. A table of Pólya-Eggenberger distribution functions for selected values of parameters concludes this chapter.

Chapter 5 deals with applications of the concepts and models discussed in Chapters 3 and 4 to various fields of scientific inquiry, and in the general framework of statistics and probability. In the first group we especially mention: applications to genetics—this section (5.2) is based on a draft prepared by Dr. W. J. Ewens; capture-recapture models (Section 5.3), including some recent applications to epidemiological research (Section 5.3.2); learning processes (Section 5.4); military applications (Section 5.5.1); some biological and computer applications (Sections 5.5.3 and 5.5.4, respectively); and finally a section (5.6) on sampling systems and procedures.

The second group of "internal" statistical and probabilistic applications includes *inter alia* a detailed discussion of the test of empty boxes in its univariate and bivariate versions, tolerance regions, decision theoretic applications, and a didactic section on Markov chains. The reader should consult the index, in which there is more detailed information on applications mentioned or discussed in this book.

Chapter 6 is of a special, advanced nature. It deals with limiting and asymptotic forms of various distributions introduced in Chapters 3 and 4. The subject of limit and asymptotic distributions is one of the most popular and time-honored topics of probability theory. Its most familiar manifestations are in the classical central limit theorems and other laws of large numbers. Most of the results presented in Chapter 6 pertain to occupancy models. There are several ways in which the reader can utilize this chapter.

For a superficial acquaintance with the subject, Section 6.1 provides an introduction and elementary results. For more detailed information on limits of occupancy distributions, Section 6.2.1 (a telescopic survey) may suffice. More persistent readers then may proceed to Sections 6.2.2 and 6.2.3 which describe results (with some proofs) on Poisson-type and normal limits, respectively, for the classical occupancy distributions.

Modified, sequential, and multivariate occupancy distributions are dealt with in some detail in Sections 6.2.4, 6.2.5, and 6.2.6, respectively. Section 6.3 discusses asymptotic and limiting results for the standard models with stochastic replacement—mainly for the Pólya-Eggenberger case. As with the occupancy distributions in Section 6.2.1, Section 6.3.1 summarizes some elementary results, while the later sections discuss more complicated cases in somewhat more detail.

The last section (6.4), written by Dr. J. Galambos, contains some of the most recent results on limit laws for exchangeable variables.

Each chapter contains a comprehensive bibliography which we have tried to make as up-to-date as possible. The detailed subject, author, and urn models indexes at the end of the volume may be helpful to readers who are interested in a particular topic or author, as we have already mentioned in connection with applications of urn models.

N.L.J.
S.K.

1

Introductory Background: Basic Concepts and Methods

1.1. SOME MATHEMATICAL FORMULAS

We give here, for convenience of reference, mathematical formulas which are used from time to time in this book. The reader should master these formulas and feel comfortable in their use before proceeding to later sections. These relationships (expressions) are purely mathematical; formulas relating to probability theory are given in Sections 1.3 and 1.4.

1.1.1. Combinatorial Formulas

We use the symbol $\binom{n}{r}$ to denote the number of different possible combinations of r from n different items. For example, given 10 different colors there are $\binom{10}{2}$ different possible pairs of colors.

The formula for $\binom{n}{r}$ is

$$\binom{n}{r} = \frac{n(n-1) \times \cdots \times 3 \times 2 \times 1}{\{r(r-1) \times \cdots \times 3 \times 2 \times 1\}\{(n-r)(n-r-1) \times \cdots \times 3 \times 2 \times 1\}}$$

$$= \frac{n!}{r!(n-r)!}, \qquad (1.1)$$

where $n!$ (called *n factorial*) is defined as

$$n! = n \times (n-1) \times \cdots \times 3 \times 2 \times 1. \tag{1.2}$$

Example 1.1. An urn contains b black and w white balls. The balls are drawn out one by one until the urn is empty. In how many distinguishable orders can this be done?

Imagine the balls set out in a line in order of their drawing. All together there are $(b + w)$ positions in the order. Distinguishable arrangements correspond to different sets of positions for the b black balls. These b positions can be chosen from the $(b + w)$ available in $\binom{b+w}{b}$ ways. This is therefore the number of distinguishable orders of drawing.

Example 1.2. The number of ways of assigning n (indistinguishable) balls to m (distinguishable) urns in such a way that each urn contains at least one ball is $\binom{n-1}{m-1}$. This can be seen by imagining the n balls set out on a line, one after the other. Assignment to the first urn is made by placing a divider between two balls and assigning all balls to the left of the divider to the first urn. (See Figure 1.1.) The process is repeated, always moving to the right. We need to place $(m - 1)$ dividers, each one occupying a different one of the $(n - 1)$ spaces between balls, and this can be done in $\binom{n-1}{m-1}$ ways.

Urn 1	Urn 2	Urn 3	. . .
• • •	• • •	• •	

FIGURE 1.1

In this case there are three balls in the first urn, three in the second, and two in the third.

Example 1.3. We now consider the number of arrangements such that the first j urns have specified numbers, $n_1, n_2, \ldots, n_j$, assigning the remaining $(n - \sum_{i=1}^{j} n_i)$ balls to the remaining $(m - j)$ urns so that each urn has at least one ball. This is

$$\omega(n_1, n_2, \ldots, n_j) = \binom{n - \sum_{i=1}^{j} n_i - 1}{m - j - 1}. \tag{1.3}$$

The total number of arrangements in which the first urn has at least $(a_1 + 1)$ balls, the second at least $(a_2 + 1), \ldots,$ and the jth at least $(a_j + 1)$ balls is then the sum of $\omega(n_1, \ldots, n_j)$ over all sets $n_1, \ldots, n_j$ for which $n_i > a_i$ $(i = 1, \ldots, j)$. This can be evaluated by supposing that we first reserve a_i balls for the ith urn $(i = 1, \ldots, j)$ and then assign the remaining $(n - \sum_{i=1}^{j} a_i)$ balls among the m urns so that there is at least one ball in every urn. The total number of arrangements is then

$$\binom{n - \sum_{i=1}^{j} a_i - 1}{m - 1}.$$

We have incidentally shown that

$$\sum_{n_1 > a_1} \cdots \sum_{n_j > a_j} \omega(n_1, n_2, \ldots, n_j) = \binom{n - \sum_{i=1}^{j} a_i - 1}{m - 1}. \tag{1.4}$$

This result is utilized in Section 2.1.

Example 1.4. Suppose we have m urns in a row. We have n balls to assign to this row $(n \leq m)$, of which n_j are of type j $(j = 1, \ldots, k; \sum_{j=1}^{k} n_j = n)$. The balls are to be placed in the urns in such a way that

(i) No urn contains more than one ball.
(ii) All balls of the same type are in neighboring urns.

In how many ways can this be done?

When the balls have been assigned there will be

(i) k blocks of $n_1, n_2, \ldots, n_k$ urns, respectively, in each of which all urns contain balls of the same type.
(ii) $(m - n)$ empty urns.

The number of ways of choosing the positions of the $(m - n)$ empty urns in the sequence of $(m - n + k)$ empty urns and blocks of occupied urns is

$$\binom{m - n + k}{m - n} = \binom{m - n + k}{k}.$$

The k blocks can be arranged in $k!$ different orders. Hence the total number of arrangements is

$$k!\binom{m - n + k}{k} = \frac{(m - n + k)!}{(m - n)!}.$$

EXERCISE 1.1. Obtain the answer to Example 1.4 if it is additionally required that there must be at least one empty urn between any two urns containing balls of different types.

We now discuss how to calculate the number of different partitions of f indistinguishable items into groups. Bloxham (1975) uses the notation s_f for the number of different partitions into at most s groups or, equivalently, into s groups some of which may be empty. Given any such partition, one can derive a partition of $(s + f)$ items into s groups, in which each group has at least one item, by adding one item to each group. Further, this will produce *all* such partitions of $(f + s)$ items into s groups. Hence

s_f = number of partitions of f items into at most s groups (each containing at least one item)
= number of partitions of f items into s groups (some of which may be empty)
= number of partitions of $(f + s)$ items into s groups (each containing at least one item).

Given a partition of $(f + s)$ items into s groups (each containing at least one item) we can construct another partition of the items by

1. Forming the first new group by taking one item from each old group.
2. Forming the second new group by taking one item from each old group which is not now empty.
3. Continuing in this way until all items are accounted for.

We now have a varying number of groups, none of which contains more than s items, and so s_f is also equal to the number of partitions of f items into any number of groups, none containing more than s items.

Numerical values of s_f can be obtained by using the recurrence relation

$$s_f = (s - 1)_f + s_{f-s}. \tag{1.5}$$

Bloxham (1975) points out that the calculation can be facilitated by using the relation

$$s_f = (s - 1)_f + (s - 1)_{f-1s} + (s - 1)_{f-2s} + \cdots, \tag{1.6}$$

which can be derived from (1.5). By this means s_f can be reduced to the sum of terms of form 1_a, and we know that $1_a = 1$ if $a \geq 0$. (Note that $1_0 = 1$, $1_a = 0$ if $a < 0$).

Example 1.5.

$$\begin{aligned} 3_4 &= 2_4 + 2_1 \\ &= (1_4 + 1_2 + 1_0) + 1_1 = 4. \end{aligned}$$

EXERCISE 1.2. Evaluate

(i) 4_3.
(ii) 3_5.
(iii) 5_3.

EXERCISE 1.3. Show that, if $s > f$, then $s_f = 1 + f_f$.

The well-known binomial expansion

$$(a+b)^m = a^m + \binom{m}{1} a^{m-1}b + \cdots + \binom{m}{m-1} ab^{m-1} + b^m$$
$$= \sum_{j=0}^{m} \binom{m}{j} a^j b^{m-j} \tag{1.7}$$

is a classic example of the use of the $\binom{n}{r}$ symbol in mathematics.

Example 1.6. There are n balls numbered $1, 2, \ldots, n$ in an urn. They are drawn out, one at a time without replacement, the number on the ball being recorded at each drawing. In how many ways can the drawings occur so that there is only one mode [i.e., denoting by a_j the number drawn jth in order, $a_1 < a_2 < \cdots < a_s$ and $a_s > a_{s+1} > \cdots > a_n$ (with $a_s = n$, of course), for some s]?

In this case, for a given s there are $(s-1)$ a_j's with $j < s$ and $(n-s)$ a_j's with $j > s$. We can select the two sets of a_j's in $\binom{n-1}{s-1}$ ways (since choice of the first set automatically defines the second set). Each set must be arranged in one specific way (increasing or decreasing). Hence the total number of suitable ways is

$$\binom{n-1}{0} + \binom{n-1}{1} + \cdots + \binom{n-1}{n-2} + \binom{n-1}{n-1} = (1+1)^{n-1} = 2^{n-1}.$$

(We assume that the arrangements $123\cdots n$ and $n(n-1)(n-2)\cdots 1$ satisfy the stated conditions. If they do not, the end terms in the above summation have to be omitted, and the number of suitable ways is $2^{n-1} - 2$.)

EXERCISE 1.4. Suppose only m $(<n)$ balls are drawn from the urn in Example 1.6. What is the number of unimodal sequences in this case?

The binomial expansion generalizes to the *multinomial expansion*

$$(b_1 + b_2 + \cdots + b_k)^m = \sum \cdots \sum \binom{m}{j_1\, j_2 \cdots j_k} b_1^{j_1} b_2^{j_2} \cdots b_k^{j_k} \tag{1.8}$$

where

(i) the summation is over all sets of nonnegative integers $(j_1, j_2, \ldots, j_k)$ such that $\sum_{i=1}^{k} j_i = m$.

(ii) $$\binom{m}{j_1\, j_2 \cdots j_k} = \frac{m!}{j_1!j_2!j_3!\cdots j_k!}. \tag{1.9}$$

The quantity in (ii) is called a *multinomial coefficient*.

It also represents the number of ways in which m (distinguishable) items can be divided into k (distinguishable) groups containing $j_1, j_2, \ldots, j_k$ items, respectively.

Example 1.7. A group of 11 persons is to be split up into three subgroups A, B, C, each to discuss a specific type of problem. A and B will have four members each; C will only have three members.

Since the subgroups are *distinguishable* (on account of their subject matter), the required number is

$$\binom{11}{4\ 4\ 3} = \frac{11!}{4!4!3!} = \frac{11 \times 10 \times 9 \times 8 \times 7 \times 6 \times 5}{4 \times 3 \times 2 \times 3 \times 2} = 11{,}550.$$

Note that, if the subgroups had been *indistinguishable* (so that only the number of subgroupings consisting of two subgroups of four members each, and one of three members, was required), the number of different groupings would be half this.

EXERCISE 1.5. In how many ways can a group of 11 persons be split up into four subgroups of 3, 3, 3, and 2 members, respectively, when the subgroups are (i) distinguishable and (ii) indistinguishable?

1.1.2. Finite Difference Formulas

Unless otherwise stated, all functions are supposed real-valued, with real arguments.

The rth *descending factorial* of x is

$$x^{(r)} = x(x-1)(x-2)\cdots(x-r+1). \tag{1.10}$$

Using this definition [and noting that $n! = n^{(r)}(n-r)!$] we have, from (1.10),

$$\binom{n}{r} = \frac{n^{(r)}}{r!} \tag{1.10$'$}$$

The rth *ascending factorial* of x is

$$x^{[r]} = x(x+1)\cdots(x+r-1). \tag{1.11}$$

Some authors use the notation

$$x^{(r,c)} = x(x + c)(x + 2c) \cdots \{x + (r - 1)c\}.$$

Since $x^{(r,c)} = c^r(x/c)^{[r]} = c^r(-x/c)^{(r)}(-1)^r$, we shall not use this particular notation.

The *displacement operator* E increases the argument of a real-valued function $f(\cdot)$ by 1, so that

$$Ef(x) = f(x + 1). \tag{1.12}$$

Repeated application of E is denoted by an appropriate power index. Thus

$$E^2f(x) = Ef(x + 1) = f(x + 2).$$

More generally, we define

$$E^h f(x) = f(x + h) \tag{1.13}$$

for any h.

The *forward difference operator* Δ is defined by

$$\Delta f(x) = f(x + 1) - f(x).$$

Noting that

$$f(x + 1) - f(x) = Ef(x) - f(x) = (E - 1)f(x),$$

we have the symbolic (or operational) identity

$$\Delta \equiv E - 1. \tag{1.14}$$

If n is an integer, then

$$\begin{aligned} \Delta^n f(x) &= (E - 1)^n f(x) \\ &= \sum_{j=0}^{n} \binom{n}{j} (-1)^j f(x + n - j). \end{aligned} \tag{1.15}$$

Note that

$$\begin{aligned} \Delta^2 f(x) &= \Delta \cdot \Delta f(x) = \Delta\{f(x + 1) - f(x)\} \\ &= \{f(x + 2) - f(x + 1)\} - \{f(x + 1) - f(x)\}. \end{aligned}$$

The quantity $\Delta^n f(x)$ is called the *nth forward difference of* $f(x)$. Applying the operator Δ to the *descending factorial* $x^{(r)}$, we obtain

$$\begin{aligned} \Delta x^{(r)} &= (x + 1)^{(r)} - x^{(r)} \\ &= x(x + 1) \cdots (x - r + 2)(x + 1 - x + r - 1) \\ &= r x^{(r-1)}. \end{aligned} \tag{1.16}$$

More generally,

$$\Delta^m x^{(r)} = r^{(m)} x^{(r-m)} \qquad (m = 1, 2, \ldots, r). \tag{1.17}$$

In particular,

$$\Delta^r x^{(r)} = r! \tag{1.18}$$

and

$$\Delta^m x^{(r)} = 0 \qquad (\text{for } m > r). \tag{1.19}$$

Rewriting (1.14) as

$$E \equiv 1 + \Delta, \tag{1.14$'$}$$

we have, for n an integer,

$$\begin{aligned} f(x+n) &= (1+\Delta)^n f(x) \\ &= \sum_{j=0}^{n} \binom{n}{j} \Delta^j f(x) = \sum_{j=0}^{n} n^{(j)} \frac{\Delta^j f(x)}{j!}. \end{aligned} \tag{1.20}$$

Taking $f(x) = x^s$ in (1.20), we have

$$(x+n)^s = \sum_{j=0}^{n} \binom{n}{j} \Delta^j x^s \tag{1.21}$$

and, in particular, replacing x by $\mathbf{0}$ and n by x,

$$x^s = \sum_{j=1}^{s} x^{(j)} \frac{\Delta^j \mathbf{0}^s}{j!}, \tag{1.22}$$

where

$$\Delta^j \mathbf{0}^s = \Delta^j x^s |_{x=0}. \tag{1.23}$$

(Note that $\Delta^0 \mathbf{0}^s = 0$ for any $s > 0$.)

A multivariate version of (1.22) is

$$\left(\sum_{i=1}^{r} x_i\right)^s = \sum_{j=0}^{s} \sum\nolimits^{(j)} \frac{x_1^{(j_1)} x_2^{(j_2)} \cdots x_r^{(j_r)}}{j_1!\, j_2! \cdots j_r!} \Delta^j \mathbf{0}^s, \tag{1.24}$$

where $\sum^{(j)}$ denotes summation over all $j_1, j_2, \ldots, j_r$ satisfying

$$j_1 + j_2 + \cdots + j_r = j.$$

Quantities of form $\Delta^n \mathbf{0}^s = \Delta^n x^s|_{x=0}$ are very useful. They are called *differences of zero*. Table 1.1 contains some values of $\Delta^n \mathbf{0}^s/n!$—the *reduced differences of zero*, or *Stirling numbers of the second kind*.

Note that (1.22) can be written

$$x^s = \sum_{j=0}^{s} \mathscr{S}_s^{(j)} x^{(j)}. \tag{1.22$'$}$$

TABLE 1.1

Stirling Numbers of the Second Kind (Reduced Differences of Zero)

$$\mathscr{S}_n^{(m)} = \Delta^m \mathbf{0}^n / m!$$

$n \backslash m$	2	3	4	5	6	7	8	9
2	1							
3	3	1						
4	7	6	1					
5	15	25	10	1				
6	31	90	65	15	1			
7	63	301	350	140	21	1		
8	127	966	1701	1050	266	28	1	
9	255	3025	7770	6951	2646	462	36	1
10	511	9330	34105	42525	22827	5880	750	45

NOTES: (i) $\mathscr{S}_n^{(m)} = 0$ for $m > n$. (ii) $\mathscr{S}_n^{(1)} = 1 = \mathscr{S}_n^{(n)}$. (iii) $\mathscr{S}_{n+1}^{(m)} = m\mathscr{S}_n^{(m)} + \mathscr{S}_n^{(m-1)}$.

The coefficients in the inverse expansion

$$x^{(s)} = \sum_{j=0}^{s} S_s^{(j)} x^j \tag{1.25}$$

are called *Stirling numbers of the first kind*. Note that $S_s^{(j)}$ is negative if $(s - j)$ is negative, and that

$$x^{[s]} = \sum_{j=0}^{s} |S_s^{(j)}| x^j. \tag{1.25$'$}$$

More extensive tables of Stirling numbers are available for $n, m \leq 25$ in Abramovitz and Stegun (1964), for $m \leq 32$ and $n \leq 100$ in Schäfer (1954) to six significant figures and for $n, m \leq 40$ in Goldberg et al. (1976). Stirling numbers of the second kind can be calculated directly by using formula (1.15). For example,

$$\begin{aligned}\frac{\Delta^5 \mathbf{0}^6}{5!} &= \frac{1}{120}\left[5^6 - \binom{5}{1}4^6 + \binom{5}{2}3^6 - \binom{5}{3}2^6 + \binom{5}{4}1^6\right] \\ &= \tfrac{1}{120}[15{,}625 - (5 \times 4096) + (10 \times 729) - (10 \times 64) + 5] \\ &= 15.\end{aligned}$$

EXERCISE 1.6.

(i) Show that $\Delta^m \mathbf{0}^{s+1} = m\{\Delta^m \mathbf{0}^s + \Delta^{m-1}\mathbf{0}^s\}$.

(ii) Show that

$$\left(\frac{d}{ds}\right)^n h(e^{As}) = A^n\left[\sum_{j=1}^{n} \frac{\Delta^j \mathbf{0}^n}{j!} e^{jAs} D^j\right] h(y)\bigg|_{y=e^{As}},$$

where $h(\cdot)$ is a function which is differentiable at least n times.

(iii) Show that if $k < s$ then

$$\sum_{j=k}^{s} \mathscr{S}_s^{(j)} S_j^{(k)} = 0.$$

[*Hint*: Use (1.25) and (1.22′) and equate coefficients.]

EXERCISE 1.7. Show that

$$\sum_{j=0}^{k} (m+j)^{(g)} = (g+1)^{-1}\{(m+h+1)^{(g+1)} - m^{(g+1)}\}.$$

[*Hint*: Use (1.16).]

Example 1.8. (Paul, 1971). As an interesting exercise in the use of finite difference symbols we derive the identity:

$$1 + \sum_{j=0}^{r-1} \binom{r}{j} S_j(n) = (n+1)^r, \tag{1.26}$$

where $S_j(n) = 1^j + 2^j + \cdots + n^j$.

From (1.22),

$$m^j = \sum_{h=0}^{\infty} m^{(h)} \frac{\Delta^h \mathbf{0}^j}{h!},$$

and so

$$S_j(n) = \sum_{m=1}^{n} m^j = \sum_{h=0}^{\infty} \left\{\sum_{m=1}^{n} m^{(h)}\right\} \frac{\Delta^h \mathbf{0}^j}{h!}$$

$$= \sum_{h=0}^{\infty} \{(n+1)^{(h+1)} - 1^{(h+1)}\} \frac{\Delta^h \mathbf{0}^j}{(h+1)!} \qquad \text{[using Exercise 1.7].}$$

Hence

$$1 + \sum_{j=0}^{r-1} \binom{r}{j} S_j(n) = 1 + \sum_{j=0}^{r-1} \binom{r}{j} \sum_{h=0}^{\infty} \{(n+1)^{(h+1)} - 1^{(h+1)}\} \frac{\Delta^h \mathbf{0}^j}{(h+1)!}$$

$$= 1 + \sum_{h=0}^{\infty} \frac{(n+1)^{(h+1)} - 1^{(h+1)}}{(h+1)!}$$

$$\times \Delta^h \left\{1 + \binom{r}{1}\mathbf{0} + \cdots + \binom{r}{r-1}\mathbf{0}^{r-1}\right\}.$$

Since

$$\Delta^h\left\{1 + \binom{r}{1}\mathbf{0} + \cdots + \binom{r}{r-1}\mathbf{0}^{r-1}\right\} = \Delta^h\{(1 + \mathbf{0})^r - \mathbf{0}^r\} = \Delta^h\Delta\mathbf{0}^r = \Delta^{h+1}\mathbf{0}^r,$$

we have

$$\begin{aligned} 1 + \sum_{j=0}^{r-1}\binom{r}{j}S_j(n) &= 1 + \sum_{h=0}^{\infty}\frac{(n+1)^{(h+1)} - 1^{(h+1)}}{(h+1)!}\Delta^{h+1}\mathbf{0}^r \\ &= 1 + \sum_{h=0}^{\infty}(n+1)^{(h)}\frac{\Delta^h\mathbf{0}^r}{h!} - \frac{1^{(1)}}{1!}\Delta\mathbf{0}^r \\ &= (n+1)^r, \end{aligned}$$

since $1^{(1)} = 1! = \Delta\mathbf{0}^r = 1$ and $1^{(g)} = 0$ for $g > 1$.

EXERCISE 1.8. There are n distinct biological functions which are distributed among $m \geq n$ different but not identifiable classes of cells in such a way that *each class must receive at least one biological function.*

Let R_m^n denote the total number of different ways of distributing n biological functions into m classes (observe that $R_m^m = 1$; $R_m^n = 0$ for $n < m$).

Show that

$$R_2^n = 2^{n-1} - 1.$$

[*Hint*: The number of ways of putting k functions into one class and $(n - k)$ into the other (for $k - 1, \ldots, n - 1$) is $\binom{n}{k}$. Assignments are essentially the same if classes are interchanged. (Rashevsky, 1955.)]

EXERCISE 1.9.

(i) Show that

$$R_{m+1}^n = \frac{1}{m+1}\left[nR_m^{n-1} + \frac{n!}{2!(n-2)!}R_m^{n-2} + \cdots + \frac{n!}{(n-m)!m!}R_m^m\right].$$

[*Hint*: Assign p elements to one class and the remaining $(n - p)$ to the other m classes for $p = 1, \ldots, n - m$ and observe that this procedure counts each possible distribution $(m + 1)$ times.]

(ii) Deduce from (i), by induction, that

$$R_m^n = \frac{(-1)^m}{m!}\Delta^m\mathbf{0}^n.$$

EXERCISE 1.10. Show that

(i) $\Delta^n\mathbf{0}^n = n!$

(ii) $\Delta^m\mathbf{0}^n = 0$, for $m > n$.

(iii) $\Delta^n 2^x = 2^x$, for all n.

For any finite functions $g(x)$, $h(x)$:

(iv) $\Delta^m(g(x)h(x)) = \sum_{j=0}^{m} \binom{m}{j} \Delta^j g(x) \Delta^{m-j} h(x)$ (with Δ^0 interpreted as 1).

(v) $\Delta^m(\mathbf{0}h(\mathbf{0})) = m\Delta^{m-1}h(\mathbf{0})$. [*Note*: Use (ii)]

Example 1.9. Show that, if h and r are integers, with $0 < h < r$, and $z_1, z_2, \ldots, z_r$ are real numbers, then

$$\sum_{j=1}^{r} z_j^h - \sum_{j<j'}^{r-1} \sum^{r} (z_j + z_{j'})^h + \sum_{j<j'<j''}^{r-2} \sum^{r-1} \sum^{r} (z_j + z_{j'} + z_{j''})^h - \cdots + (-1)^{r-1}\left(\sum_{j=1}^{r} z_j\right)^h = 0.$$

(This was suggested by W. L. Smith.)

The left-hand side of the equation is a polynomial of degree h in $z_1, z_2, \ldots, z_r$. If the z_j's are all positive integers, then this polynomial can be written as

$$(-1)^{r-1} \prod_{j=1}^{r} (E^{z_j} - 1)\mathbf{0}^h = (-1)^{r-1}\left\{\prod_{j=1}^{r} \sum_{i=0}^{z_j-1} E^i\right\}(E-1)^r\mathbf{0}^h.$$

This is zero if $r > h$, since $(E-1)^r\mathbf{0}^h = \Delta^r\mathbf{0}^h = 0$ [see Exercise 1.10(ii)].

A polynomial that is zero for all positive integer values of the arguments must be identically zero. This completes the demonstration.

Occasionally the *backward difference* operator ∇, defined by

$$\nabla f(x) = f(x) - f(x-1), \tag{1.27}$$

and the *central difference* operator δ, defined by

$$\delta f(x) = f(x + \tfrac{1}{2}) - f(x - \tfrac{1}{2}), \tag{1.28}$$

are useful.

Analogously to (1.17) we have

$$\nabla^m x^{[r]} = r^{(m)} x^{[r-m]} \tag{1.29}$$

and

$$\delta^m x^{\{r\}} = r^{(m)} x^{\{r-m\}} \tag{1.30}$$

where

$$x^{\{r\}} = \left(x - \frac{r-1}{2}\right) \cdots \left(x + \frac{r-1}{2}\right)$$

is the *rth central factorial* of x. Note that formulas (1.17), (1.29), and (1.30) are analogous to the formula

$$D^m x^r = r^{(m)} x^{r-m}, \tag{1.31}$$

where D stands for the differentiation operator d/dx.

If the function $f(x)$ can be expressed as a Taylor series, then we have the expansion

$$f(x+h) = \sum_{j=0}^{\infty} \left(\frac{h^j}{j!}\right) D^j f(x). \tag{1.32}$$

The *whole* operator acting on $f(x)$ on the right-hand side of (1.32) can be written formally as

$$\sum_{j=0}^{\infty} \left\{\frac{(hD)^j}{j!}\right\} \equiv e^{hD}. \tag{1.33}$$

Comparing (1.20) and (1.33) we have (again formally)

$$e^{hD} \equiv (1+\Delta)^h \equiv E^h \tag{1.34}$$

and, in particular,

$$e^D \equiv 1 + \Delta \equiv E.$$

Although this is only a *formal* relation among operators, it gives exact results when $f(x)$ is a polynomial of finite order, and useful approximations in other cases, especially when $D^j f(x)$ and $\Delta^j f(x)$ decrease rapidly as j increases. Note the analogy between (1.20) and (1.32).

We sometimes need to solve simple *finite difference equations* of the form

$$a_m u_{r+m} + a_{m-1} u_{r+m-1} + \cdots + a_1 u_{r+1} + a_0 u_r = g(r), \tag{1.35}$$

where $u_1, u_2, \ldots$ is an unknown sequence except for r values (usually $u_1, u_2, \ldots, u_r$), but the values of the a's are known. Symbolically we can write

$$\left(\sum_{j=0}^{m} a_j E^j\right) u_r = g(r).$$

These equations may be solved in the following way. We first seek a *general solution* (with m arbitrary constants) of the equation

$$\left(\sum_{j=0}^{m} a_j E^j\right) u_r = 0.$$

This is obtained by first finding the m roots (in θ) of the auxiliary equation

$$\sum_{j=0}^{m} a_j \theta^j = 0. \tag{1.36}$$

If these roots are all different, say $\theta_1, \theta_2, \ldots, \theta_m$, then the general solution is

$$u_r = \sum_{j=1}^{m} \alpha_j \theta_j^r, \tag{1.37}$$

where $\alpha_1, \alpha_2, \ldots, \alpha_m$ are parameters which are, for the moment arbitrary. (If there are repeated roots, then for the second, third, and so on, equal value of θ_j the corresponding α_j is multiplied by $r, r^2, \ldots$, correspondingly.)

We then seek a particular solution, with no arbitrary parameters, of (1.36). Usually we try to find as simple a particular solution as possible. If $g(r)$ is a constant, say g_0, then we can take

$$u_r = \frac{g_0}{\sum_{j=0}^{m} a_j} \qquad \left(\text{provided} \sum_{j=0}^{m} a_j \neq 0\right).$$

If $g(r)$ is a linear function of r, then u_r can be taken as a linear function of r, with appropriate coefficients.

The complete solution is then: general solution plus particular solution, with parameters α_j chosen to give required initial values, for example, of $u_1, u_2, \ldots, u_r$. (The analogy between this method and the solution of linear differential equations is noteworthy.)

Example 1.10.

$$u_{r+1} - au_r = b \qquad (a \neq 0).$$

The auxiliary equation is $\theta - a = 0$, and the root is $\theta = a$. So the general solution is $u_r = \alpha a^r$.

A particular solution is $u_r = b(1 - a)^{-1}$. So the complete solution is

$$u_r = b(1 - a)^{-1} + \alpha a^r.$$

Choosing α to give a specified value of u_0, we obtain

$$u_r = b(1 - a)^{-1} + \{u_0 - b(1 - a)^{-1}\}a^r. \tag{1.38}$$

This formula is used repeatedly in Chapter 5.

Example 1.11.

$$u_{r+2} - u_{r+1} - u_r = 0.$$

The auxiliary equation is $\theta^2 - \theta - 1 = 0$, and the roots are $\frac{1}{2}(1 \pm \sqrt{5})$. The general solution is

$$u_r = 2^{-r}[\alpha(1 + \sqrt{5})^r + \beta(1 - \sqrt{5})^r]. \tag{1.39}$$

This is also the complete solution, since $u_r = 0$ is a particular solution.

Sequences $\{u_r\}$ satisfying $u_{r+2} - r_{r+1} - u_r = 0$ have attracted considerable interest (see Reference F, 1962–). They are called *Fibonacci* sequences.

EXERCISE 1.11. Solve the difference equation

$$u_{r+2} - 5u_{r+1} + 4u_r = r,$$

given that $u_0 = 0, u_1 = 1$.

1.1.3. Miscellaneous Formulas

It is well-known that, if $g_1(t)$ and $g_2(t)$ are polynomials (possibly of infinite order) in t and $g_1(t) = g_2(t)$ for all t in some interval, then the polynomials are identical and so the coefficients of t^j are the same in each of $g_1(t)$ and $g_2(t)$.

Example 1.12. Suppose we need to find the coefficient of t^{10} in $(1 + t + t^2 + \cdots + t^{19})^7$ (see Section 1.4.3 for situations of this kind). For $|t| < 1$, this expression is also equal to

$$\left(\frac{1 - t^{20}}{1 - t}\right)^7 = (1 - t^{20})(1 - t)^{-7}.$$

The coefficient of t^{10} in the expression on the right-hand side is easily obtained. It is just the coefficient of t^{10} in the expansion of $(1 - t)^{-7}$, and this is

$$\binom{-7}{10} = \frac{7 \cdot 8 \cdot 9 \cdot 10 \cdot 11 \cdot 12 \cdot 13 \cdot 14 \cdot 15 \cdot 16}{1 \cdot 2 \cdot 3 \cdot 4 \cdot 5 \cdot 6 \cdot 7 \cdot 8 \cdot 9 \cdot 10}$$
$$= 11 \times 12 \times 13 \times 14 \times 15 \times 16/(2 \cdot 3 \cdot 4 \cdot 5 \cdot 6) = 7908.$$

We now use this method to derive a useful algebraic identity. Suppose we want to determine constants $A_1, A_2, \ldots, A_m$ such that

$$\sum_{j=1}^{m} A_j(1 - a_j t)^{-1} = \prod_{j=1}^{m} (1 - a_j t)^{-1}, \tag{1.40}$$

where $a_1, a_2, \ldots, a_m$ are m constants, no two of which are equal.

Multiplying both sides of (1.40) by $\prod_{j=1}^{m} (1 - a_j t)$, we have

$$\sum_{j=1}^{m} A_j(1 - a_1 t) \cdots (1 - a_{j-1}t)(1 - a_{j+1}t) \cdots (1 - a_m t) = 1.$$

Putting $t = a_g^{-1}$, we obtain

$$A_g = [(1 - a_1 a_g^{-1}) \cdots (1 - a_{g-1}a_g^{-1})(1 - a_{g+1}a_g^{-1}) \cdots (1 - a_m a_g^{-1})]^{-1}$$

or, formally,

$$A_g = \left.\frac{1 - a_g a^{-1}}{\prod_{j=1}^{m}(1 - a_j a^{-1})}\right|_{a=a_g}. \tag{1.41}$$

This method is known as the *method of undetermined coefficients*. It is used in Chapter 4 (Section 4.4).

Example 1.13. Suppose $m = 4$ and $a_j = j$. Then,

$$\begin{aligned}
A_1 &= \{(1 - \tfrac{2}{1})(1 - \tfrac{3}{1})(1 - \tfrac{4}{1})\}^{-1} = -\tfrac{1}{6},\\
A_2 &= \{(1 - \tfrac{1}{2})(1 - \tfrac{3}{2})(1 - \tfrac{4}{2})\}^{-1} = 4,\\
A_3 &= \{(1 - \tfrac{1}{3})(1 - \tfrac{2}{3})(1 - \tfrac{4}{3})\}^{-1} = -\tfrac{27}{2},\\
A_4 &= \{(1 - \tfrac{1}{4})(1 - \tfrac{2}{4})(1 - \tfrac{3}{4})\}^{-1} = \tfrac{32}{3},
\end{aligned}$$

and so

$$\begin{aligned}
&\{(1 - t)(1 - 2t)(1 - 3t)(1 - 4t)\}^{-1}\\
&\quad = -\frac{1}{6(1 - t)} + \frac{4}{1 - 2t} - \frac{27}{2(1 - 3t)} + \frac{32}{3(1 - 4t)}.
\end{aligned}$$

A formula used in Chapter 6 expresses the descending factorial $(n - x)^{(r)}$ as a polynomial $\sum_{j=0}^{r} a_j x^j$ in x.

If

$$f(x) = \sum_{j=0}^{r} a_j x^j,$$

then

$$f^{(s)}(0) = a_s s!$$

where $f^{(s)}(0) = D^s f(x)_{x=0}$—the value of the s-th derivative of $f(x)$ at $x = 0$.

In the present case $f(x) = (n - x)^{(r)}$, and so

$$\begin{aligned}
f^{(s)}(x) = (-1)^s (n - x)^{(r)} \sum_{i_1 \neq i_2 \neq \cdots \neq i_s}^{r-1} \cdots \sum (n - x - i_1)^{-1} (n - x - i_2)^{-1} \cdots\\
\times (n - x - i_s)^{-1}.
\end{aligned}$$

Hence

$$a_s = \frac{(-1)^s n^{(r)}}{s!} \sum_{i_1 \neq i_2 \neq \cdots \neq i_s}^{r-1} \cdots \sum \left\{\prod_{j=1}^{s} (n - i_j)\right\}^{-1}.$$

EXERCISE 1.12. Show that, if a, b, and c are positive integers, with $a + c \leq b$, then

$$\sum_{v=a}^{b-c} \binom{v}{a}\binom{b-v}{c} = \binom{b+1}{a+c+1}.$$

[*Hint*: Use $(1-x)^{-(a+1)}(1-x)^{-(b+1)} = (1-x)^{-(a+b+2)}$.]

EXERCISE 1.13. (Bizley, 1970). Show that

$$\sum_i \binom{b}{i}\binom{c}{i-d}\binom{a+i}{b+c} = \binom{a}{b-d}\binom{a+d}{c+d} \qquad (1.42a)$$

$$\sum_i \binom{b}{i}\binom{c}{d-i}\binom{a+i}{b+c} = \binom{a}{b+c-d}\binom{a-c+d}{d}, \qquad (1.42b)$$

where a, b, c, d are positive integers with $b \geq d$ in (1.42a) and $b + c \geq d$ and $a + d \geq c$ in (1.42b). [*Hint*: Use $(1+x)^m(1+y)^n[1+x(1+y)]^b = (1+x)^m(1+y)^n[(1+x)+xy]^b$. For (1.42a) put $m = a, n = c$, and consider the coefficient of $x^{a-c}y^{b-d}$.]

A simple and useful relationship is

$$\binom{m+p}{m} = \binom{m+p-1}{m} + \binom{m+p-1}{m-1}.$$

Using this formula it is possible to build up values of $\binom{r}{s}$ from values of $\binom{r-1}{s}$ in the way set out schematically below:

$r \backslash s$	0	1	2	3
1	1	1	0	0
2	1	2(=1 + 1)	1	0
3	1	3(=1 + 2)	3(=2 + 1)	1

This is called *Pascal's triangle*.

Finally, we introduce some special mathematical functions which we will need to use.

1. The *normal integral*

$$\Phi(u) = (\sqrt{2\pi})^{-1} \int_{-\infty}^{u} \exp(-\tfrac{1}{2}t^2)\, dt. \qquad (1.43)$$

The notation $\phi(u) = d\Phi/du = (\sqrt{2\pi})^{-1}\exp(-\frac{1}{2}u^2)$, representing the *standardized normal density*, is also used.

2. The *gamma function*

$$\Gamma(\alpha) = \int_0^\infty t^{\alpha-1}e^{-t}\,dt \qquad (\alpha > 0). \tag{1.44}$$

Special cases:

$$\Gamma(n) = (n-1)! \qquad \text{for } n \text{ a positive integer.}$$

$$\Gamma(\tfrac{1}{2}) = \sqrt{\pi}.$$

Recurrence relation:

$$\Gamma(\alpha + 1) = \alpha\Gamma(\alpha) \qquad (\text{for all } \alpha > 0) \tag{1.45}$$

Gauss' multiplication formula:

$$\Gamma(n\alpha) = (2\pi)^{-(n-1)/2}n^{n\alpha-1/2}\prod_{j=0}^{n-1}\Gamma(\alpha + jn^{-1}). \tag{1.46}$$

Duplication formula ($n = 2$):

$$\Gamma(2\alpha) = (2\pi)^{-1}2^{2\alpha-1/2}\Gamma(\alpha)\Gamma(\alpha + \tfrac{1}{2}). \tag{1.47}$$

3. The *incomplete gamma function*

$$\Gamma_y(\alpha) = \int_0^y t^{\alpha-1}e^{-t}\,dt \tag{1.48}$$

and the *incomplete gamma function ratio*

$$\frac{\Gamma_y(\alpha)}{\Gamma(\alpha)}. \tag{1.49}$$

Stirling's formula:

$$\Gamma(\alpha + 1) = \sqrt{2\pi}\,\alpha^{\alpha+1/2}\exp\left(-\alpha + \frac{\theta_\alpha}{12\alpha}\right) \qquad (\text{for some } \theta_\alpha; 0 < \theta_\alpha < 1). \tag{1.50}$$

4. The *beta function*

$$B(\alpha, \beta) = \int_0^1 t^{\alpha-1}(1-t)^{\beta-1}\,dt \qquad (\alpha, \beta > 0). \tag{1.51}$$

In terms of gamma functions,

$$B(\alpha, \beta) = \frac{\Gamma(\alpha)\Gamma(\beta)}{\Gamma(\alpha + \beta)}. \tag{1.52}$$

5. The *incomplete beta function*

$$B_y(\alpha, \beta) = \int_0^y t^{\alpha-1}(1-t)^{\beta-1}\,dt \tag{1.53}$$

and the *incomplete beta function ratio*

$$I_y(\alpha, \beta) = \frac{B_y(\alpha, \beta)}{B(\alpha, \beta)}. \tag{1.54}$$

6. The *Dirichlet integral of type I*, in m dimensions, is

$$\int_{t_i>0;\sum t_i<1}\cdots\int \left(1-\sum_{i=1}^m t_i\right)^{\alpha_0-1}\prod_{i=1}^m t_i^{\alpha_i-1}\,dt_1\,dt_2\cdots dt_m = \left\{\prod_{i=0}^m \Gamma(\alpha_i)\right\}\left\{\Gamma\left(\sum_{i=0}^m \alpha_i\right)\right\}^{-1}. \tag{1.55}$$

The *incomplete Dirichlet integral* is

$$\frac{\Gamma\left(\sum_{i=0}^m \alpha_i\right)}{\prod_{i=0}^m \Gamma(\alpha_i)}\int_0^{p_m}\int_0^{p_{m-1}}\cdots\int_0^{p_1}\left(1-\sum_{i=1}^m t_i\right)^{\alpha_0-1}\prod_{i=1}^m t_i^{\alpha_i-1}\,dt_1\cdots dt_m$$

$$\left(0 < p_i, \sum_{i=1}^m p_i \le 1\right). \tag{1.56}$$

The special case $p_i = p$, with $mp \le 1$, $\alpha_i = \alpha$ for all $i = 1, \ldots, m$, has the notation

$$I_p^{(m)}(\alpha, m\alpha + \alpha_0 - 1). \tag{1.57}$$

(See Sobel et al., 1975.)

7. The *hypergeometric function*

$$\begin{aligned} F(\alpha, \beta; \gamma; x) &= 1 + \frac{\alpha\beta}{\gamma}\frac{x}{1!} + \frac{\alpha(\alpha+1)\beta(\beta+1)}{\gamma(\gamma+1)}\frac{x^2}{2!} + \cdots \\ &= \sum_{j=0}^\infty \frac{\alpha^{[j]}\beta^{[j]}}{\gamma^{[j]}}\frac{x^j}{j!}. \end{aligned} \tag{1.58}$$

8. The *confluent hypergeometric function*

$$M(\alpha; \gamma; x) = \sum_{j=0}^\infty \frac{\alpha^{[j]}}{\gamma^{[j]}}\frac{x^j}{j!}. \tag{1.59}$$

9. The derivatives of the logarithm of $\Gamma(\alpha)$ are also useful, though not needed as often as the gamma function itself. The function

$$\psi(\alpha) = \frac{d}{d\alpha}\{\log \Gamma(\alpha)\} = \frac{\Gamma'(\alpha)}{\Gamma(\alpha)} \tag{1.60}$$

is called the *digamma* function (with argument α) or the *psi function*.

Similarly,

$$\psi'(\alpha) = \frac{d}{d\alpha}\{\psi(\alpha)\} = \frac{d^2}{d\alpha^2}\{\log \Gamma(\alpha)\} \tag{1.61}$$

is called the *trigamma* function and, generally,

$$\psi^{(s)}(\alpha) = \frac{d^s}{d\alpha^s}\{\psi(\alpha)\} = \frac{d^{s+1}}{d\alpha^{s+1}}\{\log \Gamma(\alpha)\} \tag{1.62}$$

is called the $(s+2)$-*gamma* function.

Extensive tables of the digamma, trigamma, tetragamma, pentagamma, and hexagamma functions are contained in Davis (1933, 1935). Shorter, but useful, tables can be found in Abramowitz and Stegun (1964) and in Luke (1975).

From the recurrence formula (1.45) for the gamma function, the recurrence formula

$$\psi(\alpha + 1) = \psi(\alpha) + \alpha^{-1} \tag{1.63}$$

for the digamma function can be derived.

Particular values are

$$\psi(1) = -\gamma = -0.577216, \qquad \psi(\tfrac{1}{2}) = -\gamma - 2\log_e 2 = -1.963510, \tag{1.64}$$

where γ is Euler's constant,

$$\gamma = \lim_{N\to\infty}\left\{\sum_{j=1}^{N} j^{-1} - \log N\right\}. \tag{1.65}$$

A very good approximate formula for $\psi(\alpha)$ is

$$\psi(\alpha) \doteqdot \log(\alpha - \tfrac{1}{2}). \tag{1.66}$$

1.2. BACKGROUND FOR URN MODELS

An urn model is constructed by imagining a number of urns, some or all containing balls of various colors. In specific cases, we consider sequences of experiments (trials) in which balls are drawn from and possibly returned to the urns according to certain rules. These rules may include requirements for the addition of balls to or removal from certain urns at various stages of the experiment. They may also possibly call for certain balls to change color according to prescribed rules.

When a ball is drawn at random from an urn containing s balls, we suppose that it is equally likely that the chosen ball can be any particular one of the s balls in the urn. The probability that a specified ball will be chosen is then s^{-1}. (See Section 1.3.)

From this simple result we are (at least in principle) able to calculate the probability of any specified outcome of any experiment (or series of experiments) of the kind just described. In some cases of course the calculations may be technically complicated, but even then we will only be using, in effect, this simple result.

We are usually interested in

1. Distributions of the number of balls of various kinds in the urns.
2. Waiting-time distributions until a specified condition, or conditions, are satisfied.

We will see that these include quite a remarkable range of important discrete distributions, and also some less common distributions. By considering limiting cases as certain parameters—for example, number of urns, proportion of balls of a certain color, number of trials—are varied, we can extend our field, including a number of continuous distributions. This is the special topic of Chapter 6, but we encounter some cases of this kind much earlier—see, for example, Section 3.2.3.

Arrangements of urns and balls, together with sampling rules, may also be used to set up models of structures of variation. An example of historical importance is due to Lexis (1876, 1903). He considered a set of urns containing different proportions of black and white balls. We plan to draw a number (say n) of balls from the urns. We may

1. Choose an urn at random and then draw a ball n times (replacing the ball each time) from the chosen urn, or
2. Choose an equal (so far as possible) number of balls from each urn.

Considering the extreme case when we have two urns, one containing only white balls and one containing only black balls, we see that method 1 leads to an unusually high variation in the number of white balls chosen (this must be either n or 0), while method 2 leads to unusually stable results (if n is even, then the number of white balls is always $\frac{1}{2}n$).

Case 1 is termed *supernormal dispersion*, and case 2 *subnormal dispersion*, as compared with the results of simple random sampling (with replacement) from a single urn containing proportions p, $(1 - p)$ of white and black balls, respectively. (See Section 2.2.1.) Extensions and variations of these models are due to Coolidge (1921).

It is rather difficult to pinpoint the first written record in which urn models appear. The situation is similar to the case of uniform (rectangular) distribution which has probably been in use far longer than can be inferred from printed records.

Usages of the notions of urn and random drawings from urns in the Old Testament and Jewish theological literature are described in Rabinovitch's (1973) monograph.

According to some sources (e.g., Heubeck, 1974) the first reference to the use of an urn model in probabilistic problems appears in the works of Huygens (1629–1695) (in the fourteenth volume of his collected works devoted to papers in probability theory.) We cannot find an explicit reference in Huygens' *Oeuvres Complètes*, though in Huygens (1665) a problem suggested by J. Hudde is solved, which could quite easily be expressed in terms of urns. Urn models are mentioned on several occasions in J. Bernoulli's (1654–1705) *Ars Conjectandi* (Book III), posthumously published in 1713. D. Bernoulli (1700–1782) often resorts to various urn models, in particular in D. Bernoulli (1768). In fact, he was the first who proposed the model now known as the Ehrenfest model (Sheynin, 1973; see also Section 4.7.1 for more details). Urn models appear in the works of de Moivre (1667–1754) and Montmort (1678–1719) (see, for example, the article by Jordan, 1920). Laplace (1749–1827) studied in detail an urn scheme also leading to the Ehrenfest model (cf. Section 4.7.1; Molina, 1930; Ondar', 1970), as well as other less well-known urn models (cf. Laplace, 1886). Urn models appear in the works of numerous nineteenth century statisticians and probabilists, for example, Quetelet (1794–1874) (cf. Stigler, 1975), Ostrogradskiĭ (1801–1862) (cf. Maistrov, 1974), and Poisson (1781–1840), to mention only three authors with quite different backgrounds. We have already noted a contribution due to Lexis (1837–1914) (see also Bortkiewicz, 1895; Chuprov (Tschuprow), 1905).

The class of urn models, widely known today as *Pólya-Eggenberger* models (see Chapter 4) seems to have been introduced as early as 1906 in the works of Markov (1856–1922). It is difficult to trace the origins of occupancy models (see Chapter 3), that is, models dealing with distributions of numbers of balls in urns, under various systems of random assignment. Sprott (1957) mentions that classical occupancy distribution appears in the work of De Moivre (1718, pp. 109–110) and others. Most probably the first widespread usage was in connection with problems in theoretical physics in the works of Clausius (1822–1888), Boltzmann (1844–1906), and Maxwell (1831–1879), in their theory of gases, which introduced probabilistic concepts into physics (see Section 1.3.4). A combinatorial background for this work was developed in the works of Whitworth (1840–1905) and MacMahon (1854–1929) to whom many of the classical results used in modern combinatorics of partitions, arrangements, and combinations are due. Earlier contributors to the development of combinatorics include Leibniz (1646–1716)—in particular in his *Dissertatio de Arte Combinatoria* (1666)—and even earlier medieval and premedieval writers (see Maistrov, 1974, pp. 34–38).

We are indebted to Dr. Ivo Schneider for information on references to Bernoulli, Boltzmann, and Leibniz.

EXERCISE 1.14. (Hudde and Huygens, 1657–1665). A has an urn containing two white balls and one black ball. B has an urn containing white balls and black balls (only) in proportions $p:(1 - p)$.

A and B, in turn, draw balls at random from their respective urns and replace them. When either player draws a white ball, he or she takes all the stake money; when a black ball is drawn, the player contributes 1 ducat to the stake money.

If A has the first draw, with no stake money, what should be the proportion (p) of white balls in B's urn for the game to be fair, assuming it concludes with (1) the first, (2) the second drawing of a white ball.

[*Hint*: Case 2 can be thought of as two successive games like case 1, but with an uncertain first player in the second game (depending on whether A or B drew the first white ball, so ending the first game).]

EXERCISE 1.15. (D. Bernoulli, 1766–77). An urn contains $2n$ balls, of which n are black and n are white. Each set of n balls of the same color is numbered $1, 2, \ldots, n$. $(2n - r)$ balls are chosen at random and removed from the urn. What is the distribution of X, the number of pairs (one white and one black) of balls inscribed with the same number, remaining in the urn?

Show that $E[X] = \frac{1}{2}r(r - 1)(2n - 1)^{-1}$. (cf. Sec. 1.4.2).

EXERCISE 1.16. Bernoulli used the model in Exercise 1.15 to represent the duration of marriage. Explain how the model does this.

Extend the model to represent possibly polygamous and polyandrous societies.

EXERCISE 1.17. (D. Bernoulli, 1769). We have three urns A, B, and C containing n white, n red, and n black balls, respectively.

We extract, simultaneously and at random, one ball from each of A, B, and C and move them to urns B, C, and A, respectively. Find the expected numbers of white balls in each of the three urns after m successive operations of this kind.

[*Hint*: This may be done by introducing the vector $\mathbf{E}'_m = (E_{mA}, E_{mB}, E_{mC})$ and showing that

(i) $\mathbf{E}_{m+1} = \mathbf{C}\mathbf{E}_m + \mathbf{E}_m$ with

$$\mathbf{C} = n^{-1}\begin{pmatrix} -1 & 0 & 1 \\ 1 & -1 & 0 \\ 0 & 1 & -1 \end{pmatrix}.$$

(ii) If

$$\mathbf{A} = \begin{pmatrix} 0 & 0 & 1 \\ 1 & 0 & 0 \\ 0 & 1 & 0 \end{pmatrix},$$

then $\mathbf{A}^2 = \mathbf{I}$.]

$$\left[\textit{Answer}: \mathbf{E}'_m = n\{(1 - n^{-1})\mathbf{I} + n^{-1}\mathbf{A}\}^m \begin{pmatrix} 1 \\ 1 \\ 1 \end{pmatrix}.\right]$$

ALTERNATIVE. At each operation each ball has a probability of moving (once) equal to $1/n$. The probability a ball in urn A will be in urn A after m operations is

$$(1 - n^{-1})^m + \binom{m}{3}(n^{-1})^3(1 - n^{-1})^{m-3} + \binom{m}{6}(n^{-1})^6(1 - n^{-1})^{m-6} + \cdots$$

(i.e., the probability that the number of moves is a multiple of 3). Hence the expected number of white balls in urn A after m operations is n times this quantity.

Similarly, the expected number of white balls in urn B is

$$n\left[\binom{m}{1}(n^{-1})(1 - n^{-1})^{m-1} + \binom{m}{4}(n^{-1})^4(1 - n^{-1})^{m-4} + \cdots\right].$$

References for Section 1.2.

Bernoulli, D. (1768) De usu algorithmi infinitesimalis in arte conjectandi specimen, *Novi Comment. Acad. Sci. Imp. Petropolitanae*, **12**.

Bernoulli, J. (1713) *Ars Conjectandi*, Impensis Thurnisiorum, Fratrum, Basileae; (see also *Die Werke von Jakob Bernoulli*, Vol. 3, revised by B. L. van der Waerden, Naturforschend Gesellschaft, Basel: Birkhauser Verlag, 1975.)

Boltzmann, L. (1877) Über die Beziehung zwischen dem zweiten Hauptsätze der mechanischen Warmetheorie und der Wahrscheinlichkeitsrechnung respektive den Sätzen über das Warmegleichgewicht, *Wien. Ber.*, **76**, 373–435.

Bortkiewicz, L. von (1894) Kritische Betrachtung zur Theoretischen Statistik; Artikel I, *Jahr. Nationalökonomie Stat., Ser. 3*, **8**, 641–680.

Clausius, R. (1849). Über die Natur derjenigen Bestandtheile der Erdatmosphäre durch welche die Lichtreflexion in derselben bewirkt wird, *Ann. Phys. Chem.*, **76**, 161–188.

Coolidge, J. L. (1921) The dispersion of observations, *Bull. Am. Math. Soc.*, **27**, 439–442.

DeMoivre, A. (1718) *Doctrine of Chances*, London: H. Woodfall. Reprinted by Chelsea, New York, 1965.

Heubeck, K. (1974) Urnenmodelle und ihre Anwendung in der Versicherungsmathematik, *Bl. Deut. Ges. Versicherungsmath.*, **11** (3), 371–429.

Huygens, C. I. (1665) Appendix to *Van Rekeningh in Spelen van Geluk* (1657) (with French translation) in *Oeuvres Complètes de Christian Huygens*, Vol. 14, The Hague: Nijhoff, 1920.

Jordan, C. (1920) On the Montmort-Moivre Problem, *Acta Litt. Sci.*, **1**, 144–147.

Jordan, K. (1973) *Treatise on Probability Theory*, Budapest: Hungarian Academy of Sciences.

Laplace, P. S. *Théorie Analytique des Probabilités*, Paris: Courcier. 1st ed., 1812; 2nd ed., 1814, 3rd ed., 1820; also *Oeuvres Complètes*, Vol. 7, (1886), Livre 2, pp. 413–414.

Leibniz, G. W. (1666) Dissertatio de arte combinatoria, in *Mathematische Schriften*, Vol. 5, Hildesheim and New York: Georg Olms Verlag, 1971.

Lexis, W. (1876) Das Geschlechstverhältnis der Geborenen and Wahrscheinlichkeitsrechnung, *Jahrb. Nat. Oekol. Stat. Jena*, **27**.

Lexis, W. (1903) *Abhandlungen zür Theorie der Bevölkerungs und Moralstatistik*, Jena: G. Fischer.

MacMahon, P. A. (1915) *Combinatory Analysis*, Vol. 1, Cambridge: Cambridge University Press.

Maistrov, L. (1974) *Probability Theory, A Historical Sketch* (translated and edited by S. Kotz), New York: Academic Press.

Markov, A. A. (1951) *Izbrannye Trudy* (*Selected Works*), Vol. I, Moscow: Academy of Sciences of the USSR.

Maxwell, J. C. (1875) Paper presented before the London Chemical Society.

Molina, E. C. (1930) Some comments on Laplace's *Théorie Analytique*, *Bull. Am. Math. Soc.*, **36**, 369–371.

Ondar', H. O. (1970) On a paper by Steklov, *Isotor. Metod. Estestv. Nauk*, **9**, 262–266 (in Russian).

Ostrogradskiĭ, M. V. (1961) *Polnoe Sobraniye Trudov* (*Complete Collection of Works*), Vol. 3, Kiev: Academy of Sciences of the Ukrainian SSR.

Quetelet, L.-A.-J. (1846) *Lettres à S.A.R. le Duc Régnant de Saxe-Coburg et de Gotha*, Brussels: M. Hayez.

Rabinovitch, N. L. (1973) *Probability and Statistical Inference in Ancient and Medieval Jewish Literature*, Toronto: University of Toronto Press.

Sheynin, O. (1973) D. Bernoulli's work in probability, *RETE*, **1**, 273–300.

Sprott, D. A. (1957) Probability distributions associated with distinct hits on targets, *Bull. Math. Biophys.*, **19**, 163–170.

Stigler, S. M. (1975) The transition from point to distribution estimation, Invited paper No. 94, 39th Session of the International Statistics Institute, Warsaw.

Tschuprow, A. (1905) Die Aufgaben der Theorie der Statistik, *Jahrb. Gesetzg. Verwalt. Volkswirt.*, **29**, Part 2.

Whitworth, W. A. (1886) *Choice and Chance*, 4th ed., Cambridge: Deighton Bell.

William Allen Whitworth and a Hundred Years of Probability (1967) (Discussion opened by J. O. Irwin). *J. R. Stat. Soc. Ser. A*, **130**, 147–175.

1.3. PROBABILITY

1.3.1. Definitions and Basic Formulas

We use "probabilities" to estimate long-run relative frequencies of the occurrence of events in repeated trials performed under the same conditions (circumstances). Relationships among probabilities follow the pattern of relationships among relative frequencies. We thus have (with $\Pr[E]$ denoting "the probability of the event E")

$$0 \le \Pr[E] \le 1. \tag{1.67}$$

$\Pr[E] = 0$ if the event E is impossible (never happens); $\Pr[E] = 1$ if the event E is certain (always happens). If E_1 and E_2 are two events, each of which may be observed at each of a sequence of trials, we can define the *compound events*:

1. "At least one of E_1 and E_2"—this is the *union* or *logical sum* of E_1 and E_2, and may be written $E_1 \cup E_2$.
2. "Both E_1 and E_2"—this is the *intersection* or *logical product* of E_1 and E_2 and may be written $E_1 \cap E_2$.

These definitions may be extended in a natural way to more than two events. Thus

1. $E_1 \cup E_2 \cup E_3$, or $\bigcup_{j=1}^{3} E_j$, means "at least one of E_1, E_2, and E_3."
2. $E_1 \cap E_2 \cap E_3$, or $\bigcap_{j=1}^{3} E_j$, means "all of E_1, E_2, and E_3."

Also, as an example of a more complicated compound event we can have

$$(E_1 \cup E_2) \cap (E_3 \cup E_4),$$

"which means at least one of E_1 and E_2 *and* at least one of E_3 and E_4."

From the corresponding relationship among relative frequencies, it is straightforward to deduce that

$$\Pr[E_1 \cup E_2] = \Pr[E_1] + \Pr[E_2] - \Pr[E_1 \cap E_2]. \tag{1.68}$$

An important generalization of this formula is given in (1.88).

If E_1 and E_2 cannot occur simultaneously—for example, if E_j is the event of a six-sided die falling with the j face uppermost ($j = 1, 2$)—then the event $E_1 \cap E_2$ is impossible and (1.68) becomes

$$\Pr[E_1 \cup E_2] = \Pr[E_1] + \Pr[E_2]. \tag{1.69}$$

(The probability of the logical sum equals the sum of the probabilities.)

The natural extension of (1.69) to k *mutually exclusive* (disjoint) events,

$$\Pr[E_1 \cup E_2 \cup \cdots \cup E_k] = \Pr[E_1] + \Pr[E_2] + \cdots + \Pr[E_k]$$

or

$$\Pr\left[\bigcup_{j=1}^{k} E_j\right] = \sum_{j=1}^{k} \Pr[E_j] \tag{1.70}$$

is also valid. It can be established from (1.69) or from (1.88).

The *conditional probability* of the event E_2 given the event E_1 (which should represent the long-run relative frequency of E_2 in those trials in which E_1 occurs) is denoted by $\Pr[E_2 | E_1]$.

We have the relationship

$$\Pr[E_1 \cap E_2] = \Pr[E_1]\Pr[E_2 | E_1]. \tag{1.71}$$

This can be extended to

$$\Pr[E_1 \cap E_2 \cap E_3] = \Pr[E_1]\Pr[E_2 | E_1]\Pr[E_3 | E_1 \cap E_2] \tag{1.72}$$

and to similar formulas for more than three events. Note that, for any k, the events may be taken in $k!$ different orders, each giving a different formula for $\Pr[\bigcap_{j=1}^{k} E_j]$. For example, for $k = 3$, there are six formulas for

$$\Pr[E_1 \cap E_2 \cap E_3],$$

of which (1.72) is one. Also, (1.71) can be expanded to

$$\Pr[E_1 \cap E_2] = \Pr[E_1]\Pr[E_2 | E_1] = \Pr[E_2]\Pr[E_1 | E_2]. \tag{1.73}$$

The event E_2 is *independent* of E_1 if

$$\Pr[E_2 | E_1] = \Pr[E_2], \tag{1.74}$$

and (1.73) becomes in this case

$$\Pr[E_1 \cap E_2] = \Pr[E_1] \times \Pr[E_2]. \tag{1.73$'$}$$

Provided $\Pr[E_2] \neq 0$ we see from (1.73) that (1.74) implies

$$\Pr[E_1 | E_2] = \Pr[E_1],$$

that is, E_1 *is independent of* E_2. Therefore, with the exception of cases in which the probability of one (or both) of the events is zero, the property of independence is mutual.

Example 1.14. Show that, in general, mutually exclusive events are not independent.

If E_1 and E_2 are mutually exclusive, $\Pr[E_2 | E_1] = 0$ and, unless $\Pr[E_2] = 0$, they cannot be independent.

The *complement* of an event E is just "not E" and is conventionally denoted by the symbol $\bar{E}$. If $\bar{E}$ occurs, then E does not occur, and conversely. Since either E or $\bar{E}$ *must* occur, the event $(E \cup \bar{E})$ is certain, and so

$$\Pr[E \cup \bar{E}] = 1.$$

Since E and $\bar{E}$ are clearly mutually exclusive, we can use formula (1.69) and obtain

$$\Pr[E] + \Pr[\bar{E}] = 1 \tag{1.75}$$

or

$$\Pr[\bar{E}] = 1 - \Pr[E]. \tag{1.75$'$}$$

Formula (1.75) can be generalized in the following way. If the events $E_1, E_2, \ldots, E_k$ are mutually exclusive and also *exhaustive* (i.e., one of them *must* happen at each trial), then

$$\Pr[E_1] + \Pr[E_2] + \cdots + \Pr[E_k] = 1;$$

that is,

$$\sum_{j=1}^{k} \Pr[E_j] = 1. \tag{1.76}$$

For example, if we represent by E_j the event that the uppermost face of a six-sided die shows a j ($j = 1, 2, \ldots, 6$), these six events are mutually exclusive. They are also exhaustive (provided "uppermost face" is identifiable). Hence

$$\Pr[E_1] + \Pr[E_2] + \cdots + \Pr[E_6] = 1.$$

If we further suppose that the probabilities of each of the events $E_1, \ldots, E_k$ are all equal (equally likely), then (1.76) leads to the conclusion that their common value is k^{-1}.

In the special case $k = 6$ the common value is $\frac{1}{6}$.

EXERCISE 1.18. (Bernoulli, 1713, Part III). An urn contains one white and one black ball. A prize is offered to three players, A, B, and C, who will draw balls in turn from the urn (and replace them). The first player to draw the white ball will win the prize; if no player draws the white ball, the prize will be withdrawn. What are each player's chances of winning the prize? [*Answer*: A, $\frac{1}{2}$; B, $\frac{1}{4}$; C, $\frac{1}{8}$; probability of withdrawal of prize $\frac{1}{8}$.]

EXERCISE 1.19. We have two urns, one white (W) and one black (B). Urn W contains proportions $(1 - b)$ of white balls and b of black balls. Urn B contains proportions a of white and $(1 - a)$ of black balls.

Initially, a ball is chosen from a third urn which contains proportions c of white and $(1 - c)$ of black balls. Then the first stage of the experiment consists of drawing a ball from the third urn and then drawing a ball from urn B or W according to whether the first ball drawn is black or white. The balls are then returned to the urns from which they were drawn.

At each subsequent stage a ball is drawn from urn W or B according to whether the last ball drawn was white or black and then returned to the urn from which it was drawn.

What is the probability P_r that the ball selected at the rth stage ($r > 1$) will be white?

$$\left[\textit{Answer}\text{: } P_r = \frac{a}{a+b} + (1-b)^r\left(c - \frac{a}{a+b}\right).\right]$$

[*Hint*: Express P_r in terms of P_{r-1} and solve the resultant difference equation, with $P_0 = c$.]

This model is used in Chapter 5.

1.3.2. Inclusion-Exclusion Principle

Suppose we have an urn containing t balls, of which t_1 have color C_1 (only) on them, t_{12} have colors C_1 and C_2 (and no others) on them, and generally $t_{\alpha_1 \cdots \alpha_k} = t_{\boldsymbol{\alpha}}$ have colors $C_{\alpha_1}, C_{\alpha_2}, \ldots, C_{\alpha_k}$ (and no others) appearing on them. Clearly, $t_{\boldsymbol{\alpha}}$ is unchanged by any permutation of the α's. We denote by ϕ the total number of different colors. Then the total number of balls t satisfies

$$t = t_0 + \sum_{j=1}^{\phi} t_j + \sum_{i<j}^{\phi-1} \sum^{\phi} t_{ij} + \cdots + t_{12\cdots\phi}, \tag{1.77}$$

where t_0 is the number of balls that are "colorless," that is, contain none of the specified colors. We choose a ball at random from the urn. And we denote by E_α the event that the color C_α is on the chosen ball (possibly in combination with other colors). This event will occur if the chosen ball is one of those having C_α in its set of colors. The number of such balls is

$$r_\alpha = t_\alpha + \sum_{j \neq \alpha} t_{\alpha j} + \sum_{\substack{i<j \\ i,j \neq \alpha}} \sum t_{\alpha ij} + \cdots + t_{12\ldots\phi} \tag{1.78}$$

and

$$P_\alpha = \Pr[E_\alpha] = \frac{r_\alpha}{t}. \tag{1.79}$$

A natural extension of (1.78) is obtained by replacing α by $\boldsymbol{\alpha} = (\alpha_1, \alpha_2, \ldots, \alpha_k)$. Then the number of balls for which the compound event $\bigcap_{j=1}^{k} E_{\alpha_j}$ occurs is

$$r_{\boldsymbol{\alpha}} = t_{\boldsymbol{\alpha}} + \sum_{j \notin \boldsymbol{\alpha}} t_{\boldsymbol{\alpha} j} + \sum_{\substack{i<j \\ i,j \notin \boldsymbol{\alpha}}} \sum t_{\boldsymbol{\alpha} ij} + \cdots + t_{12\ldots\phi} \tag{1.80}$$

and

$$P_{\alpha} = \Pr\left[\bigcap_{j=1}^{k} E_{\alpha j}\right] = \frac{r_{\alpha}}{t}. \tag{1.81}$$

There are $1 + (\phi - 1) + \binom{\phi - 1}{2} + \cdots + 1 = 2^{\phi - 1}$ terms in (1.78); there are $2^{\phi - k}$ terms in (1.80).

The probability that *exactly* k different colors will be on the chosen ball is

$$P_{[k]} = \sum_{\alpha} \cdots \sum \frac{t_{\alpha}}{t} = \frac{\tau_k}{t}, \tag{1.82}$$

where the summation is over all possible $\binom{\phi}{k}$ sets $(\alpha_1, \ldots, \alpha_k)$, and $\tau_k \equiv \sum_{\alpha} \cdots \sum t_{\alpha}$. Similarly, we define $\rho_k = \sum_{\alpha} \cdots \sum r_{\alpha}$.

In order to express $P_{[k]}$ in terms of $\Pr\{\bigcap_{j=1}^{k} E_{\alpha j}\}$, and so on, we must express the τ's in terms of the ρ's. From (1.80), summing over all sets $(\alpha_1, \ldots, \alpha_k)$, we obtain

$$\rho_k = \tau_k + \binom{k+1}{k}\tau_{k+1} + \binom{k+2}{k}\tau_{k+2} + \cdots + \binom{\phi}{k}\tau_{\phi} \qquad (k = 1, 2, \ldots, \phi). \tag{1.83}$$

In order to solve (1.83) for the τ's we note that (1.83) means that ρ_k is the coefficient of x^k in the expansion of

$$\sum_{h=1}^{\phi} (1 + x)^h \tau_h,$$

and so

$$\sum_{h=0}^{\phi} x^h \rho_h = \sum_{h=0}^{\phi} (1 + x)^h \tau_h$$

(with $\rho_0 = \tau_0$). Putting $x = y - 1$, we see that

$$\sum_{h=1}^{\phi} y^h \tau_h = \sum_{h=1}^{\phi} (y - 1)^h \rho_h,$$

and so

$$\begin{aligned} \tau_k &= \text{the coefficient of } y^k \text{ in the expansion of } \sum_{h=1}^{\phi} (y-1)^h \rho_h \\ &= \rho_k - \binom{k+1}{k}\rho_{k+1} + \binom{k+2}{k}\rho_{k+2} - \cdots + (-1)^{\phi-k}\binom{\phi}{k}\rho_{\phi}, \end{aligned} \tag{1.84}$$

whence

$$P_{[k]} = S_k - \binom{k+1}{k} S_{k+1} + \binom{k+2}{k} S_{k+2} - \cdots + (-1)^{\phi-k} \binom{\phi}{k} S_\phi, \quad (1.85)$$

where

$$S_j = \sum_{\boldsymbol{\alpha}(j)} \cdots \sum p_{\boldsymbol{\alpha}(j)}, \quad (1.86)$$

summation being over all possible $\binom{\phi}{j}$ sets $(\alpha_1, \ldots, \alpha_j)$.

EXERCISE 1.20. The probability that *at least k* different colors are on the chosen ball is

$$P_k = \sum_{h=k}^{\phi} P_{[h]}.$$

From (1.85), show that

$$P_k = S_k - \binom{k}{k-1} S_{k+1} + \binom{k+1}{k-1} S_{k+2} - \cdots + (-1)^{\phi-k} \binom{\phi-1}{k-1} S_\phi. \quad (1.87)$$

EXERCISE 1.21. Show that the probability that at least one of *k specified* colors is on the chosen ball is

$$S'_1 - S'_2 + S'_3 - \cdots + (-1)^{k-1} S'_k, \quad (1.88)$$

where

$$S'_j = \sum_{\boldsymbol{\alpha}(j)}{}' \cdots \sum{}' p_{\boldsymbol{\alpha}(j)} \qquad (j \leq k),$$

the summation being restricted to all sets $(\alpha_1, \ldots, \alpha_j)$, with each α_i corresponding to one of the k specified colors. (This is known as *Waring's formula.*)

By identifying events appropriately, show how (1.88) generalizes (1.68).

Formulas (1.85), (1.87), and (1.88) are generally called *inclusion-exclusion* formulas. This is because the value of $P_{[k]}$ must lie between the successive sums (as more terms are included) on the right-hand side of (1.85). In turn, this is because the sums of the right-hand side, multiplied by the total number of balls t represent a set of balls alternatively including, and included among, the set for which exactly k colors are on the chosen ball.

An exhaustive discussion of the inclusion-exclusion principle and its uses can be found in Takacs (1967).

EXERCISE 1.22. Assuming that each person is equally likely to have any one of 365 days in the year as his or her birthday, what is the probability that among n persons chosen at random at least r have the same birthday? [*Hint*: Use (1.87).]

(Exercise 1.22 is a form of the classical *birthday problem.* It can be expressed in terms of urns and balls by identifying days with urns and persons with balls. In terms of the text of this section, however, we identify individuals with colors and days with balls. There is, in this model, just one urn.)

EXERCISE 1.23. Each of m urns is assigned a ball, which may be any one of $k(\geq m)$ different colors. In how many ways can this be done, so that balls of exactly μ different colors are among the m balls chosen? Balls are indistinguishable, but urns are distinguishable. [*Answer*: $(\Delta^{\mu}\mathbf{0}^{m}/\mu!)k^{(\mu)}$.] [*Note.* It is possible to obtain this result by noting that the total number of ways of assigning balls to the urns is k^m, and this may be split up according to the exact number ($\mu = 1, 2, \ldots, m$) of balls of different colors. Then use equation (1.22). See Matschinski (1962).]

Example 1.14a. n balls are assigned at random to the squares on a checkerboard of r rows and m columns. No square can be assigned more than one ball and $n < mr$. What is the probability that exactly k columns are "filled" (i.e. have all r squares occupied by a ball)?

If g columns are filled there are $(m-g)r$ remaining positions for the remaining $(n-gr)$ balls. The probability that g *specified* columns (at least) are filled is therefore

$$\binom{mr}{n}^{-1}\binom{(m-g)r}{n-gr}.$$

There are $\binom{m}{g}$ ways of selecting the g specified columns. Applying formula (1.85), the required formula is

$$\binom{mr}{n}^{-1}\sum_{g=k}^{m}(-1)^{g-k}\binom{g}{k}\binom{n}{g}\binom{(m-g)r}{n-gr}$$

From (1.87) the probability that there are *at least* k full columns is

$$\binom{mr}{n}^{-1}\sum_{g=k}^{m}(-1)^{g-k}\binom{g-1}{k-1}\binom{m}{g}\binom{(m-g)r}{n-gr}.$$

$\left(\text{Note: } \binom{A}{B} = 0 \text{ if } B < 0 \text{ or } B > A.\right)$

Grimson (1976) has discussed the application of this result to problems in clinical trials and health surveys (R. C. Grimson, *Combinatorial Problems*

Arising in the Health Sciences, Technical Report, Department of Biostatistics, University of North Carolina, 1976).

Example 1.14b (Engen (1976)). Given that $0 < b_i < n < N$ $(i = 1, 2, \ldots, k)$ where n, N and a_i are integers, show that

$$\sum_{r=1}^{n}(-1)^{r+1}\frac{\binom{n}{r}}{\binom{N}{r}}\sum_{i=1}^{k}\binom{b_i}{r} = k - \sum_{i=1}^{k}\frac{(N-b_i)^{(n)}}{N^{(n)}}.$$

It suffices to show that

$$\sum_{r=1}^{n}(-1)^{r+1}\frac{\binom{n}{r}}{\binom{N}{r}}\binom{b_i}{r} = 1 - \frac{(N-b_i)^{(n)}}{N^{(n)}}$$

Suppose we have an urn containing b_i black and $(N - b_i)$ white balls. Balls are drawn randomly one at a time, without replacement, n balls in all being drawn. Let E_j denote the event that the j-th ball drawn is black. The probability that at least one of the n chosen balls is black is

$$\Pr\left[\bigcup_{j=1}^{k} E_j\right] = 1 - \frac{\binom{N-b_i}{n}}{\binom{N}{n}}$$

$$= 1 - \frac{(N-b_i)^{(n)}}{N^{(n)}} \qquad \text{(zero if } N - b_i < n\text{)}.$$

The probability that a black ball is drawn at any r specified drawings (j_1-th, $\ldots$, j_r-th) is

$$\Pr\left[\bigcap_{i=1}^{r} E_{j_i}\right] = \frac{\binom{b_i}{r}}{\binom{N}{r}}.$$

The required result follows from (1.87).

1.3.3. Bayes' Theorem

Suppose we have m urns $U_1, \ldots, U_m$, in which the proportions of black balls are $\pi_1, \pi_2, \ldots, \pi_m$. By drawing from *another* $[(m + 1)\text{th}]$ urn we can choose one of the m urns with probabilities $P_1, P_2, \ldots, P_m$. From the urn so chosen, a ball is drawn and found to be black. What is the probability that the chosen urn is U_j?

Denote the choice of a black ball by B. Then $\Pr[B|U_j] = \pi_j$, and so $\Pr[B \cap U_j] = P[U_j] \times P[B|U_j] = P_j\pi_j$. We want to find $\Pr[U_j|B] = \Pr[B \cap U_j]/\Pr[B]$. Since

$$\begin{aligned}\Pr[B] &= \Pr[U_1]\Pr[B|U_1] + \Pr[U_2]\Pr[B|U_2] + \cdots + \Pr[U_m]\Pr[B|U_m] \\ &= P_1\pi_1 + P_2\pi_2 + \cdots + P_m\pi_m \\ &= \sum_{i=1}^{m} P_i\pi_i, \end{aligned} \tag{1.89a}$$

we have

$$\Pr[U_j|B] = \frac{P_j\pi_j}{\sum_{i=1}^{m} P_i\pi_i}.$$

This result is known as *Bayes' theorem*. It is usually expressed in the form

$$\Pr[H_j|E] = \frac{\Pr[H_j]\Pr[E|H_j]}{\sum_{i=1}^{m} \Pr[H_i]\Pr[E|H_i]}. \tag{1.89b}$$

where E is thought of as an (observed) event, and $H_1, H_2, \ldots, H_m$ as hypotheses concerning the possible circumstances under which the event E may be observed to occur. $\Pr[H_j]$ is called the *prior probability* of $H_j (j = 1, \ldots, m)$. $\Pr[H_j|E]$ is called the *posterior probability* of H_j (after observing E) $(j = 1, \ldots, m)$. Formula (1.89a) is known as the *total probability formula.*

Note that in order to use (1.89) it is necessary that

1. $\sum_{i=1}^{m} \Pr[H_i] = 1$ (the hypotheses must be *exhaustive*), and
2. Each of $\Pr[H_1], \ldots, \Pr[H_m]$ must be known.

Suppose that an urn contains N balls of which S are black. We take $n\ (<N)$ balls at random from the urn without replacement and observe that k of them are black. What is the probability that the next ball drawn from the urn will also be black? To answer this question we need to introduce a prior distribution for a *random variable* S with $p_s = \Pr[S = s]$ $(s = 0, 1, \ldots, N)$ and $\sum_{s=0}^{N} p_s = 1$ (cf. Section 1.4.1). The posterior distribution of S, given

that k black balls have been observed among n taken from the urn (without replacement), is, using (1.89b):

$$\Pr[S = s|k, n] = \frac{p_s\binom{s}{k}\binom{N-s}{n-k}}{\sum_{j=0}^{N} p_j\binom{j}{k}\binom{N-j}{n-k}}. \tag{1.90}$$

[We identify the hypothesis H_s with the event $(S = s)$.]

In particular, if, $p_0 = p_1 = \cdots = p_N = (N + 1)^{-1}$, then

$$\Pr[S = s|k, n] = \frac{\binom{s}{k}\binom{N-s}{n-k}}{\sum_{j=0}^{N} \binom{j}{k}\binom{N-j}{n-k}}.$$

The denominator is equal to the coefficient of $x^k y^{n-k}$ in the expansion of

$$\sum_{j=0}^{N} (1 + x)^j(1 + y)^{N-j} = \frac{(1 + x)^{N+1} - (1 + y)^{N+1}}{(1 + x) - (1 + y)}$$

$$= (x - y)^{-1}\sum_{j=1}^{N+1}\binom{N+1}{j}(x^j - y^j)$$

$$= \sum_{j=1}^{N+1}\binom{N+1}{j}(x^{j-1} + x^{j-2}y + \cdots + xy^{j-2} + y^{j-1}).$$

The term in $x^k y^{n-k}$ appears only in the $(j = n + 1)$ term of this last summation. Hence

$$\sum_j \binom{j}{k}\binom{N-j}{n-k} = \binom{N+1}{n+1} \tag{1.91}$$

and

$$\Pr[S = s|k, n] = \frac{\binom{s}{k}\binom{N-s}{n-k}}{\binom{N+1}{n+1}}.$$

The probability that the next $[(n+1)\text{th}]$ ball to be drawn will be black is

$$\sum_s \left(\frac{s-k}{N-n}\right)\Pr[S=s\,|\,k,n] = \frac{\sum_s (s-k)\binom{s}{k}\binom{N-s}{n-k}}{(N-n)\binom{N+1}{n+1}}$$

$$= \frac{(k+1)\sum_s \binom{s}{k+1}\binom{N-s}{(n+1)-(k+1)}}{(n+2)\binom{N+1}{n+2}}$$

$$= \frac{k+1}{n+2} \qquad \text{[using (1.91)].}$$

If black balls have been observed at every one of the n drawings, then the probability that the next ball will be black is $(n+1)/(n+2)$. This result is well known as *Laplace's law of succession*. It is discussed in Laplace (1812).

EXERCISE 1.24. An urn contains 15 white balls. A second urn contains 15 black balls. Five balls are taken from one urn (chosen at random) and placed in the other.

Two balls are taken simultaneously from an urn chosen at random. They are both the same color. What is the probability that the other urn contains 10 balls?

1.3.4. Distinguishability

When enumerating the different ways in which n balls can be assigned to m urns, it is important to be know whether (1) the balls and (2) the urns are or are not distinguishable. The number of *distinguishably* different ways of assignment depends on these points.

As an example, suppose we have three balls and two urns. If the balls (urns) are distinguishable, we can number them 1 through 3 (1 through 2) and we have the following eight different assignments (*ball* numbers are shown):

Assignment	Urn 1	Urn 2
A	—	1,2,3
B	1,2,3	—
C	1	2,3
D	2,3	1
E	2	1,3
F	1,3	2
G	3	1,2
H	1,2	3

However, if urns 1 and 2 are indistinguishable, then assignments A and B are the same, and so are C and D, E and F, and G and H; so there are only four distinguishable assignments. If in addition the balls are indistinguishable, there are only *two* different distinguishable assignments: A and B, a 3–0 split, and C through H, a 2–1 split. If the urns are distinguishable but not the balls, there are four distinguishable arrangements. These are

	Number of Balls in	
Assignment	Urn 1	Urn 2
A	0	3
B	3	0
C,E,G,	1	2
D,F,H	2	1

Streefkerk (1975) further includes conditions on whether occupied urns should be "neighbors" or not (the urns being supposed set out in a line), while Collings (1966) and Wright (1967) allow for the *order* in which the balls are assigned to the urns, when the balls are distinguishable. [The last two authors discuss the problem in terms of railway sidings (instead of urns) and trucks (instead of balls).] They further consider separately the case when some of the urns may be empty.

For reference we give Table 1.2, based on the work of Collings and Wright.

TABLE 1.2

Numbers of Arrangements

	Balls			
	Distinguishable		Not distinguishable	
Urns	All cases	No empty urns	All cases	No empty urns
Distinguishable	m^n	$\Delta^m \mathbf{0}^n$	$\binom{n+m-1}{m-1}$	$\binom{n-1}{m-1}$
Not distinguishable	$\sum_{j=1}^{m} \frac{\Delta^j \mathbf{0}^n}{j!}$	$\frac{\Delta^m \mathbf{0}^n}{m!}$	$p_m(n)$	$q_m(n)$

Wright (1967) has shown that $p_m(n) = q_m(n - m)$ and

$$q_k(n) = \frac{\{n + \frac{1}{4}m(m+1)\}^{m-1}}{m!(m-1)} - \frac{(m+1)(2m+1)n + \{\frac{1}{4}m(m+1)\}^{m-3}}{144(m-1)!(m-3)!}$$

$$+ 0(n^{m-5}) \qquad (\text{for } m \geq 7).$$

We urge the reader to verify such table entries as needed. We do this for one entry in the following example.

Example 1.15. We verify the entry for both balls and urns distinguishable, using all m urns. When $n_1, n_2, \ldots, n_m$ are the numbers of balls assigned to the first, second, ..., mth urns, respectively, the number of possible assignments is

$$\binom{n}{n_1 n_2 \cdots n_m} = \frac{n!}{\prod_{i=1}^{m} n_i!}.$$

The table entry should therefore be the sum of these quantities over all sets of nonnegative integers $(n_1, n_2, \ldots, n_m)$, such that $n_j > 0$ for all j and $n_1 + n_2 + \cdots + n_m = n$.

If the restriction $n_j > 0$ is removed, we have

$$S = \sum \cdots \sum \frac{n!}{\prod_{i=1}^{m} n_i!} = \underbrace{(1 + 1 + \cdots + 1)^n}_{m \text{ terms}} = m^n.$$

(This, incidentally, verifies the table entry for "all cases" with distinguishable balls and urns.)

If r specified n's are equal to zero (and perhaps some others), the sum S is equal to

$$\underbrace{(1 + 1 + \cdots + 1)^n}_{(m-r) \text{ terms}} = (m - r)^n.$$

There are $\binom{m}{r}$ ways of selecting the n's to be specified equal to zero. Hence by the inclusion-exclusion formula (1.88) the table entry should be

$$m^n - \binom{m}{1}(m-1)^n + \binom{m}{2}(m-2)^n - \cdots = \Delta^m \mathbf{0}^n$$

[cf. (1.15) and (1.23)].

The three well-known systems of physical statistics, *Maxwell–Boltzmann* (M–B), *Bose–Einstein* (B–E) (Bose, 1924; Einstein, 1924, 1925), and *Fermi–*

Dirac (F–D), all ascribe equal probabilities to each distinguishable member of certain classes of events. They differ in regard to which members are regarded as being distinguishable.

The M–B system, which corresponds to classical probability theory, regards every member as distinguishable from every other member. In the B–E system urns are distinguishable but not balls. This is also true in the F–D system, but here no urn can contain more than one ball.

For the B–E system we find from Table 1.2 that there are $\binom{n+m-1}{m-1}$ possible distinguishable arrangements, and so each has probability

$$\binom{n+m-1}{m-1}^{-1}. \tag{1.92}$$

It is possible of course to consider B–E statistics under the further restriction that each urn must contain at least one ball. In this case, using Table 1.2, we find there are $\binom{n-1}{m-1}$ distinguishable arrangements, each with probability

$$\binom{n-1}{m-1}^{-1}.$$

In the F–D system we must have $n \leq m$, and the number of distinguishable arrangements is the number of ways of choosing the n occupied urns, which is $\binom{n}{m}$. The probability of each arrangement is $\binom{n}{m}^{-1}$.

Example 1.16. A B-E version of the birthday problem (Exercise 1.22) would be obtained if it were supposed that all possible *configurations* of birthdays were equally likely. For example, with two persons A and B, the configurations

Person	Birthday		Person	Birthday		Person	Birthday
A	Jan. 1	and	A	Jan. 1	or	A	Jan. 2
B	Jan. 1		B	Jan. 2		B	Jan. 1

are regarded as equally likely. (In Exercise 1.22, the stated conditions indicate that M–B statistics are appropriate. We would then assign, for example, probabilities of the two above configurations in the ratio $(365)^{-2}:2(365)^{-2} = 1:2$.)

EXERCISE 1.25. Suppose that we have five urns and three balls. Obtain the distribution of distinguishable assignments of the balls to the urns on

(i) The M–B system.
(ii) The B–E system.
(iii) The F–D system.

EXERCISE 1.26. Using B–E statistics, show that the probability that a specified j urns contain at least $(a + 1)$ balls each, given that each urn contains at least one ball is

$$\binom{n-1}{m-1}^{-1}\binom{n-ja-1}{m-1}.$$

[*Hint*: Subtract a balls from each of the j specified urns and consider the resultant situation.]

1.3.5. Ballot Problems

Suppose we have a balls in an urn with numbers $b_1, b_2, \ldots, b_a$ (each zero or a positive integer) inscribed on them, and $b = b_1 + b_2 + \cdots + b_a \leq a$. The balls are drawn one at a time from the urn (without replacement). What is the probability that after each drawing the sum of numbers on the drawn balls will be less than the number of balls drawn?

It is remarkable that the probability sought is simply $1 - ba^{-1}$. The probability does not depend on the individual numbers $b_1, b_2, \ldots, b_a$, but only on their sum b.

To establish this result we first note that it is certainly true when $b = a$. The probability given by the formula is then $1 - aa^{-1} = 0$ and, indeed, the condition (that the sum of numbers is less than the number of balls) is certainly violated when all the balls have been drawn from the urn, for then the two quantities are *equal*.

This means that we can assume $b < a$. We use a process of induction. We first note that the result is certainly true for $a = 1$, because in this case we must have $b = 0$, there is just one ball, and it has zero on it; the condition is certainly satisfied, while the formula gives a probability of $1 - 0{\cdot}1^{-1} = 1$ for this sure event.

Now suppose the result established for all $a < a_0$. We can subdivide all sequences of drawings with a_0 balls into a_0 subsets, according to the number of the ball drawn last. These subsets are equally likely (common probability a_0^{-1}). Consider the subset obtained by omitting b_i. This is now a sequence of $(a_0 - 1)$ balls with $b_{(i)} = b - b_i \leq a_0 - 1$. (Note that, since $b < a_0$, we have $b \leq a_0 - 1$, and so $b_{(i)} = b - b_i \leq a_0 - 1$.) Since the result is established for $a = a_0 - 1$, the proportion of the subset satisfying the condition is

$$1 - b_{(i)}(a_0 - 1)^{-1} = 1 - (b - b_i)(a_0 - 1)^{-1}.$$

Hence the probability that the condition is satisfied when $a = a_0$ is

$$\begin{aligned}\sum_{i=1}^{a_0} a_0^{-1}\{1 - (b - b_i)(a_0 - 1)^{-1}\} &= \sum_{i=1}^{a_0} a_0^{-1}(a_0 - 1)^{-1}(a_0 - 1 - b + b_i)\\ &= a_0^{-1}(a_0 - 1)^{-1}\{a_0(a_0 - 1) - a_0 b + b\}\\ &= 1 - ba_0^{-1},\end{aligned}$$

which is the required formula.

By means of this result we can solve the classical ballot problem, which may be stated in the following way (Dvoretzky and Motzkin, 1947; Takacs, 1962a,b):

There are two candidates, A and B, in an election. A receives a votes and B receives b votes, with $a \geq b$. If the votes are counted one at a time, and each possible order of votes (given the totals a and b) is equally likely, what is the probability that A is always ahead in the voting?

In order to see the equivalence of the two problems we identify b_j as the number of votes for B between the jth and $(j + 1)$th votes for A. (b_a is the number of votes for B following the last vote for A.) The probability that A is always ahead is

$$\Pr[\text{first vote is for A}] \cdot (1 - ba^{-1}) = \frac{a}{a + b} \cdot \frac{a - b}{a} = \frac{a - b}{a + b}.$$

A straightforward extension of this result is obtained by considering the probability that, from the Lth vote onward, A is always at least L votes ahead of B ($L \leq a - b$). Narayana (1959) (see also Steck, 1974) shows that this probability is

$$\frac{a!(a + b - L)!(a - b - L + 1)!}{(a + b)!(a - L + 1)!}.$$

EXERCISE 1.27. An urn contains 50 white balls, 20 black balls, and 10 orange balls. The balls are drawn randomly, one at a time without replacement, until the urn is empty. What is the probability that after every drawing

(i) The number of white balls drawn exceeds the total number of black and orange balls drawn, *and*

(ii) The number of black balls drawn exceeds the number of orange balls drawn?

[*Answer*: $\frac{1}{12}$.]

EXERCISE 1.28. An urn is filled with $(2n + 1)$ balls, each ball being equally likely to be black or white (but no other color).

The $(2n + 1)$ balls are then drawn out of the urn, one at a time without replacement. Show that the probability that, among the balls drawn from the urn at any stage, the majority color is always the same (and equality does not occur at any stage) is $2^{-2n}\binom{2n}{n}$. [*Answer*: $1/\sqrt{\pi n}$.]

Find an asymptotic expression for this quantity, as $n \to \infty$.

Express this problem in terms of ballots.

EXERCISE 1.29. An urn contains n_1 white balls and n_2 black balls $(n_1 \geq n_2 + a)$. The balls are drawn from the urn one at a time without replacement. What is the probability that [except for the first $(a - 1)$ draws] the number of white balls drawn is always at least a in excess of the number of black balls drawn? [*Answer*: $[n_1^{(a)} - n_1^{(a-1)}n_2]/(n_1 + n_2)^{(a)}$.]

Example 1.17. The following example, discussed by Oakley and Perry (1965) (see also Gardner, 1969) is an interesting application of difference equations (see Section 1.1.2).

An urn contains b (>0) black and w (>0) white balls. Then

(i) A ball is drawn from the urn and discarded.

(ii) Balls are drawn, one at a time, from the urn and discarded so long as they are of the same color as that drawn in step (i).

(iii) When a ball is drawn that is of color opposite that drawn in step (i), it is returned to the urn, and we return to step (i).

This is continued until the urn is empty. It is required to show that the probability that the last ball drawn is black is equal to $\frac{1}{2}$, whatever the values of b and w.

Let $P(b, w)$ be the required probability. Then, by considering the process up to the first return to step (i), we find

$$P(b, w) = \sum_{h=1}^{b} \frac{b^{(h)}}{(b + w)^{(h)}}\left(1 - \frac{b - h}{b + w - h}\right)P(b - h, w)$$

$$+ \sum_{j=1}^{w} \frac{w^{(j)}}{(b + w)^{(j)}}\left(1 - \frac{w - j}{b + w - j}\left(1 - \frac{w - j}{b + w - j}\right) P(b, w - j)\right)$$

$$= \sum_{h=1}^{b-1} \left\{\frac{b^{(h)}}{(b + w)^{(h)}} - \frac{b^{(h+1)}}{(b + w)^{(h+1)}}\right\}P(b - h, w)$$

$$+ \sum_{j=1}^{w-1} \left\{\frac{w^{(j)}}{(b + w)^{(j)}} - \frac{w^{(j+1)}}{(b + w)^{(j+1)}}\right\}P(b, w - j) + \frac{w!b!}{(b + w)!}.$$

[Note that $P(0, w) = 0$; $P(b, 0) = 1$.]

If $P(b, w) = P$ for all $b > 0$ and $w > 0$, then, from this formula,

$$P = \left\{\frac{b}{b+w} - \frac{b!w!}{(b+w)!} + \frac{w}{b+w} - \frac{w!b!}{(b+w)!}\right\} P + \frac{w!b!}{(b+w)!},$$

whence $P = \frac{1}{2}$, which indeed does not depend on b or w.

[It is easy to establish, by direct argument, that

$$\begin{aligned} P(1, w) &= \frac{w}{w+1}\,\frac{w-1}{w}\left(1 + \frac{1}{w}\right)\frac{w-2}{w-1}\left(1 + \frac{1}{w-1}\right)\cdots \\ &= \frac{w}{w+1}\,\frac{(w-1)(w+1)}{w^2}\,\frac{(w-2)w}{(w-1)^2}\cdots\frac{1.3}{2^2} = \frac{1}{2} \end{aligned}$$

and, similarly, $P(b, 1) = \frac{1}{2}$.]

[Note that, *if* $P(b, w)$ does not depend on b (>0) or w (>0), its value *must* be $\frac{1}{2}$, since $P(b, b) = \frac{1}{2}$ by symmetry.]

EXERCISE 1.30. Try to extend the above result to the case when there are balls of k different colors in the urn and show that the probability that the last ball remaining has a specific color is no longer independent of the initial constitution of the urn. [There is an examination of this problem in an unpublished report by Downton (1967).]

1.4. RANDOM VARIABLES

1.4.1. Definitions—Independence

In Section 1.2 we introduced the concept of probability related to the relative frequency of occurrence of an event E in a sequence of repeated trials. We now consider a situation in which a certain character is measured or observed at each trial. If we represent the observed value by, say X, then we can define the event E to be "X does not exceed x," where x is a fixed number. Provided the probability of this event ($\Pr[X \le x]$) exists for every x, X is called a *random variable*, and the function of x,

$$F_X(x) = \Pr[X \le x], \tag{1.93}$$

is called its *cumulative distribution function* (cdf). It is clear from (1.93) that

$$F_X(x) \le 1 \tag{1.94}$$

and, if $x' < x''$,

$$F_X(x') \leq F_X(x'') \tag{1.95}$$

[i.e., $F_X(\cdot)$ is a nondecreasing function of the argument].

The function $\Pr[X > x] = 1 - F_X(x)$ is called the *survival distribution function* of X.

As examples of quantities that may be represented by random variables we mention

1. The number j on the uppermost face of a six-sided die after it has been thrown.
2. The weight of a cigarette.
3. The effective lifetime of an electric light bulb in a system where a bulb is automatically replaced after a fixed period of time, say τ.

The repeated trials in these cases are (1) successive throws of the die, (2) individual cigarettes chosen from current production and (3) individual light bulbs used in the system, respectively.

(It should not be assumed automatically that any event in any kind of sequence of repeated trials can be represented by a random variable, though one can use this representation in *trying* to construct an adequate model.)

We note that, though all three quantities may be represented by random variables, there are certain differences among them.

In case (1) the only values that can be taken are the six values 1, 2, 3, 4, 5, 6. In case (2) it is reasonable to suppose that the weight may take any value over some interval. In principal this implies an infinity of possible values, though actual measurements are restricted by the accuracy of the measuring instruments. In fact, therefore, we still have situations similar to case 1, although the number of possible values is usually much larger and the probability of the random variable taking any particular value is very small.

In case 3 there is a range of values $(0 < X < \tau)$ over which the situation described in case 2 applies, but it is likely there is a sizeable probability that the value τ will be observed (when a bulb is replaced before failure) and quite possibly a sizable probability that the value zero will be observed (when a bulb fails the first time it is used).

It is found convenient to distinguish the random variables used in these three cases as (1) discrete, (2) continuous, and (3) mixed.

A *discrete* random variable is supposed to take only isolated values with nonzero probabilities. In many cases there is only a finite number of such values, but there may be an infinity of them (as, for example, in the Poisson distribution—see Section 2.2.3). It is essential, however, that the possible be enumerable, that is, it must be possible to arrange them all in a sequence so that the rth number is identifiable.

For such random variables, the cumulative distribution function ($F_X(x)$) is a step function.

Example 1.18. For case 1, for a six-sided die with equal probabilities $\frac{1}{6}$ for the j side to be uppermost, with X representing the number on the uppermost side after a throw, we have

$$F_X(x) = \begin{cases} 0 & (x < 1) \\ \frac{1}{6} & (1 \leq x < 2), \\ \frac{1}{3} & (2 \leq x < 3), \\ \frac{1}{2} & (3 \leq x < 4), \\ \frac{2}{3} & (4 \leq x < 5), \\ \frac{5}{6} & (5 \leq x < 6), \\ 1 & (6 \leq x) \end{cases} \tag{1.96}$$

Figure 1.2 shows why this is called a *step function*.

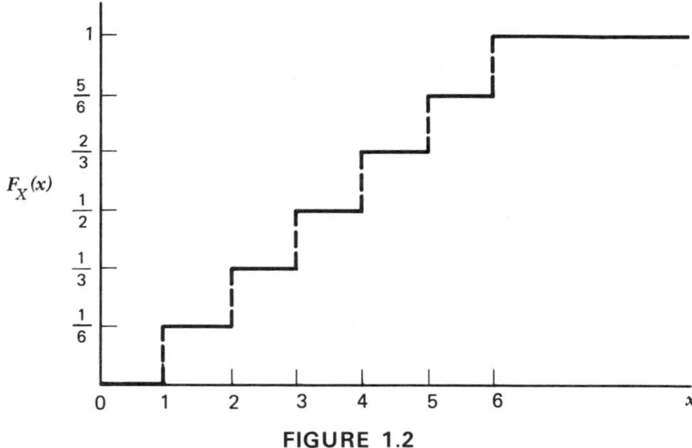

FIGURE 1.2

A discrete distribution.

An (*absolutely*) *continuous* random variable is one for which the cumulative distribution function can be expressed as the integral of a function $p_X(x)$:

$$F_X(x) = \int_{-\infty}^{x} p_X(t)\, dt. \tag{1.97}$$

In the cases we discuss in this book

$$p_X(x) = \frac{dF_X(x)}{dx}. \tag{1.98}$$

This function of x is called the *density function* of X. An essential property of continuous random variables is that, if X is such a variable, then

$$\Pr[X = x] = 0 \tag{1.99}$$

for any value of x. Despite the fact that (1.99) holds, we can still calculate (for any $\alpha < \beta$)

$$\Pr[\alpha < X < \beta] = F_X(\beta) - F_X(\alpha) = \int_\alpha^\beta p_X(x)\,dx. \tag{1.100}$$

Some random variables are *mixed* in the sense that they take on certain values, say x_i, with nonzero probabilities, that is, $\Pr[X = x_i] > 0$ (as for discrete random variables), but over ranges of other values behave as continuous random variables. In case 3 such a random variable would be part of an appropriate model. Possible forms for the $F_X(x)$ functions for continuous and mixed random variables are sketched in Figure 1.3*a* and *b*.

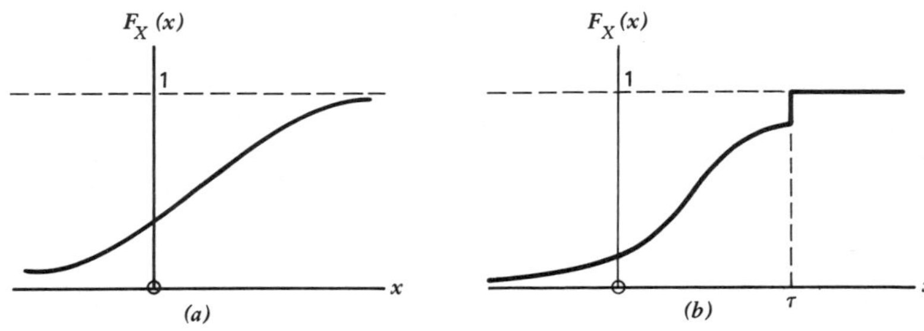

FIGURE 1.3

(a) A continuous distribution; (b) a mixed distribution.

It is noted that, in Example 1.13,

$$\lim_{x \to -\infty} F_X(x) = 0 \qquad \text{and} \qquad \lim_{x \to \infty} F_X(x) = 1. \tag{1.101}$$

The curves sketched in Figure 1.3 also indicate the validity of these relations. Although it is possible to construct random variables with distributions for which (1.101) is not valid, we consider only distributions for which it is valid. (These are called *proper* distributions.)

We also need to use the concept of *joint* and *conditional* distributions. The joint cumulative distribution function of m random variables $X_1, X_2, \ldots, X_m$ is

$$F_{X_1,\ldots,X_m}(x_1, \ldots, x_m) = \Pr\left[\bigcap_{j=1}^{m} (X_j \le x_j)\right]. \tag{1.102}$$

The left-hand side can be abbreviated to $F_{\mathbf{X}}(\mathbf{x})$. These distributions are called *multivariate*. If $m = 2$ they are *bi*variate, if $m = 3$, *tri*variate, and so on. Just as univariate distributions can be continuous, discrete, or mixed, so multivariate distributions can be of different types. For a discrete multivariate distribution there is only a countable set of combinations of values x_{1i}, $x_{2i}, \ldots, x_{mi}$ for which $\Pr[\bigcap_{j=1}^{m} (X_{ji} = x_{ji})] > 0$.

For a continuous multivariate distribution we have

$$F_{\mathbf{X}}(\mathbf{x}) = \int_{-\infty}^{x_m} \int_{-\infty}^{x_{m-1}} \cdots \int_{-\infty}^{x_1} p_{\mathbf{X}}(\mathbf{x})\, dx_1\, dx_2 \cdots dx_m, \tag{1.103}$$

with (in our cases)

$$p_{\mathbf{X}}(\mathbf{x}) = \frac{\partial^m F_{\mathbf{X}}(\mathbf{x})}{\partial x_1\, \partial x_2 \cdots \partial x_m}. \tag{1.104}$$

The conditional joint distribution of $X_1, X_2, \ldots, X_m$ given $Y_1, Y_2, \ldots, Y_q$ is just the joint distribution of the random variables $X_1, \ldots, X_m$ when other random variables $Y_1, Y_2, \ldots, Y_q$ have specified values. The set $X_1, X_2, \ldots, X_m$ is *independent* of the set $Y_1, Y_2, \ldots, Y_q$ if this conditional joint distribution is the same as the overall (unconditioned) joint distribution of $X_1, X_2, \ldots, X_m$ (so that the joint distribution of the X's is *unaffected* by the values of the Y's.) Formally, if $\mathbf{X}$ is independent of $\mathbf{Y}$,

$$F_{\mathbf{X}|\mathbf{Y}}(\mathbf{x} | \mathbf{y}) = F_{\mathbf{X}}(\mathbf{x}), \tag{1.105}$$

whence in this case

$$F_{\mathbf{X},\mathbf{Y}}(\mathbf{x}, \mathbf{y}) = F_{\mathbf{X}}(\mathbf{x}) F_{\mathbf{Y}}(\mathbf{y}). \tag{1.106}$$

In particular, X_1 is independent of X_2 if the events $(X_1 \le x_1)$ and $(X_2 \le x_2)$ are independent *for all* x_1 *and all* x_2. Equation (1.64) specializes, in this case, to

$$F_{X_1, X_2}(x_1, x_2) = F_{X_1}(x_1) F_{X_2}(x_2). \tag{1.107}$$

If X_1 and X_2 are each continuous, then there is a similar relationship,

$$p_{X_1, X_2}(x_1, x_2) = p_{X_1}(x_1) p_{X_2}(x_2), \tag{1.108}$$

among the density functions.

EXERCISE 1.31. If X_1, X_2 are mutually independent and X_1 is continuous, while X_2 has the discrete distribution

$$\Pr[X_2 = 0] = p, \qquad \Pr[X_2 = 1] = 1 - p \qquad (0 < p < 1),$$

show that $Y = X_1 X_2$ has a mixed distribution.

EXERCISE 1.32. If X_1, X_2 have the same distributions as in Exercise 1.31, show that the distribution of $Z = X_1(X_2 + 1)$ is continuous.

EXERCISE 1.33. W_1 and W_2 each have the distribution (1.96). What is the value of $F_{W_1+W_2}(4)$? [*Answer*: $\frac{1}{6}$.]

If every subset of the variables $X_1, X_2, \ldots, X_m$ is independent of any disjoint subset of these variables, the set of variables $X_1, X_2, \ldots, X_m$ is called a *mutually independent set*. Note that satisfaction of the conditions "X_i and X_j are mutually independent, for any pair (i, j)" does not ensure that $X_1, X_2, \ldots, X_m$ is a mutually independent set.

Example 1.19. Consider the set X_1, X_2, X_3 with the following distribution:

x_1	0	0	0	0	1	1	1	1
x_2	0	0	1	1	0	0	1	1
x_3	0	1	0	1	0	1	0	1
$\Pr\left[\bigcap_{j=1}^{3}(X_j = x_j)\right]$	.1	.15	.15	.1	.05	.2	.2	.05.

$$\Pr[X_j = 0] = \Pr[X_j = 1] = .5 \qquad (\text{for } j = 1, 2, 3)$$

and

$$\Pr[X_{j'} = 0 | X_j = 0] = .5 \qquad (\text{for } j \neq j'),$$

But, for example,

$$\Pr[X_1 = 0 | (X_2 = 0) \cap (X_3 = 0)] = \frac{.1}{.15} = \frac{2}{3} \neq .5.$$

1.4.2. Expected Values and Related Quantities

The *arithmetic mean* of q quantities $g_1, g_2, \ldots, g_q$ is

$$\bar{g} = q^{-1} \sum_{i=1}^{q} g_i. \tag{1.109}$$

Denoting the measured values of a random variable X in a sequence of n trials by $X_1, \ldots, X_n$ the arithmetic mean

$$\bar{X}_n = n^{-1} \sum_{i=1}^{n} X_i$$

usually appears to tend to a limiting value, say ξ, as n increases without limit. If X is a discrete random variable, this limiting value (if it exists) will be

$$\xi = \sum_j x_j \Pr[X = x_j]; \tag{1.110}$$

if X is a continuous random variable,

$$\xi = \int_{-\infty}^{\infty} x p_X(x)\, dx. \tag{1.111}$$

The number ξ is called the *expected value* of X. It is conventionally written $E[X]$. Note that the expected value need not be a value of X that can actually be observed.

Another formula for the expected value, which applies to a random variable X that takes only the values 0, 1, 2, . . . , is

$$E[X] = \sum_{j=0}^{\infty} \Pr[X > j]. \tag{1.110$'$}$$

Analogously, for a nonnegative random variable X,

$$E[X] = \int_{0}^{\infty} \{1 - F_X(x)\}\, dx. \tag{1.111$'$}$$

Example 1.20. If $\Pr[X = j] = \frac{1}{6}$ for $j = 1, 2, 3, 4, 5, 6$ (see Example 1.18), then

$$E[X] = (\tfrac{1}{6} \times 1) + (\tfrac{1}{6} \times 2) + \cdots + (\tfrac{1}{6} \times 6) = 3\tfrac{1}{2},$$

but

$$\Pr[X = 3\tfrac{1}{2}] = 0.$$

If a is a constant, then

$$E[aX] = aE[X]. \tag{1.112}$$

For any two random variables X and Y (whether independent or not),

$$E[X + Y] = E[X] + E[Y] \tag{1.113}$$

and, more generally,

$$E[Y_1 + Y_2 + \cdots + Y_k] = E[Y_1] + E[Y_2] + \cdots + E[Y_k], \tag{1.114}$$

that is,

$$E\left[\sum_{i=1}^{k} Y_i\right] = \sum_{i=1}^{k} E[Y_i]. \tag{1.114$'$}$$

Combining (1.112) and (1.114) we obtain the formula

$$E\left[\sum_{i=1}^{k} a_i Y_i\right] = \sum_{i=1}^{k} a_i E[Y_i]. \tag{1.115}$$

There is a formula similar to (1.113) for the expected value of the product of two random variables,

$$E[XY] = E[X]E[Y], \tag{1.116}$$

which is valid *if X and Y are mutually independent*. This formula *may* be valid even if X and Y are not independent, but in general it is not then valid.

If $Y_1, Y_2, \ldots, Y_k$ are a mutually independent set, then (1.116) generalizes to

$$E\left[\prod_{i=1}^{k} Y_i\right] = \prod_{i=1}^{k} E[Y_i]. \tag{1.117}$$

Example 1.21. Urn 1 contains five white balls and five black balls. Urn 2 contains three white balls and seven black balls.

A ball is chosen at random (i.e., so that each ball in the urn has an equal probability of being chosen) from each of the urns and placed in the other urn. What is the expected number of white balls in urn 1 after this has been done if

(i) The balls are chosen simultaneously?
(ii) The ball from urn 1 is chosen first and placed in urn 2 before the ball from urn 2 is chosen?

Solution. (i) Let Y_j denote the number of white balls chosen from urn j ($j = 1, 2$). Then

$$\begin{aligned}
&\Pr[Y_1 = 1] = \tfrac{5}{10} = .5, \\
&\Pr[Y_2 = 0] = .5, \qquad E[Y_1] = (1 \times .5) + (0 \times .5) = .5
\end{aligned}$$

$$\begin{aligned}
&\Pr[Y_2 = 1] = \tfrac{3}{10} = .3, \\
&\Pr[Y_2 = 0] = .7, \qquad E[Y_2] = (1 \times .3) + (0 \times .7) = .3.
\end{aligned}$$

The number of white balls in urn 1 after the exchange is

$$(\text{Number originally in urn 1}) + Y_2 - Y_1 = 5 + Y_2 - Y_1.$$

The expected number is $5 + E[Y_2] - E[Y_1] = 4.8$.

(ii) With the same notation as in part (i), $E[Y_1] = .5$ as before, but we now have

$$\begin{aligned}
&\Pr[Y_2 = 1 \mid Y_1 = 1] = \tfrac{4}{11}, \qquad \Pr[Y_2 = 0 \mid Y_1 = 1] = \tfrac{7}{11} \\
&\Pr[Y_2 = 1 \mid Y_1 = 0] = \tfrac{3}{11}, \qquad \Pr[Y_2 = 0 \mid Y_1 = 0] = \tfrac{8}{11},
\end{aligned}$$

whence

$$\begin{aligned}
&\Pr[Y_2 = 1] = (\tfrac{4}{11} \times \tfrac{1}{2}) + (\tfrac{3}{11} \times \tfrac{1}{2}) = \tfrac{7}{22}, \\
&\Pr[Y_2 = 0] = 1 - \Pr[Y_2 = 1] = \tfrac{15}{22}
\end{aligned}$$

and

$$E[Y_2] = \tfrac{7}{22}.$$

The expected number of white balls in urn 1 is

$$5 + E[Y_2] - E[Y_1] = 5 + \tfrac{7}{22} - \tfrac{1}{2} = 4\tfrac{9}{11} = 4.82 \text{ (approximately)}.$$

EXERCISE 1.34. Show that, if n balls are assigned to m ($>n$), urns the expected number of balls assigned to a specified urn is nm^{-1} under *each* of the systems, M–B, B–E, and F–D. [*Hint*: Use (1.114).]

The expected value of a function $g(X)$ of a random variable X is defined by direct analogy with (1.110) and (1.111) as

$$E[g(X)] = \sum_j g(x_j)\Pr[X = x_j] \tag{1.118}$$

or

$$E[g(X)] = \int_{-\infty}^{\infty} g(x)p_X(x)\,dx. \tag{1.119}$$

In particular we note the *moments* of X. The *rth moment of X about θ* is $E[(X - \theta)^r]$. Putting $\theta = E[X]$, we obtain the *rth central moment of X*,

$$\mu_r = E[\{X - E(X)\}^r]. \tag{1.120}$$

Putting $\theta = 0$, we obtain the *rth moment of X* (sometimes called the *rth crude moment*),

$$\mu'_r = E[X^r]. \tag{1.121}$$

The notations μ_r, μ'_r are almost universally employed. We also have the *rth absolute central moment of X*,

$$\nu_r = E[|X - E[X]|^r] \tag{1.122}$$

and the *rth absolute (crude) moment of X*

$$\nu'_r = E[|X|^r]. \tag{1.123}$$

Clearly, if r is an even integer,

$$\mu_r = \nu_r \qquad \text{and} \qquad \mu'_r = \nu'_r.$$

The *rth (descending) factorial moment of X*,

$$\mu_{(r)} = E[X^{(r)}], \tag{1.124}$$

is also useful sometimes, as is the *rth ascending factorial moment*,

$$\mu_{[r]} = E[X^{[r]}] = E[X(X+1)\cdots(X+r-1)]. \qquad (1.125)$$

(Note that in (1.121) through (1.125) the *subscripts* of the μ's and ν's correspond to the *indices* of the function in $E[\cdot]$.)

EXERCISE 1.35. Show that, if X is a random variable taking only values 0, 1, 2, ..., then

$$E[X^{[2]}] = E[X(X+1)] = 2\sum_{r=0}^{\infty} T_r^{(1)},$$

where $T_r^{(1)} = \sum_{j=r}^{\infty} \Pr[X > j]$.

(For a positive integer k there is a similar expression for $E[X^{[k]}]$ in terms of repeated summations on $\Pr[X > j]$.)

EXERCISE 1.36. Show that, if X is a nonnegative random variable,

$$E[X^k] = k\int_0^{\infty} x^{k-1}\{1 - F_X(x)\}\,dx,$$

provided the integral converges.

Relationships among the moments can be deduced by using (1.112) and (1.114). These include

$$\mu_r = \sum_{j=0}^{r-2}(-1)^j\binom{r}{j}\mu'_j\mu_1'^{r-j} + (-1)^{r-1}(r-1)\mu_1'^r, \qquad (1.126)$$

$$\mu'_r = \begin{cases} \displaystyle\sum_{j=0}^{r}\binom{r}{j}\mu_j\mu_j'^{r-j} = \mu_1'^r + \sum_{j=1}^{r}\binom{r}{j}\mu_j\mu_1'^r & (1.127\text{a}) \\ \displaystyle\sum_{j=0}^{r}\left(\frac{\Delta^j\mathbf{0}^r}{j!}\right)\mu_{(j)} \qquad \text{[cf. (1.22)]}, & (1.127\text{b}) \end{cases}$$

$$\mu_{(r)} = \sum_{j=1}^{r} S_r^{(j)}\mu'_j \qquad \text{[cf. (1.25)]}. \qquad (1.128)$$

In particular, from (1.126),

$$\begin{aligned} \mu_2 &= \mu'_2 - \mu_1'^2, \\ \mu_3 &= \mu'_3 - 3\mu'_2\mu'_1 + 2\mu_1'^3, \\ \mu_4 &= \mu'_4 - 4\mu'_3\mu'_1 + 6\mu'_2\mu_1'^2 - 3\mu_1'^4. \end{aligned} \qquad (1.129)$$

If it is desired to identify the random variable to which the moments relate, we write $\mu_r(X)$, $\mu'_r(Y)$, $\mu_{(r)}(Z)$, and so on.

An alternative notation for $\mu_2(X)$ is var(X)—the *variance* of X. The (positive) square root of the variance—a quantity of the same dimensions as the measurements of X—is called the *standard deviation of* X. It is often denoted by $\sigma(X)$, so that

$$\text{var}(X) = \{\sigma(X)\}^2.$$

Note that, if α, β are constants, then $\text{var}(\alpha + \beta X) = \beta^2 \text{ var}(X)$. In particular, by choosing $\beta = \{\sigma(X)\}^{-1}$ and $\alpha = -E[X]/\sigma(X)$ we see that the *standardized variable*; $X' = \{X - E[X]\}/\sqrt{\text{var}(X)}$ has expected value 0 and standard deviation 1.

The following result is used later in this book:

Chebyshev's Inequality

If X has a finite expected value $E[X]$, then, for any $\lambda > 0$,

$$\Pr[|X - E[X]| \leq \lambda\{\mu_r(X)\}^{1/r}] \geq 1 - \lambda^{-r}. \qquad (1.130a)$$

The most commonly used case is with $r = 2$. In terms of the standardized variate X' we have

$$\Pr[|X'| \leq \lambda] \leq 1 - \lambda^{-2}. \qquad (1.130b)$$

The third and fourth central moments (μ_3, μ_4) are not often used on their own in describing a distribution. They are usually employed in certain ratios called *shape factors*. These are

$$\sqrt{\beta_1} = \alpha_3 = \frac{\mu_3}{\mu_2^{3/2}}, \qquad \beta_2 = \alpha_4 = \frac{\mu_4}{\mu_2^2}. \qquad (1.131)$$

Note that the powers of μ_2 in the denominators have been chosen so that the ratios are unaffected by linear transformations of the variables. This is because, if $Y = \alpha + \beta X$, then $E[Y] = \alpha + \beta E[X]$ and

$$Y - E[Y] = \beta(X - E[X]).$$

$\sqrt{\beta_1}$ is a measure of *skewness*; β_2 is a measure of *kurtosis*.

Let us now consider the evaluation of the variance $[\mu_2(Z)]$ of the linear function $\sum_{i=1}^{k} a_i Y_i = Z$. As we have already seen,

$$E[Z] = \sum_{i=1}^{k} a_i E[Y_i]. \qquad (1.132)$$

Now

$$
\begin{aligned}
\operatorname{var}(Z) = E[\{Z - E[Z]\}^2] &= E\left[\left\{\sum_{i=1}^{k} a_i(Y_i - E[Y_i])\right\}^2\right] \\
&= E\left[\sum_{i=1}^{k} a_i^2(Y_i - E[Y_i])^2\right. \\
&\qquad \left. + 2\sum_{i<i'}\sum a_i a_{i'}(Y_i - E[Y_i])(Y_j - E[Y_j])\right] \\
&= \sum_{i=1}^{k} a_i^2 E[(Y_i - E[Y_i])^2] \\
&\qquad + 2\sum_{i<i'}\sum a_i a_{i'} E[(Y_i - E[Y_i])(Y_{i'} - E[Y_{i'}])].
\end{aligned}
\tag{1.133}
$$

We know that $E[(Y_i - E[Y_i])^2] = \mu_2(Y_i) = \operatorname{var}(Y_i)$. As yet we do not have a name for the expected value of the product of deviations of Y_i, $Y_{i'}$ from their respective expected values. We call this the *covariance* of Y_i and $Y_{i'}$ and write

$$\operatorname{cov}(Y_i, Y_{i'}) = E[(Y_i - E[Y_i])(Y_{i'} - E[Y_{i'}])]. \tag{1.134}$$

Using this notation we obtain, from (1.133),

$$\operatorname{var}\left(\sum_{i=1}^{k} a_i Y_i\right) = \sum_{i=1}^{k} a_i^2 \operatorname{var}(Y_i) + 2\sum_{i<i'}\sum a_i a_{i'} \operatorname{cov}(Y_i, Y_{i'}). \tag{1.135}$$

The *correlation coefficient* between Y_i and $Y_{i'}$ is

$$\operatorname{corr}(Y_i, Y_{i'}) = \frac{\operatorname{cov}(Y_i, Y_{i'})}{\{\operatorname{var}(Y_i)\operatorname{var}(Y_{i'})\}^{1/2}}. \tag{1.136}$$

Using the notation $\rho_{ii'} = \operatorname{corr}(Y_i, Y_{i'})$, $\sigma_i^2 = \operatorname{var}(Y_i)$, (1.135) can be written

$$\operatorname{var}\left(\sum_{i=1}^{k} a_i Y_i\right) = \sum_{i=1}^{k} a_i^2\sigma_i^2 + 2\sum_{i<i'}\sum a_i a_{i'} \rho_{ii'}\sigma_i\sigma_{i'}. \tag{1.136$'$}$$

Note that

$$\operatorname{var}(Y_1 \pm Y_2) = \sigma_1^2 + \sigma_2^2 \pm 2\rho_{12}\sigma_1\sigma_2. \tag{1.137}$$

EXERCISE 1.37. Show that

$$\operatorname{corr}(\alpha_1 + \beta_1 Y_1, \alpha_2 + \beta_2 Y_2) = \operatorname{sgn}(\beta_1\beta_2)\operatorname{corr}(Y_1, Y_2).$$

Hence show that the correlation between the standardized variables Y'_1, Y'_2 equals $\text{corr}(Y_1, Y_2)$.

From (1.24),

$$E\left[\left(\sum_{i=1}^{r} X_i\right)^k\right] = \sum_{j=0}^{k} \sum^{(j)} E[X_1^{(j_1)} X_2^{(j_2)} \cdots X_r^{(j_r)}] \frac{\Delta^j \mathbf{0}^k}{j_1! j_2! \cdots j_r!},$$

where $\sum^{(j)}$ denotes summation over $j_1, j_2, \ldots, j_r$, with $j_1 + \cdots + j_r = j$. If $X_1, X_2, \ldots, X_r$ are random variables which take only values 0 and 1, then $X_j^{(\alpha)}$ is identically 0 for any α greater than 1, and so

$$\begin{aligned} E\left[\left(\sum_{i=1}^{r} X_i\right)^k\right] &= \sum_{j=0}^{k} \left\{ \sum_{i_1 \neq i_2 \cdots \neq i_j} \cdots \sum E[X_{i_1}^{(1)} X_{i_2}^{(1)} \cdots X_{i_j}^{(1)}] \right\} \Delta^j \mathbf{0}^k \\ &= \sum_{j=1}^{s} \left\{ \sum_{i_1 \neq i_2 \neq \cdots \neq i_j} \cdots \sum E[X_{i_1} X_{i_2} \cdots X_{i_j}] \right\} \Delta^j \mathbf{0}^k. \end{aligned} \tag{1.138}$$

(Note that $\Delta^0 \mathbf{0}^k = 0$.)

1.4.3. Generating Functions

We frequently use *generating functions* in later chapters. Many generating functions are power series:

$$G(s) = a_0 + a_1 s + a_2 s^2 + \cdots = \sum_{j=0}^{\infty} a_j s^j. \tag{1.139}$$

The argument s of the function $G(s)$ plays only an auxiliary role; the function is used as a source of information on the coefficients $a_0, a_1, a_2, \ldots$. [Seal (1949) gives an illuminating historical discussion.]

The representation (1.139) can be used only if, for values of s in a sufficiently small interval, say $|s| \leq s_0$, the right-hand side of (1.139) converges uniformly. Its utility resides in the fact that in such a case the representation of $G(s)$ as a power series is *unique*. This means that, once we obtain $G(s)$, then—if we can expand it in a power series—we can reveal the values $a_0, a_1, a_2, \ldots$ which are "hidden" in $G(s)$.

For a discrete distribution taking only nonnegative integer values (0, 1, 2, ...), with $\Pr[X = j] = P_j$ $(j = 0, 1, \ldots)$, the *probability generating function* (pgf) is

$$P(s) = \sum_{j=0}^{\infty} p_j s^j. \tag{1.140}$$

This series will certainly converge uniformly for $|s| \leq 1$. If the distribution is *proper* [see (1.101)], then

$$P(1) = \sum_{j=0}^{\infty} p_j = 1.$$

Differentiating (1.140) with respect to s we obtain

$$P'(1) = \sum_{j=0}^{\infty} jp_j = E[X] \tag{1.141}$$

(provided the expected value exists).

Repeated differentiation (and subsequently putting s equal to 1) yields

$$P^{(r)}(1) = \sum_{j=0}^{\infty} j^{(r)}p_j = E[X^{(r)}] = \mu_{(r)}(X). \tag{1.142}$$

If $G_j(s)$ denote the pgf's of X_j ($j = 1, 2$) (each variable taking only values 0, 1, 2, ...) and X_1 *and* X_2 *are mutually independent*, then the generating function of $(X_1 + X_2)$ is

$$G_1(s)G_2(s). \tag{1.143}$$

The results extend directly to the sum of k *mutually independent random variables* (each taking only integral values 0, 1, 2, ...).

Example 1.22. A ball is drawn at random from an urn containing six balls numbered 1, 2, 3, 4, 5, 6. The number on the ball is noted, and it is replaced in the urn. This procedure is repeated four times. What is the probability that the sum of the four noted numbers equals 20?

If it is assumed that each ball is equally likely to be drawn, the pgf of X_j, the number on the ball drawn on the jth occasion, is

$$\begin{aligned} G_j(s) &= \tfrac{1}{6}s^1 + \tfrac{1}{6}s^2 + \tfrac{1}{6}s^3 + \tfrac{1}{6}s^4 + \tfrac{1}{6}s^5 + \tfrac{1}{6}s^6 \\ &= \frac{1}{6}\frac{s(1 - s^6)}{1 - s}. \end{aligned}$$

Note that (1) this value does not depend on j, and (2) the distribution of X_j is the same as that in Example 1.18, where the results of throwing a six-sided die were considered. The pgf of the total of the noted numbers $(X_1 + X_2 + X_3 + X_4)$ is

$$G_1(s)G_2(s)G_3(s)G_4(s) = 6^{-4}s^4(1 - s^6)^4(1 - s)^{-4}.$$

$\Pr[X_1 + X_2 + X_3 + X_4 = 20]$ is the coefficient of s^{20} in the expansion of the right-hand side. This equals

$$\begin{aligned}
&6^{-4} \times [\text{coefficient of } s^{16} \text{ in } (1 - s^6)^4(1 - s)^{-4}] \\
&\quad = 6^{-4} \times [\{\text{coefficient of } s^{16} \text{ in } (1 - s)^{-4}\} \\
&\qquad -4\{\text{coefficient of } s^{10} \text{ in } (1 - s)^{-4}\} \\
&\qquad +6\{\text{coefficient of } s^{4} \text{ in } (1 - s)^{-4}\}] \\
&\quad = 6^{-4}\left[\binom{19}{16} - 4\binom{13}{10} + 6\binom{7}{4}\right] \\
&\quad = 6^{-4}[969 - (4 \times 286) + (6 \times 35)] = \frac{35}{1296}.
\end{aligned}$$

EXERCISE 1.38. In Example 1.22, what is the probability that the sum of the four noted numbers equals

(i) 21?
(ii) 7?

In many problems it is possible to effect explicit evaluation of probabilities by identifying required values with the coefficients of specified powers of a dummy variable t in the expansion of appropriate expressions, that is, generating functions.

Suppose we have m urns, with the jth urn containing a_j balls numbered $0, 1, 2, \ldots, a_j - 1$. A ball is drawn at random from each urn. The numbers on the chosen balls are $X_1, X_2, \ldots, X_m$, respectively. What is the probability that

$$X_1 + X_2 + \cdots + X_m = c,$$

where c is an integer between 0 and $(\sum_{j=1}^{m} a_j - m)$ inclusive?

The number of ways of drawing the balls so that the sum of the numbers on them equals c is the coefficient of t^c in the expansion of

$$\prod_{j=1}^{m}(1 + t + t^2 + \cdots + t^{a_j - 1}) = (1 - t)^{-m}\prod_{j=1}^{m}(1 - t^{a_j}). \tag{1.144}$$

The total number of ways of drawing the balls is $\prod_{j=1}^{m} a_j$, and the required probability is the ratio of the two numbers.

As a particular case suppose that $m = 4$, $a_1 = a_2 = 5$, $a_3 = a_4 = 10$, and $c = 10$. We then seek the coefficient of t^{10} in the expansion of

$$(1 - t)^{-4}(1 - t^5)^2(1 - t^{10})^2.$$

Since $(1 - t^5)^2(1 - t^{10})^2 = 1 - 2t^5 - t^{10} +$ (higher powers of t), the required coefficient is

$$b_{10} - 2b_5 - b_0 = \binom{13}{10} - 2\binom{8}{5} - 1 = 286 - 112 - 1 = 173,$$

where $b_g =$ coefficient of t^g in the expansion of $(1 - t)^{-4} = \binom{g + 3}{g}$. Hence the required probability is $173(5^2 \times 10^2)^{-1} = 173/2500 = .0692$.

This example may be generalized as follows. What is the probability that

$$c_1X_1 + c_2X_2 + \cdots + c_mX_m = c, \tag{1.145}$$

where $c_1, c_2, \ldots, c_m, c$ are given integers?

The total number of possible (equally likely) combinations of values of $X_1, X_2, \ldots, X_m$ is $a_1a_2 \cdots a_m = \prod_{j=1}^m a_j$. The number of combinations for which (1.145) is satisfied is the coefficient of t^c in the expansion of

$$(1 + t^{c_1} + t^{2c_1} + \cdots + t^{(a_1 - 1)c_1}) \cdots (1 + t^{c_m} + t^{2c_m} + \cdots + t^{(a_m - 1)c_m})$$

$$= \prod_{j=1}^m \left(\frac{1 - t^{a_jc_j}}{1 - t^{c_j}}\right). \tag{1.146}$$

In the particular case when $c_1 = c_2 = \cdots = c_m = 1$, we require the coefficient of t^c in the expansion of

$$(1 - t)^{-m}\prod_{j=1}^m (1 - t^{a_j}). \qquad \text{[cf. (1.144)].}$$

The concept of a *partition* (dividing up) of a positive integer n is sometimes useful. A partition of n into s parts is a set of positive integers $a_1, a_2, \ldots, a_s$ such that $a_1 + a_2 + \cdots + a_s = n$. The number ($s$) of parts is the *weight* of the partition. It is customary to arrange that $a_1 \leq a_2 \leq \cdots \leq a_s$, but we often wish to distinguish between, say (112), (121), and (211), as partitions of 4. In such cases we use the term *ordered partition*.

Example 1.23 (Garside, 1971). Suppose we have r numbered balls in an urn and draw them out one at a time, replacing them after each draw. Let $N_1, N_2, \ldots, N_r$ denote the number of times the balls numbered $1, 2, \ldots, r$, respectively, appear in n draws, and let $N'_1 \leq N'_2 \leq \cdots \leq N'_r$ be the corresponding *order statistics*. The set of numbers $N'_1, N'_2, \ldots, N'_r$ is called a *configuration*.

How many different configurations are there (i) altogether, and (ii) with $N'_r = q$ (a specified value)?

(i) The total number of configurations is the number of ordered partitions of n with up to (and including) r parts. This is the coefficient of t^n in the expansion of $\prod_{j=1}^r (1 - t^j)^{-1}$.

(ii) This is the number of ordered partitions of $n - q$ with up to (and including) $(r - 1)$ parts, and no part exceeding q. It is the coefficient of t^{n-q} in the expansion of

$$\prod_{j=1}^{r-1} \left(\frac{1 - t^{j(q+1)}}{1 - t^j} \right).$$

The following exercises can be solved by this kind of approach.

EXERCISE 1.39. Obtain a general formula for the distribution of the sum of the numbers obtained when one ball is drawn from each of m urns, each of which contains n balls numbered $0, 1, \ldots, n - 1$. (This is a special case of the m-fold convolution of a discrete rectangular distribution.)

EXERCISE 1.40. Show that the probability that the sum of the numbers thrown in the first three throws of an unbiased six-sided die equals the sum of the next two throws is equal to

$$6^{-5} \times \left[\text{coefficient of } t^9 \text{ in } \left(\frac{1 - t^6}{1 - t} \right)^5 \right].$$

EXERCISE 1.41. Generalize the formula given in Exercise 1.39.

For the determination of moments it is often more convenient to use *moment generating functions* if they exist in a suitable form.

The moment generating function of a random variable X is the expected value of $\exp(Xs)$, where s as before is a dummy variable. The *formal* relation

$$\phi_X(s) = E[\exp(Xs)] = E\left[1 + Xs + \frac{1}{2!} X^2 s^2 + \cdots \right]$$

$$= 1 + \mu'_1 s + \frac{1}{2!} \mu'_2 s^2 + \cdots \tag{1.147}$$

leads us to expect that the coefficient of $s^r/r!$ in the expansion of $\phi_X(s)$ will be equal to $\mu'_r(X)$. For this to be so we need the following conditions

(i) $\phi_X(s)$ can be expressed as a power series in s.
(ii) This series is uniformly convergent for $0 < |s| < s_0$ (for some s_0).

If the range of variation of X is finite, these conditions are certainly satisfied.

The *central moment generating function* is

$$\phi_{X - E[X]}(s) = E[\exp\{X - E[X]\}s] = \phi_X(s)\exp\{-E[X]s\}$$

$$= 1 + \frac{1}{2!} \mu_2 s^2 + \frac{1}{3!} \mu_3 s^3 + \cdots. \tag{1.148}$$

This is easily calculated, once $\phi_X(s)$ is known. Under conditions (i) and (ii) the coefficient of $(s^r/r!)$ in the expansion is $\mu_r(X)$, the rth central moment of X [cf. (1.120)].

If X_1 and X_2 are mutually independent, so are $\exp(X_1 s)$ and $\exp(X_2 s)$, and so

$$\phi_{X_1+X_2}(s) = E[(X_1 + X_2)s] = E(X_1 s)E(X_2 s) = \phi_{X_1}(s)\phi_{X_2}(s). \quad (1.149)$$

Hence

$$\log_e \phi_{X_1+X_2}(s) = \log_e \phi_{X_1}(s) + \log_e \phi_{X_2}(s)$$

and, if we denote the coefficient of $(s^r/r!)$ in the expansion of $\log_e \phi_X(s)$ as a power series in s by $\kappa_r(X)$ (it being supposed that the series converges uniformly for $0 < |s| < s_0$), then

$$\kappa_r(X_1 + X_2) = \kappa_r(X_1) + \kappa_r(X_2), \quad (1.150)$$

if X_1 and X_2 are mutually independent.

Since

$$\phi_X(s) = \phi_{X-E(X)}(s)\exp(E[X]s),$$

we see that

$$\psi_X(s) = \log_e[\phi_X(s)] = E[X]s + \log_e[\phi_{X-E[X]}(s)]. \quad (1.151)$$

From (1.151), the coefficient of s is $E[X]$, that is, $\kappa_1(X) = E[X]$, and [also from (1.151)] all the other coefficients are functions of the central moments $\mu_r(X)$ [and *not* of $E[X] = \mu'_1(X)$]. In fact, we find

$$\kappa_2 = \mu_2, \qquad \kappa_3 = \mu_3, \qquad \kappa_4 = \mu_4 - 3\mu_2^2, \qquad \kappa_5 = \mu_5 - 10\mu_3\mu_2 \cdots \quad (1.152)$$

Equation (1.150) generalizes straightforwardly to

$$\kappa_r(X_1 + X_2 + \cdots + X_n) = \kappa_r(X_1) + \kappa_r(X_2) + \cdots + \kappa_r(X_n), \quad (1.153)$$

provided the n random variables $X_1, X_2, \ldots, X_n$ are mutually independent.

On account of the property (1.153), $\kappa_r(X)$ is called the *rth cumulant* of X, and $\psi_X(s)$, as defined in (1.151), is called the *cumulant generating function* of X.

Note that, if a is a constant,

$$\phi_{aX}(s) = E[\exp(Xas)] = \phi_X(as), \quad (1.154)$$

hence

$$\psi_{aX}(s) = \psi_X(as). \quad (1.155)$$

It follows that

$$\kappa_r(aX) = a^r\kappa_r(X). \quad (1.156)$$

Example 1.24. If $X_1, X_2, \ldots, X_n$ are mutually independent and have common moments, hence common cumulants, so that

$$\kappa_r(X_j) = \kappa_r \qquad (j = 1, 2, \ldots, n),$$

then

$$\kappa_r(X_1 + X_2 + \cdots + X_n) = n\kappa_r. \tag{1.157}$$

The arithmetic mean $\bar{X} = n^{-1} \sum_{j=1}^{n} X_j$ has the rth cumulant

$$\kappa_r(\bar{X}) = n\kappa_r\left(\frac{X_j}{n}\right) = \frac{\kappa_r}{n^{r-1}} \qquad \text{[from (1.156)].}$$

In particular,

$$\text{var}(\bar{X}) = \mu_2(\bar{X}) = \frac{\kappa_2}{n} = \frac{\mu_2(X)}{n} \tag{1.158}$$

$$\mu_3(\bar{X}) = \frac{\kappa_3}{n^2} = \frac{\mu_3(X)}{n^2}; \qquad \alpha_3(\bar{X}) = \frac{\alpha_3(X)}{\sqrt{n}} \tag{1.159}$$

$$\mu_4(\bar{X}) - 3\{\mu_2(\bar{X})\}^2 = \frac{\mu_4(X) - 3\{\mu_2(X)\}^2}{n^3},$$

whence

$$\begin{aligned} \mu_4(\bar{X}) &= \mu_4(X)n^{-3} + 3\{\mu_2(X)\}^2(n-1)n^{-3} \\ \alpha_4(\bar{X}) &= \alpha_4(X)n^{-1} + 3(1 - n^{-1}) \\ &= 3 + (\alpha_4(X) - 3)n^{-1}. \end{aligned} \tag{1.160}$$

Note that, if $\alpha_3(X) = 0$, then also $\alpha_3(\bar{X}) = 0$; if $\alpha_4(X) = 3$, then also $\alpha_4(\bar{X}) = 3$, for all n.

For a discrete variable X, taking values 0, 1, 2, ..., we define the *factorial moment generating function*:

$$M(s) \Rightarrow \phi_X^*(s) \tag{1.161}$$

If the greatest value X can take is m, then $\mu_{(j)} = 0$ for $j > m$, and the series on the right-hand side of (1.161) terminates at $j = m$.

In such a case we have

$$\sum_{r=0}^{m} p_r(1 + s)^r = \sum_{j=0}^{m} \frac{\mu_{(j)}}{j!} s^j,$$

where $p_r = \Pr[X = r]$. Putting $1 + s = t$, we have

$$\sum_{r=0}^{m} p_r t^r = \sum_{j=0}^{m} \frac{\mu_{(j)}}{j!} (t - 1)^j.$$

Equating the coefficients of t^h, we find

$$p_h = \sum_{j=h}^{m} \frac{\mu_{(j)}}{j!}(-1)^{j-h}\binom{j}{j-h} = \sum_{j=h}^{m}(-1)^{j-h}\frac{\mu_{(j)}}{(j-h)!h!}$$
$$= \frac{1}{h!}\sum_{j=0}^{m-h}(-1)^j \frac{\mu_{(j+h)}}{j!}. \tag{1.162}$$

Equation (1.162) enables us to calculate the probability distribution from the factorial moments. It is a special kind of *inversion formula*.

By analogy with (1.151) the function

$$\psi_X^*(s) = \log_e \phi_X^*(s) = \sum_{j=1}^{\infty} \frac{\kappa_{(j)}(X)}{j!} s^j \tag{1.163}$$

is called the *factorial cumulant generating function* of X, and $\kappa_{(j)}(X)$, the coefficient of $(s^j/j!)$ in the expansion of $\psi_X^*(s)$, is the jth *factorial cumulant* of X. Analogously to (1.127b) and (1.128) we have

$$\kappa_r = \sum_{j=1}^{r} \left(\frac{\Delta^j \mathbf{0}^r}{j!}\right)\kappa_{(j)} \tag{1.164a}$$

$$\kappa_{(r)} = \sum_{j=1}^{r} S_r^{(j)}\kappa_j. \tag{1.164b}$$

Generating functions can also be defined for multivariate distributions. We discuss bivariate distributions only—extension to more variables is straightforward, although sometimes calculations are cumbersome.

If X_1, X_2 each can take only the values 0, 1, 2, ..., with

$$\Pr[(X_1 = i) \cap (X_2 = j)] = p_{ij}$$

and

$$\sum_i \sum_j p_{ij} = 1,$$

then the *joint probability generating function* of X_1 and X_2 is

$$P(s_1, s_2) = \sum_{i=0}^{\infty}\sum_{j=0}^{\infty} p_{ij} s_1^i s_2^j.$$

Clearly, $P(1, 1) = 1$,

$$\left.\frac{\partial P}{\partial s_h}\right|_{\mathbf{s}=\mathbf{1}} = E[X_h]$$

and, generally,

$$\left.\frac{\partial^{r_1+r_2} P}{\partial s_1^{r_1}\,\partial s_2^{r_2}}\right|_{\mathbf{s}=\mathbf{1}} = E[X_1^{(r_1)} X_2^{(r_2)}]. \tag{1.165}$$

The *joint central moment generating function* of X_1 and X_2 is

$$[\exp\{-s_1 E[X_1] - s_2 E[X_2]\}] \cdot E[\exp\{s_1 X_1 + s_2 X_2\}].$$

The coefficient of $s_1^{r_1} s_2^{r_2}/r_1!r_2!$ in the expansion of this quantity is the *mixed central moment*,

$$\mu_{r_1, r_2}(X_1, X_2) = E[\{X_1 - E[X_1]\}^{r_1}\{X_2 - E[X_2]\}^{r_2}]. \tag{1.166a}$$

The notion of generating function of a distribution can be extended to *generating function of a family of distributions* in certain cases. A simple example arises when the members of the family of distributions can be indexed by the natural numbers 0, 1, 2 . . . .

Denoting the corresponding pgf's by $G_0(s)$, $G_1(s)$, $G_2(s)$, . . . we construct an *extended* pgf (epgf):

$$G(s, z) = \sum_{j=0}^{\infty} z^j G_j(s) \tag{1.166b}$$

Provided conditions like (i) and (ii) stated in the definition of the moment generating function (equation 1.147) are satisfied, we see that from $G(s, z)$ we can find $G_j(s)$ (and hence the corresponding distribution) as the coefficient of z^j in the expansion of $G(s, z)$.

These kinds of pgf's are analogous to bivariate generating functions defined above with one of the variates replaced by a dummy variate indexing the distributions. (The analogy is not complete, because the coefficient of $z^j s^i$ is $\Pr[X = i]$ evaluated for the jth distribution—say $p_i^{(j)}$—and the condition $\sum_j \sum_i p_i^{(j)} = 1$ is not satisfied (in fact the sum diverges).

Many modifications of (1.166b) can be used, for mathematical convenience. For example z^j may be replaced by $z^j/j!$ or even [see (3.16)] by $(mz)^j/j!$, where m is a parameter in a particular problem.

Furthermore, we also use generating functions that define the cumulative distribution function rather than individual probability. Such a generating function is

$$H(s) = \sum_{j=0}^{\infty} \Pr[X \le j] s^j. \tag{1.167}$$

For two variables we have the joint generating function

$$H(s_1, s_2) = \sum_{j_1=0}^{\infty} \sum_{j_2=0}^{\infty} \Pr[(X_1 \le j_1) \cap (X_2 \le j_2)] s_1^{j_1} s_2^{j_2}, \tag{1.168}$$

and so on. (Of course, s^j may be replaced by $s^j/j!$ or some other quantity.)

EXERCISE 1.42. Show that, for a discrete distribution taking only nonnegative integer values,

$$P(s) = (1 - s)H(s),$$

where $P(s)$ is defined in (1.140).

Alternatively to (1.167), we can define a generating function in terms of the survival distribution function, taking

$$Q(s) = \sum_{j=0}^{\infty} \Pr[X > j]s^j.$$

EXERCISE 1.43. Show that

(i) $Q(1) = E[X]$.
(ii) $Q'(1) = \frac{1}{2}E[X(X - 1)]$.

(*Hint*: Write $\Pr[X > j] = \sum_{i=1}^{\infty} \Pr[X = j + i]$.)

(iii) From (i) and (ii) deduce a formula for var(X) in terms of $Q(1)$ and $Q'(1)$.

1.4.4. Patterns in Repeated Drawings

Suppose we have an urn containing a proportion p of white (W) balls and $q = (1 - p)$ of black (B) balls. We draw one ball at a time from the urn, returning the ball to the urn after each drawing. Sequences of colors of drawn balls can be represented as

WBBWWBWWWB....

The probability of observing the above sequence of colors in the first 10 balls drawn is

$$p \times q \times q \times p \times p \times q \times p \times p \times p \times q = p^6q^4.$$

We can use formula (1.73′) here, on the assumption that the results of successive drawings are mutually independent.

Generally, the probability that there will be w white and $(n - w)$ black balls, *in a specified order*, in the first n drawings is $p^w q^{n-w}$.

We consider the occurrence of specified *patterns* in the sequences of colors. A pattern is a sequence such as WBBBW, WWW, WBWB, and so on. We are interested in the distribution of the number N of drawings needed to produce a specified pattern. We introduce the generating function

$$F(s) = \sum_{n=1}^{\infty} F_n s^n,$$

where F_n is the probability that the specified pattern will be *first* completed on the nth drawing.

We also introduce another generating function

$$R(s) = \sum_{n=1}^{\infty} R_n s^n.$$

where R_n is the probability that the pattern will be completed *as a recurrent event* on the nth drawing. For this to be so we suppose the series to start again after each completion of the pattern, so that no previous results can contribute to the next completion.

As an example suppose the pattern is BWB. In the sequence

WWBWBWBBWWBBWB,

the *first completion* of BWB is on the fifth drawing; the pattern *is completed* on the fifth, seventh, and fourteenth drawings; the pattern is completed as a *recurrent event* on the fifth and fourteenth drawings. (The seventh drawing does not count, since the BWB pattern completed there uses, as its first B, the results of the fifth drawing.)

Consider now a pattern of *length* g (i.e., containing g letters). If on the nth drawing the pattern is completed, this *may* correspond to completion as a recurrent event. If it does not, then the pattern must have been completed as a recurrent event at some point (and only one point) on its length. Points at which this could happen are called *critical points*. The following are some examples.

Pattern	Critical Points at Drawings
WWWW	1st, 2nd, 3rd, 4th
WBWB	2nd, 4th
WBBB	4th
BWWB	1st, 4th

Note that (1) there is always a critical point at the end of the pattern, and (2) the subpattern from the beginning to a critical point is identical with the subpattern of the same length at the end of the pattern. Denote the number of critical points by k, and the numbers of white and black balls up to the ith critical point by w_i, b_i, respectively, with $w_1 \leq w_2 \leq \cdots \leq w_k$, $b_1 \leq b_2 \leq \cdots \leq b_k$. Clearly $w_k + b_k = g$.

The probability of completion of the pattern on the $n(\geq g)$th drawing (whether as a recurrent event or not) is just $p^{w_k}q^{b_k}$. The event can be split up according to which of the k critical points corresponds to completion as a recurrent event. We have, for $n = g, g+1, g+2, \ldots,$

$$\begin{aligned} p^{w_k}q^{b_k} &= R_{n-g+w_1+b_1}p^{w_k-w_1}q^{b_k-b_1} + R_{n-g+w_2+b_2}p^{w_k-w_2}q^{b_k-b_2} + \cdots + R_n \\ &= \sum_{i=1}^{k} R_{n-g+w_i+b_i}p^{w_k-w_i}q^{b_k-b_i}. \end{aligned}$$

Multiplying by s^n and summing with respect to n from g to ∞,

$$\frac{s^g p^{w_k} q^{b_k}}{1-s} = \left\{\sum_{i=1}^{k} s^{g-w_i-b_i} p^{w_k-w_i} q^{b_k-b_i}\right\} R(s),$$

whence

$$R(s) = \frac{1}{(1-s)\sum_{i=1}^{k} s^{-w_i-b_i} p^{-w_i} q^{-b_i}}. \tag{1.169}$$

When the pattern is completed as a recurrent event, this must be for the first, second, or ..., time. Hence

$$\begin{aligned} R(s) &= F(s) + \{F(s)\}^2 + \cdots \\ &= \frac{F(s)}{1-F(s)} \qquad [\text{taking } |s| < 1, \text{ so } 0 \le F(s) < 1], \end{aligned}$$

so

$$F(s) = \frac{R(s)}{1+R(s)}, \tag{1.170}$$

and, using (1.169),

$$F(s) = \frac{1}{1+(1-s)\sum_{i=1}^{k} s^{-w_i-b_i} p^{-w_i} q^{-b_i}}. \tag{1.171}$$

To find the expected value of the corresponding distribution we evaluate

$$\begin{aligned} F'(s) &= \left\{1 + (1-s)\sum_{i=1}^{k} s^{-w_i-b_i} p^{-w_i} q^{-b_i}\right\}^{-2} \\ &\quad \times \left\{\sum_{i=1}^{k} s^{-w_i-b_i} p^{-w_i} q^{-b_i} + (1-s)\sum_{i=1}^{k} (w_i+b_i) s^{-w_i-b_i} p^{-w_i} q^{-b_i}\right\} \end{aligned}$$

and

$$E[N] = F'(1) = \sum_{i=1}^{k} p^{-w_i} q^{-b_i} \qquad [\text{using } (1.141)].$$

The method described above was given by Bizley (1962).

EXERCISE 1.44. Show that the variance of the distribution (of number of draws needed) is

$$\left(\sum_{i=1}^{k} p^{-w_i} q^{-b_i}\right)\left(1 + \sum_{i=1}^{k} p^{-w_i} q^{-b_i}\right) - 2\sum_{i=1}^{k} (w_i+b_i) p^{-w_i} q^{-b_i}.$$

[*Hint*: Use (1.142) with $r = 2$.]

EXERCISE 1.45. Suppose there are balls of k different colors $\mathscr{C}_1, \mathscr{C}_2, \ldots, \mathscr{C}_m$ and that the proportion of balls of color $\mathscr{C}_j$ is p_j ($j = 1, \ldots, m$; $\sum_{j=1}^m p_j = 1$). Defining the critical points of a pattern as before, let a_{ji} denote the number of $\mathscr{C}_j$'s up to and including the ith critical point ($i = 1, \ldots, k$).

(i) Derive the corresponding generating function $F(s)$, and

(ii) Show that the expected number of drawings needed to complete the pattern for the first time is

$$\sum_{i=1}^{k}\left(\prod_{j=1}^{m} p_j^{-a_{ij}}\right).$$

$$\left(\textit{Answer for (i)}: F(s) = \left[1 + (1 - s)\sum_{i=1}^{k} s^{-\Sigma a_{ji}}\left\{\prod_{j=1}^{m} p_j^{-a_{ji}}\right\}\right]^{-1}\right).$$

EXERCISE 1.46. An urn contains five balls, one marked A, one B, one C, one D, and one R. A ball is drawn at random from the urn, the letter on it noted, and the ball returned to the urn. What is the expected number of times this must be done to obtain the sequence ABRACADABRA? [*Answer*: $5^4 + 5^{11}$.]

EXERCISE 1.47. One ball is to be added to the urn in Exercise 1.47. Which of the five letters (A, B, C, D, R) should be on it to minimize the expected number of draws needed to obtain the sequence ABRACADABRA?

EXERCISE 1.48. Given two urns, I and II, each containing five balls marked as in Exercise 1.46, what is the probability that the number of draws N_1 needed to obtain the sequence ABC in draws from urn I equals the number N_2 needed to obtain BBB in draws from urn II?

We conclude by deriving an interesting special result.

Suppose we are observing a sequence as described in Exercise 1.45. We stop the sequence at an arbitrary point and observe the pattern of the last g colors observed. There are m^g possible patterns; the probability of obtaining a pattern containing a_{jk} colors C_j ($j = 1, \ldots, m$) with $\sum_{j=1}^m a_{jk} = g$ is

$$\prod_{j=1}^{m} p_j^{a_{jk}}.$$

We then start the sequence again and wish to calculate the expected number of draws needed to repeat the pattern just observed, in the new sequence (i.e., ignoring the results of any previous drawings).

The required expected value is clearly the sum of

$$\left\{\prod_{j=1}^{m} p_j^{a_{jk}}\right\}\left\{\sum_{i=1}^{k}\left(\prod_{j=1}^{m} p_j^{-a_{ji}}\right)\right\} \tag{1.172}$$

over all m^g possible patterns.

In each expression (1.172) there is a term ($i = k$)

$$\left(\prod_{j=1}^{m} p_j^{a_{jk}}\right)\left(\prod_{j=1}^{m} p_j^{-a_{jk}}\right) = 1, \tag{1.173}$$

and there are m^g expressions (1.172) in the formula for the expected value. In all, the terms (1.173) contribute m^g to the expected value.

The remaining terms are of form $\prod_{j=1}^{m} p_j^{a_{jk}-a_{ji}}$. When taking all patterns with a critical point on the h ($<g$)th drawing in the pattern, these contribute $(\sum_{j=1}^{m} p_j)^{g-h} = 1$ to the sum. This is so for $h = 1, 2, \ldots, g - 1$. (We have already shown that, for $h = g$, the contribution is m^g.) The expected value is therefore $m^g + g - 1$. Note that this is an integer and does not depend on the p_j's.

It can be shown (Johnson, 1976) that the variance of the number of draws needed to repeat the pattern is

$$2\sum\left(\prod_{j=1}^{m} p_j^{a_{jk}}\right)\left\{\sum_{i=1}^{k} p_j^{-a_{ji}}\right\}^2 - m^{2g} - (4g - 3)m^g - 2(g - 1)^2, \tag{1.174}$$

the first summation being over all patterns of length g.

It is possible to express (1.174) in terms of the inverse power sums

$$\phi_h = \sum_{j=1}^{m} p_j^{-h}:$$

(i) For $g = 1$, the variance is $2\phi_1 - m(m + 1)$.
(ii) For $g = 2$, the variance is $2(\phi_1^2 + 2\phi_1) - m^4 - 5m^2 + 2m - 2$
(iii) For $g = 3$, the variance is $2\{\phi_1^3 + 2\phi_2 + (2m + 1)\phi_1\} - m^6 - 9m^3 + 6m - 8$.

EXERCISE 1.49. Verify that, for $g = 4$, the variance is

$$2\{\phi_1^4 + 2\phi_3 + 2\phi_1^2 + \phi_2 + 2(m^2 + 1)\phi_1\} - m^8 - 13m^4 + 2m^2 + 6m - 14.$$

References

Abramovitz, M. and Stegun, I. A. (Eds.) (1964) *Handbook of Mathematical Functions*, National Bureau of Standards, Applied Mathematics Series, No. 55, Washington, D.C.: U.S. Government Printing Office.

Abramson, M. (1970) Certain distributions of unlike objects into cells, *Math. Mag.*, **41**, 214–218.

Bernoulli, J. (1713) *Ars Conjectandi*, Impensis Thurnisiorum, Fratrun, Basileae.

Bizley, M. T. L. (1962) Patterns in repeated trials, *J. Inst. Actu., London*, **88**, 360–366.

Bizley, M. T. L. (1970) A generalization of Nanjundiah's identity, *Amer. Math. Mthly*, **77**, 863–865.

Bloxham, M. (1975) Letter to the editor: Pascal generalizations, *Am. Stat.*, **29** (1), 67.

Bose, S. N. (1924) Planck's law and the hypothesis of light quanta, *Z. Phys.*, **26**, 178–181 (translated by A. Einstein into German).

Collings, S. N. (1966) Number of arrangements, *Math. Gaz.*, **50**, 287–289.

Davis, H. T. (1933, 1935) *Tables of the Higher Mathematical Functions*, 2 vols., Bloomington, Ind.: Principia Press.

Downton, F. (1967) *A problem of sampling from an urn*; Department of Mathematics and Statistics, University of Birmingham (preprint).

Dvoretsky, A. and Motzkin, T. (1947) A problem of arrangements, *Duke Math. J.*, **14**, 305–313.

Einstein, A. (1924, 1925) Quanten theorie des idealen Gases, 1 and 2, *Preuss. Akad. Wiss. Phys.-Math. Kl. Sitzungsber.*, 261–267 (1924); 3–14 (1925).

Engen, S. (1976) A note on estimation of the species-area curve, *J. Cons. Int. Explor. Mer*, **36** (3), 286–288.

F (1962–) *Fibonacci Quart.*, published by the Fibonacci Association, St. Mary's College, Calif.

Gardner, M. (1969) Mathematical games, *Sci. Am.*, **220**(4), 124–126; **220**(5) 118–124.

Garside, G. R. (1971) A recursive approach to raffles with replacement, *Math. Gaz.*, **55**, 41–43.

Goldberg, K., Leighton, F. T., Newman, M. and Zuckerman, S. L. (1976) Tables of Binomial coefficients and Stirling numbers, *Journal of Res., Nat. Bur. Stand.* **80B**, 99–171.

Johnson, N. L. (1968) Repetitions, *Am. Math. Mon.*, **75**, 382–383.

Johnson, N. L. (1976) A return to repetitions, in *Essays in Probability and Statistics*, Ogawa Volume (Ed: S. Ikeda) Tokyo: Shinko Tsusho, 635–644.

Laplace, P. S. (1812) *Theorie Analytique des Probabilités*, 1st ed., Paris: Courcier.

Luke, Y. L. (1975) *Mathematical Functions and Their Approximations*, New York: Academic Press.

Matschinski, M. (1962) Nombre des séries à k places occupées par des objets appartenant à m classes quelconques prises parmi n classes fixées, *C. R. Acad. Sci.* (*Paris*), 3799–3801.

Narayana, T. V. (1959) A partial order and its applications to probability theory, *Sankhya*, **21**, 91–98.

Oakley, B. E. and Perry, R. L. (1965) A sampling process, *Math. Gaz.*, **49**, 42–44.

Paul, J. L. (1971) On the sum of the kth powers of the first n integers, *Am. Math. Non.*, **78**, 271–273.

Rashevsky, N. (1955) Note on a combinatorial problem in topological biology, *Bull. Math. Biophys.*, **17**, 45–50.

Riordan, J. (1964) The enumeration of election returns by number of lead positions, *Ann. Math. Stat.*, **35**, 369–379.

Schäfer, W. (1954) Das Mütungsproblem der Besetzungs-Verteilung, *Mitteilungsbl. Math. Stat.*, **5**, 1–38.

Seal, H. L. (1949) The historical development of the use of generating functions in probability theory, *Ver. Schweiz-Versicherungsmath.*, **49**, 209–228.

Sobel, M., Uppuluri, V. R. R. and Frankowski, K. (1975) *Dirichlet Distributions Type I*, Department of Statistics, University of Minnesota (preprint of Tables with an introduction).

Steck, G. P. (1974) Evaluation of some Steck determinants with applications, *Comm. Stat.*, **3**, 121–138.

Streefkerk, H. (1975) Van N ballen in drie, vier of vijf kastjes, *Nieuw. Tidschr. Wisk.*, **62**(3), 279–287.

Takacs, L. (1962a) A generalization of the ballot problem and its application in the theory of queues, *J. Am. Stat. Assoc.*, **57**, 327–337.

Takacs, L. (1962b) Ballot problems, *Z. Wahrscheinlichkeitstheorie*, **1**, 154–158.

Takacs, L. (1967) On a method of inclusion and exclusion, *J. Am. Stat. Assoc.*, **62**, 102–113.

Wright, E. M. (1967) Number of arrangements, *Math. Gaz.*, **51**, 305–307.

2

Some Special Distributions (Mostly via Urn Models)

2.1. INTRODUCTION

We have already seen, in Chapter 1, how urn models can be used in developing some basic ideas in probability theory. In this chapter we begin a systematic development of distribution theory using urn models. We obtain the most important distributions used in statistics in this way.

By their nature, urn models give rise only to discrete distributions. By the use of limiting cases, however, we can reach many important continuous distributions. Although we leave detailed study of limiting cases to Chapter 6, we make free use of elementary limiting processes in this and the succeeding chapters to extend the scope of our results.

A simple discrete distribution is obtained by supposing we have k balls, numbered $a_1, a_2, \ldots, a_k$ (with no two a's equal), in an urn and take a ball at random from the urn. The distribution of the random variable X representing the number on the chosen ball is

$$\Pr[X = a_j] = k^{-1} \qquad (j = 1, 2, \ldots, k).$$

If we take the a's to be equally spaced [e.g., $a_j = a + (j - 1)h$], we obtain a *discrete rectangular distribution*. Further specialization, by taking $a = 0$, $h = 1$ gives a distribution with

$$\Pr[X = j] = k^{-1} \qquad [j = 0, 1, \ldots, (k - 1)]. \tag{2.1}$$

The probability generating function (see Section 1.4.3) for this distribution is

$$k^{-1}\sum_{j=0}^{k-1} s^j = k^{-1}\frac{1-s^k}{1-s}. \tag{2.2}$$

EXERCISE 2.1. Obtain a general formula for the distribution of the sum S of the numbers obtained when one ball is drawn from each of m urns, each of which contains n balls numbered $0, 1, \ldots, n-1$. (This is a special case of the m-fold convolution of a discrete rectangular distribution.) [*Hint*: Use the method of undetermined coefficients (Section 1.1.3).]

$$\left[\textit{Answer}: \Pr[S = \alpha n + \theta] = \sum_{j=0}^{\min(\alpha, m)} (-1)^j \binom{m}{j}\binom{m + \theta + (\alpha - j)n - 1}{m-1}.\right]$$

2.2. BINOMIAL, NORMAL, POISSON, GAMMA, AND BETA DISTRIBUTIONS

2.2.1. Binomial Distributions

Suppose we have m balls in an urn, of which m' are white and m'' are black. If we draw a ball at random from the urn, each of the m balls is equally likely to be drawn. For each individual ball the probability of being drawn is m^{-1}. These events are mutually exclusive, since we draw only one ball. Hence formula (1.70) can be used, and the probability of drawing a white ball (W) is m'/m [since this corresponds to the occurrence of (at least) one event out of m' mutually exclusive events, each of which has probability m^{-1}].

Similarly, the probability of drawing a black ball (B) is $1 - m'/m = m''/m$.

If the drawn ball is *replaced* and then a ball again chosen at random from the urn, the results of the two draws can be represented as one of four mutually exclusive possibilities:

WW, with probability $(m'/m)^2$.
WB, with probability $(m'/m)(m''/m) = m'm''/m^2$.
BW, with probability $(m''/m)(m'/m) = m'm''/m^2$.
BB, with probability $(m''/m)^2$.

For convenience we write $m'/m = p$ and $m''/m = 1 - p = q$. Then the four above probabilities are, in order, p^2, pq, qp, and q^2.

If we do not take account of order of occurrence, but only of *number of times* of occurrence X of white balls, we have

$$\Pr[X = 2] = p^2$$
$$\Pr[X = 1] = 2pq$$
$$\Pr[X = 0] = q^2.$$

From (1.96) we recognize these values as representing the distribution of the (discrete) random variable X.

Suppose now that the process of drawing a ball, observing its color, and replacing it, is repeated n times.

Each series of drawings in which white balls are drawn on x occasions, and black balls on $(n - x)$ occasions has the probability

$$p^x q^{n-x}.$$

The number of different series with exactly x white and $(n - x)$ black drawings is equal to the number of ways of choosing the x drawings (out of n in all) where white is observed. This is $\binom{n}{x}$ (see Section 1.1.1).

Hence the distribution of X, the number of white balls drawn, is

$$\Pr[X = x] = \binom{n}{x} p^x q^{n-x} \qquad (x = 0, 1, \ldots, n). \tag{2.3}$$

The probabilities in (2.3) are the terms of the binomial expansion of $(q + p)^x$ [see (1.7)]. For this reason the distribution is called a *binomial distribution.*

Since $q + p = 1$, it is clear that the sum of $\Pr[X = x]$ over $x = 0, 1, \ldots, n$ is equal to 1.

A series of independent trials in which the probability of an event remains constant may be called a *binomial series*; the name *Bernoulli series* is also used, after J. Bernoulli who first derived the binomial distribution (see also Section 1.2).

A useful formula is

$$\Pr[X \geq k] = I_p(k, n - k + 1) \qquad \text{(for integer } k\text{)}, \tag{2.4}$$

where $I_p(\cdot, \cdot)$ is the incomplete beta function ratio defined in (1.54).

The rth factorial moment of X is

$$\begin{aligned} \mu_{(r)}(X) = E[X^{(r)}] &= \sum_{j=0}^{n} j^{(r)} \binom{n}{j} p^j q^{n-j} \\ &= n^{(r)} p^r \sum_{i=0}^{n-r} \binom{n-r}{i} p^i q^{n-r-i} \\ &= n^{(r)} p^r (q + p)^{n-r} \\ &= n^{(r)} p^r. \end{aligned} \tag{2.5}$$

In particular, we find

$$E[X] = np, \tag{2.6a}$$

$$\operatorname{var}(X) = npq, \tag{2.6b}$$

$$\alpha_3(X) = \frac{q - p}{\sqrt{npq}}, \tag{2.6c}$$

$$\alpha_4(X) = 3 + \frac{1 - 6pq}{npq}. \tag{2.6d}$$

By considering the ratio

$$\frac{\Pr[X = x + 1]}{\Pr[X = x]} = \frac{n - x}{x + 1} \frac{p}{q}$$

we find that there is a mode at x equal to the integer part of np (if np is an integer, then $\Pr[X = np + 1] = \Pr[X = np]$, with $\Pr[X = x]$ increasing for $x \leq np$ and decreasing for $x \geq np + 1$).

Example 2.1. Among many possible variants on the binomial distribution, we discuss in a little detail one proposed by Consul and Mittal (1975). These authors wished, in particular, to construct a model in which the player (the person drawing balls from the urn) would be able to influence the results by choice of a suitable strategy.

They consider a situation with four urns, initially containing white and black balls as set out below:

	Number of Balls	
Urn	White	Black
A	a	b
B	—	b
C	a	—
D	a	b

The player is allowed to select an integer, say k, from $0, 1, 2, \ldots, n$. Once this integer has been selected, $(n - k)\theta$ white balls are added to each of B and D, and $k\theta$ black balls are added to each of C and D, so that the constitution of the urns is now:

Urn	Number of Balls: White	Black
A	a	b
B	$(n-k)\theta$	b
C	a	$k\theta$
D	$a+(n-k)\theta$	$b+k\theta$

(θ is a known positive integer.)

The player now draws at random a ball from urn A. If he draws a white ball, then he draws a ball from urn B; if he draws a black ball first, then he draws a ball from urn C. If the two balls so far drawn are of *different* colors, he then draws n balls *with* replacement from urn D. If they are not of different colors, the player's sequence of drawings terminates.

The player wins if he (1) gets so far as to make the drawing from urn D *and* (2) draws exactly k white balls among the n taken from urn D. (Remember that the number k was originally chosen by the player.)

The probability that the player gets so far as to make the n drawings from urn D is

$$\frac{a}{(a+b)}\cdot\frac{b}{(b+(n-k)\theta)}+\frac{b}{(a+b)}\cdot\frac{a}{(a+k\theta)}$$
$$=\frac{ab(a+b+n\theta)}{(a+b)(a+k\theta)\{b+(n-k)\theta\}}.$$

Since the proportion of white balls in urn D is $(a+k\theta)(a+b+n\theta)^{-1}$ and drawing is with replacement, the probability of obtaining k white balls in n draws is given by (2.3), with x replaced by k and p by $(a+k\theta)/(a+b+n\theta)$, which is

$$\binom{n}{k}(a+k\theta)^k\{b+(n-k)\theta\}^{n-k}(a+b+n\theta)^{-n},$$

and the probability of winning is therefore

$$P_k=\binom{n}{k}\frac{ab}{a+b}\frac{1}{a+b+n\theta}\left\{\frac{a+k\theta}{a+b+n\theta}\right\}^{k-1}\left\{\frac{b+(n-k)\theta}{a+b+n\theta}\right\}^{n-k-1}. \tag{2.7}$$

It is remarkable that $P_0 + P_1 + \cdots + P_n = 1$, and so (2.7), can be used to define a distribution, called the *quasi-binomial distribution* by Consul and Mittal (1975).

EXERCISE 2.2 (Consul, 1974). Show that the same probability of winning as that in Example 2.1 is obtained in the following situation. There are two urns, urn I containing a white balls, and urn II, b black balls. The player chooses an integer k ($0 \le k \le n$), and then $k\theta$ black balls are added to each of the two urns, while $(n - k)\theta$ white balls are added to urn II.

The player then draws at random one ball from urn I. If it is black, he loses. If it is white he takes a sample of size n (with replacement) from the balls in urn II. He wins if there are exactly k white balls among the n chosen.

We now consider two very important limiting cases.

2.2.2. Normal (or Gaussian) Distributions

The variable

$$X' = \frac{X - np}{\sqrt{npq}}$$

is a standardized binomial variable (see (2.3), (2.6a) and (2.6b)). *Laplace's theorem* states that

$$\lim_{n\to\infty} \Pr[\alpha < X' < \beta] = (\sqrt{2\pi})^{-1} \int_\alpha^\beta e^{-(1/2)u^2}\, du = \Phi(\beta) - \Phi(\alpha), \tag{2.8}$$

where $\Phi(\cdot)$ is defined in (1.43). This means that the limiting distribution of X' as the sample size $n \to \infty$ is that of a random variable U for which

$$\Pr[U < u] = \Phi(u). \tag{2.9}$$

This distribution is a *unit normal* (or *Gaussian*) *distribution*. The variable U is called a *unit normal variable*. It is in fact standardized. One might expect this, because X' is standardized (for every n), but it does not follow automatically.

EXERCISE 2.3. Show that the rth moment of U about zero is

$$\mu_r'(X) = \begin{cases} 0 & \text{if } r \text{ is odd} \\ \pi^{-\frac{1}{2}} 2^{(\frac{1}{2})r}\Gamma(\frac{1}{2}(r+1)) & \text{(if } r \text{ is even).} \end{cases} \tag{2.10}$$

Using the duplication formula (1.47), or otherwise, show that for r even $\mu_r' = (r-1)(r-3)\cdots 5 \cdot 3 \cdot 1$.

Hence show that

$$E[U] = 0, \qquad \text{var}[U] = 1, \qquad \alpha_3(U) = 0, \qquad \alpha_4(U) = 3. \tag{2.11}$$

The family of normal distributions is generated by the random variables

$$Y = \eta + \sigma U \qquad (\sigma > 0).$$

The distribution of Y has the same shape as that of U, but $E[Y] = \eta$ and $\text{var}(Y) = \sigma^2$.

All cumulants [see (1.152)] of order greater than 2 are zero. Normal distributions are the only ones with this property, so it *characterizes* normality. This result is used in Section 6.2.3.

2.2.3. Poisson Distributions

If, in (2.3), n is increased and p decreased in such a way that $np = \theta$ remains constant, the limiting value of (2.3) is

$$\frac{e^{-\theta}\theta^x}{x!}.$$

The distribution of a random variable X with

$$\Pr[X = x] = \frac{e^{-\theta}\theta^x}{x!} \qquad (x = 0, 1, \ldots) \tag{2.12}$$

is called a *Poisson distribution. X* is called a *Poisson variable.*

The limiting process corresponds to increasing m, the number of balls in the urn, while keeping the number of white balls m' constant (and so decreasing $p = m'/m$) and simultaneously increasing the number of draws n in the same proportion.

EXERCISE 2.4. For a random variable X with the Poisson distribution (2.12) show that

$$\mu_{(r)}(X) = \theta^r, \tag{2.13}$$

hence that

$$\mu'_r(X) = \sum_{j=0}^{r} \frac{\Delta^j \mathbf{0}^r}{j!} \theta^r.$$

EXERCISE 2.5. Show that $\kappa_r(X) = \theta$ for all r.

2.2.4. Gamma and Beta Distributions

If $U_1, U_2, \ldots, U_\nu$ are ν independent unit normal variables (i.e., each with distribution 2.9), then the sum of their squares

$$Y = \sum_{j=1}^{\nu} U_j^2$$

has the density function

$$p_Y(y) = \{2^{(1/2)\nu}\Gamma(\tfrac{1}{2}\nu)\}^{-1} y^{(1/2)\nu - 1} e^{-(1/2)y} \qquad (y > 0). \tag{2.14}$$

This is called a χ^2 *distribution with v degrees of freedom*. The χ^2 distributions belong to a family of distributions with density functions

$$p_Y(y) = \{\beta^\alpha \Gamma(\alpha)\}^{-1} y^{\alpha-1} e^{-y/\beta} \qquad (y > 0; \alpha, \beta > 0). \tag{2.15}$$

Note that 2α is not necessarily an integer. For such functions the cumulative distribution function is

$$\begin{aligned} F_Y(y) &= \{\beta^\alpha \Gamma(\alpha)\}^{-1} \int_0^y t^{\alpha-1} e^{-t/\beta}\, dt \\ &= I_\alpha\left(\frac{y}{\beta}\right), \end{aligned} \tag{2.16}$$

where $I_\alpha(\cdot)$ is the incomplete gamma function ratio [see (1.49)]. These distributions are known as *gamma distributions*.

In the special case $\alpha = 1$, the density function is

$$p_Y(y) = \beta^{-1} e^{-y/\beta} \qquad (0 < y). \tag{2.17}$$

This distribution is an *exponential distribution*. The χ^2 distribution with 2 degrees of freedom is an exponential distribution with $\beta = 2$.

Note that, if Y has the distribution (2.17), then

$$\Pr[Y > y] = e^{-y/\beta}. \tag{2.18}$$

Since, for any $\eta > 0$,

$$\begin{aligned} \Pr[Y > \eta + y \mid Y > \eta] &= \frac{\Pr[Y > \eta + y]}{\Pr[Y > \eta]} \\ &= e^{-y/\beta}, \end{aligned} \tag{2.19}$$

we see that the conditional distribution of $Y - \eta$, given Y is greater than η, is the same as the unconditional distribution of Y.

If Y_1 and Y_2 are independent gamma variables with parameters α_1, β and α_2, β, respectively, then

$$Z = \frac{Y_1}{Y_1 + Y_2}$$

has the density function

$$p_Z(z) = \frac{1}{B(\alpha_1, \alpha_2)} z^{\alpha_1 - 1}(1 - z)^{\alpha_2 - 1} \qquad (0 < z < 1). \tag{2.20}$$

We have

$$F_Z(z) = \frac{1}{B(\alpha_1, \alpha_2)} \int_0^z t^{\alpha_1 - 1}(1 - t)^{\alpha_2 - 1}\, dt$$

$$= I_z(\alpha_1, \alpha_2), \tag{2.21}$$

where $I_z(\cdot, \cdot)$ is the incomplete beta function ratio defined in (1.54).

These distributions are known as (*standard*) *beta distributions*.

EXERCISE 2.6.

(i) Show that for a gamma distribution with parameters α and β the rth moment about zero is

$$\beta^r \alpha^{[r]} \qquad (r \text{ is a positive integer}). \tag{2.22}$$

(ii) Using (i), show that the mean and variance of a χ^2 with v degrees of freedom are v, $2v$, respectively.

(iii) Obtain expressions for $\sqrt{\beta_1}$ and β_2 (see (1.131)) of a χ^2 with v degrees of freedom.

EXERCISE 2.7.

(i) Show that for a beta distribution with parameters α_1 and α_2 the rth moment about zero is

$$\frac{\Gamma(\alpha_1 + r)}{\Gamma(\alpha_1)} \frac{\Gamma(\alpha_1 + \alpha_2)}{\Gamma(\alpha_1 + \alpha_2 + r)} = \frac{\alpha_1^{[r]}}{(\alpha_1 + \alpha_2)^{[r]}}. \tag{2.23}$$

(ii) Obtain an expression for the variance of this distribution.

EXERCISE 2.8. Show that if X has the gamma distribution (2.15) with $\beta = 1$ and α a positive integer, and Y has the Poisson distribution (2.12), then

$$\Pr[X > \theta] = \Pr[Y < \alpha].$$

2.3. HYPERGEOMETRIC DISTRIBUTIONS

The system of sampling described in Section 2.2 is called *sampling with replacement*. If at each stage the ball drawn is not returned to the urn, the procedure is called *sampling without replacement*. The distribution of X is no longer binomial, as in (2.3). In order to find the distribution of X (the number of white balls) we have to find the probability that a *subset* of *n balls* chosen at random from the m balls in the urn will contain exactly x white balls. There are $\binom{m}{n}$ possible subsets, and they each have the same probability of being chosen. This common probability is therefore $\binom{m}{n}^{-1}$.

Of these $\binom{m}{n}$ subsets, each containing n balls in all, how many contain exactly x white balls? There are $\binom{m'}{x}$ ways of choosing x white balls from the m' white balls in the urn, and $\binom{m''}{n-x}$ ways of choosing the remaining $(n-x)$ black balls in the subset from the m'' black balls in the urn. So the number of subsets containing exactly x white balls is

$$\binom{m'}{x}\binom{m''}{n-x},$$

and

$$\Pr[X = x] = \frac{\binom{m'}{x}\binom{m''}{n-x}}{\binom{m}{n}}. \tag{2.24}$$

This formula is valid for all integer values of x such that, for each symbol of form $\binom{a}{b}$ in (2.24), $a \geq b$. This means that

$$0 \leq x \leq m' \qquad \text{and} \qquad 0 \leq n - x \leq m'';$$

that is,

$$\max(0, n - m'') \leq x \leq \min(m', n).$$

Example 2.2. The following application of hypergeometric distributions is due to Murty (1975). [The problem was posed (orally) by L. F. Yancey and R. J. Reynolds.]

An urn contains n balls, and each ball has one number on it. Suppose that there are n_r numbers which each appear on exactly r of the balls. For example, if there are $n = 7$ balls numbered 1, 1, 2, 2, 3, 3, 4, then $n_1 = 1$, $n_2 = 3$. It is supposed that no number can appear on more than k balls, so $n_r = 0$ for $r > k$. It is required to estimate $n_1, n_2, \ldots, n_k$.

We note that, since each ball has one number on it,

$$\sum_{j=1}^{k} j n_j = n. \tag{2.25}$$

We take a random sample of m balls (without replacement) and note which numbers appear on exactly t of the chosen balls. Suppose there are M_t such numbers ($t = 1, 2, \ldots, k$). Analogously to (2.15) the sample values satisfy the equation

$$\sum_{t=1}^{k} tM_t = m.$$

Then the conditional probability that a number will appear on t balls in the sample, given that it appears on r balls in the population, is

$$P_{rt} = \begin{cases} 0 & \text{if} \quad t > r, \\ \dfrac{\dbinom{r}{t}\dbinom{n-r}{m-t}}{\dbinom{n}{m}} & \text{if} \quad t \le r. \end{cases}$$

Now

$$E[M_t] = \sum_{r=1}^{k} n_r P_{rt},$$

and so the solutions $\hat{N}_1, \hat{N}_2, \ldots, \hat{N}_k$ of the system of linear equations

$$\sum_{r=1}^{k} \hat{N}_r P_{rt} = M_t \qquad (t = 1, \ldots, k)$$

will be unbiased estimators of $n_1, n_2, \ldots, n_r$, respectively.

EXERCISE 2.9. Show that for distribution (2.24) the rth (descending) factorial moment of X is

$$E[X^{(r)}] = \frac{n^{(r)}m'^{(r)}}{m^{(r)}}. \tag{2.26}$$

Hence show that

$$E[X] = \frac{nm'}{m} = np \qquad \left(\text{where } p = \frac{m'}{m} \quad \text{and} \quad q = 1 - p\right)$$

$$\operatorname{var}[X] = \frac{m-n}{m-1}npq \tag{2.27}$$

$$\alpha_3(X) = \frac{q-p}{\left[\left(\dfrac{m-n}{m-1}\right)npq\right]^{\frac{1}{2}}} \cdot \frac{m-2n}{m-2}.$$

EXERCISE 2.10. An urn contains m white balls and n black balls. A sample of two balls is chosen without replacement. Under what conditions is the probability that both the chosen balls will be of the same color greater than $\frac{1}{2}$? [*Answer*: $(m - n)^2 > m + n$.]

V'ndev (1974) has also considered this problem when r (>2) balls are chosen in the sample.

Example 2.3. (S. Bernard, 1976) An urn contains b_0 black and w_0 white balls. An "operation" consists of adding r black balls and then withdrawing at random (without replacement) r balls from the $r + b_0 + w_0$ in the urn.

It is required to find the distribution of the number of white balls, W_t say, remaining in the urn after t operations.

We first note that after each operation the number of balls in the urn is $b_0 + w_0$, and W_t cannot exceed w_0, since no white balls are ever added to the urn.

After one operation it is easy to see that

$$\begin{aligned}\Pr(k; w_0) = \Pr[W_1 = k] &= \Pr[\text{choosing } (r-k) \text{ white balls in a sample of } r \text{ from a mixture of } w_0 \text{ white balls and } (b_0 + r) \text{ black balls}] \\ &= \frac{\binom{w_0}{r-k}\binom{b_0+r}{k}}{\binom{w_0+b_0+r}{r}} \qquad (k = 0, 1, \ldots, w_0)\end{aligned}$$

[using (2.24)].

The recurrence formula

$$\Pr[W_t = k] = \sum_{j=k}^{w_0} \Pr[W_{t-1} = j]\Pr(k; j) \qquad (k = 0, 1, \ldots, w_0)$$

can be used repeatedly to obtain values of $\Pr[W_t = k]$ for general t and r.

Using the formula (2.26) for the first two factorial moments of the hypergeometric distribution we see that

$$E[W_1] = \frac{rw_0}{w_0 + b_0 + r},$$

$$E[W_1(W_1 - 1)] = \frac{r(r-1)w_0(w_0-1)}{(w_0 + b_0 + r)(w_0 + b_0 + r - 1)}.$$

The recurrence formula shows that

$$E[W_t] = E\left[\frac{rW_{t-1}}{w_0 + b_0 + r}\right] = \frac{r}{w_0 + b_0 + r} E[W_{t-1}] = \left(\frac{r}{w_0 + b_0 + r}\right)^t w_0$$

$$E[W_t(W_t - 1)] = \frac{r(r-1)}{(w_0 + b_0 + r)(w_0 + b_0 + r - 1)} E[W_{t-1}(W_{t-1} - 1)]$$

$$= \left\{\frac{r(r-1)}{(w_0 + b_0 + r)(w_0 + b_0 + r - 1)}\right\}^t w_0(w_0 - 1).$$

Generally,

$$E(W_t^{(s)}) = \left\{\frac{r^{(s)}}{(w_0 + b_0 + r)^{(s)}}\right\}^t w_0^{(s)}.$$

Bernard (1976) applied this model to studies where radionucleides (white balls) are injected into animals and are subject to continual attenuation while being replenished by a fairly constant flow of stable nucleides (black balls).

2.4. NEGATIVE BINOMIAL DISTRIBUTIONS

Suppose we have an urn containing np white and $n(1-p)$ black balls. We draw balls at random one at a time from the urn and replace them. What is the distribution of the number N of drawings needed to draw k white balls.

$$\Pr[N = j] = \Pr[(k-1) \text{ white balls among first } (j-1) \text{ drawings}]$$

$$\times \Pr[\text{white ball at } j\text{th drawing}]$$

$$= \binom{j-1}{k-1} p^{k-1}(1-p)^{j-k} p$$

$$= \binom{j-1}{k-1} p^k (1-p)^{j-k} \qquad (j = k, k+1, \ldots). \tag{2.28}$$

Note that $\Pr[N = j]$ is the coefficient of t^j in the expansion of

$$\left(\frac{p}{1-p}\right)^k \left(1 - \frac{t}{1-p}\right)^k$$

in powers of t. This is a negative binomial expansion, and the distribution is called a *negative binomial distribution*. It is conventional to express the distribution in terms of parameters $Q = p^{-1}, P = (1 - p)/p$ (so that $Q - P = 1$). Then

$$\Pr[N = j] = \binom{j-1}{k-1} P^{-k} \left(\frac{P}{Q}\right)^j, \qquad (2.28')$$

and this is the coefficient of t^j in the expansion of

$$P^{-k}(1 - PQ^{-1}t)^{-k} = (Q - Pt)^{-k}.$$

The rth (descending) factorial moment of N is

$$E[N^{(r)}] = k^{[r]}P^r. \qquad (2.29)$$

[Note that it is the *ascending* factorial $k^{[r]}$ and not the descending factorial $k^{(r)}$ that appears in (2.29).]

The negative binomial distribution, as derived above, is an example of a *waiting-time* distribution. We encounter several other waiting-time distributions later in this book (see, e.g., Section 3.2.3). From our derivation it is necessary for k to be an integer. However, formula (2.29) defines a proper distribution for *any* positive number k. In order to obtain a variable taking all (nonnegative) integer values we may use

$$X = N - k$$

as the variate in place of N. The random variable X takes values 0, 1, 2, ...

In the special case $k = 1$ we have

$$\Pr[N = j] = p(1 - p)^{j-1}. \qquad (2.30)$$

These terms form a geometric series [ratio $(1 - p)$], and the distribution is called a *geometric distribution*.

EXERCISE 2.11. An urn contains m white and n black balls. Balls are drawn one at a time *until* a white ball appears. The balls are replaced after each drawing. What is the expected value and the variance of the number X of balls chosen? [*Answer*: $E(X) = 1 + n/m$; $\text{var}(X) = [n(m + n)]/m^2$.]

2.5. NEGATIVE HYPERGEOMETRIC DISTRIBUTIONS

Suppose we have b black balls and w white balls in an urn. We draw balls one at a time from the urn, without replacement. What is the distribution of the number R of draws needed to obtain a prespecified number k of black balls?

Using an argument similar to that employed for the negative binomial distribution we find that

$$\begin{aligned}\Pr[R = r] &= \Pr[(k-1) \text{ black balls in the first } (r-1) \text{ draws}]\\ &\quad \times \Pr[\text{drawing a black ball from an urn containing } (b-k+1) \text{ black and } (w-r+k) \text{ white balls}]\\ &= \binom{r-1}{k-1} \frac{b(b-1)\cdots(b-k+2)w(w-1)\cdots(w-r+k-1)}{(b+w)(b+w-1)\cdots(b+w-r+2)}\\ &\quad \times \frac{b-k+1}{b+w-r+1}\\ &= \binom{r-1}{k-1} \frac{b^{(k)}w^{(r-k)}}{(b+w)^{(r)}} \qquad (r = k, k+1, \ldots).\end{aligned} \tag{2.31}$$

The distribution of R is a *negative hypergeometric distribution* with parameters b, w, k.

The αth *ascending* factorial moment of R is

$$\mu_{[\alpha]}(R) = E[R^{[\alpha]}] = \frac{k^{[\alpha]}(b+w+1)^{[\alpha]}}{(b+1)^{[\alpha]}}. \tag{2.32}$$

In particular,

$$E[R] = \frac{k(b+w+1)}{b+1}$$

$$\operatorname{var}[R] = k(b+w+1)w(b+1-k)(b+1)^{-2}(b+2)^{-1}. \tag{2.33}$$

EXERCISE 2.12. For the case $b = 5$, $w = 5$, $k = 2$, calculate the distribution function of R. Compare it with the negative binomial distribution with $p = \frac{1}{2}$ (i.e., $Q = 2$, $P = 1$), $k = 2$.

2.6. POWER SERIES AND FACTORIAL SERIES DISTRIBUTIONS

The following brief summaries are given for reference purposes.

2.6.1. Power Series Distributions

Any discrete distribution that can be written in the form

$$\Pr[X = j] = \frac{a_j \theta^j}{f(\theta)} \qquad (j = 0, 1, \ldots), \tag{2.34}$$

where the a_j's are nonnegative constants and $f(\theta) = \sum_{j=0}^{\infty} a_j\theta^j$ is finite, is a *power series distribution.* Patil (1962) uses the term *generalized power series* distribution for this distribution.

The probability generating function corresponding to (2.34) is

$$E[s^X] = \frac{f(s\theta)}{f(\theta)}. \tag{2.35}$$

EXERCISE 2.13.

(i) Show that the factorial moment generating function corresponding to (2.34) is $f((s+1)\theta)/f(\theta)$.

(ii) Obtain expressions for the expected value and variance of this quantity in terms of $f(\theta)$ and its derivatives.
[*Answer*: $f'(\theta)/f(\theta)$, $[f(\theta)\{f'(\theta) + f''(\theta)\} - \{f'(\theta)\}^2]/\{f(\theta)\}^2$.]

(iii) Show that

$$\mu_{(r)}(X) = \frac{f^{(r)}(\theta)}{f(\theta)}. \tag{2.36}$$

where $f^{(r)}$ denotes the rth derivative.

The class of power series distributions includes Poisson, binomial, and negative binomial distributions.

If

$$f(\theta) = \sum_{j=0}^{\infty} \frac{\theta^j}{j!} D^j f(\mathbf{0}) \qquad \text{[see (1.32)]},$$

then

$$a_j = \frac{1}{j!} D^j f(\mathbf{0}).$$

If this is so,

$$E[g(X)] = \sum_{j=0}^{\infty} \frac{\theta^j}{j!} \frac{g(j)D^j f(\mathbf{0})}{f(\theta)}.$$

Putting $g(X) = \{D^X g_1(\mathbf{0})\}/\{D^X f(\mathbf{0})\}$, we have

$$E[g(X)] = \frac{g_1(\theta)}{f(\theta)}, \tag{2.37}$$

provided

$$\sum_{j=0}^{\infty} \frac{\theta^j}{j!} D^j g_1(\mathbf{0}) = g_1(\theta).$$

Taking $g_1(\theta) = h(\theta)f(\theta)$, we see that

$$\frac{D^X(h(\mathbf{0})f(\mathbf{0}))}{D^X(f(\mathbf{0}))}$$

is an unbiased estimator of $h(\theta)$ [provided $\sum_{j=0}^{\infty} (\theta^j/j!)D^j(h(\mathbf{0})f(\mathbf{0})) = h(\theta)f(\theta)$].

2.6.2. Factorial Series Distributions

Any proper discrete distribution that can be written in the form

$$\Pr[X = j] = \frac{\theta^{(j)}}{j!}\frac{\Delta^j f(\mathbf{0})}{f(\theta)} \qquad (j = 0, 1, 2, \ldots) \tag{2.38}$$

is called a *factorial series distribution* (e.g., Berg, 1974, 1975).

The fact that (2.38) represents a proper distribution implies that

1. $f(\theta) = \sum_{j=0}^{\infty} \frac{\theta^{(j)}}{j!} \Delta^j f(\mathbf{0})$ (2.39)

[cf. (1.20); (2.38) is certainly satisfied if $f(\theta)$ is a polynomial in θ, or θ is an integer].

2. $\theta^{(j)}\Delta^j f(\mathbf{0}) \geq 0$ for all $j = 0, 1, 2, \ldots$.
3. If θ is a positive integer, then $\Pr[X = j] = 0$ for all $j > \theta$.

Note that, since

$$E[g(X)] = \sum_{j=0}^{\infty} \frac{\theta^{(j)}}{j!}\frac{g(j)\Delta^j f(\mathbf{0})}{f(\theta)},$$

by taking $g(X) = \{\Delta^X g_1(\mathbf{0})\}/\{\Delta^X f(\mathbf{0})\}$, we have

$$E\left[\frac{\Delta^X g_1(\mathbf{0})}{\Delta^X f(\mathbf{0})}\right] = \frac{g_1(\theta)}{f(\theta)} \tag{2.40}$$

[provided $g_1(\theta) = \sum_{j=0}^{\infty} (\theta^{(j)}/j!)\Delta^j g_1(\mathbf{0})$].

Hence

$$\frac{\Delta^X(h(\mathbf{0})f(\mathbf{0}))}{\Delta^X f(\mathbf{0})}$$

is an unbiased estimator of $h(\theta)$.

The probability generating function corresponding to (2.38) can be written

$$\frac{\sum_{j=0}^{\infty} (\theta^{(j)}/j!)(s\Delta)^j f(\mathbf{0})}{f(\theta)} = \frac{(1 + s\Delta)^\theta f(\mathbf{0})}{f(\theta)}. \tag{2.41}$$

By repeated differentiation with respect to s (and putting $s = 1$) we obtain the following expression for the rth (descending) factorial moment of X:

$$\mu_{(r)}(X) = \frac{\theta^{(r)}\Delta^r f(\theta - r)}{f(\theta)}. \tag{2.42}$$

EXERCISE 2.14. Show that the factorial moment generating function corresponding to (2.38) is

$$\frac{\{1 + (1 + s)\Delta\}^\theta f(\mathbf{0})}{f(\theta)}.$$

2.6.3. Inverse Factorial Series Distributions; Waring Distributions

Distributions that can be written in the form

$$\Pr[X = j] = \frac{a_j}{\theta^{[j+1]}} f(\theta), \tag{2.43}$$

with

$$f(\theta) = \int_0^1 t^{\theta - 1}\phi(t)\, dt$$

and

$$\phi(t) = \sum_{j=0}^{\infty} \left(\frac{a_j}{j!}\right)(1 - t)^j,$$

it being supposed that the series converges, are called *inverse factorial distributions.* They are also called *Irwin distributions*, after J. O. Irwin (1963, 1965, 1975) who has studied them very thoroughly.

Distributions of this class for which $a_j = a^{[j]}$ $(a > 0)$ are called *Waring distributions.* In this case we take $\theta > a$ and find $f(\theta) = \theta - a$, so that (2.43) becomes

$$\Pr[X = j] = \frac{(\theta - a)a^{[j]}}{\theta^{[j+1]}}. \tag{2.44}$$

The probability generating function corresponding to (2.44) is

$$(1 - a\theta^{-1})F(a, 1; \theta + 1; s), \tag{2.45}$$

where $F(\cdot)$ is the hypergeometric function defined in (1.58). Waring distributions are thus in the general class of hypergeometric distributions. The rth factorial moment is

$$\mu_{(r)}(X) = \frac{r!a^{[r]}}{(\theta - a - 1)^{(r)}}. \tag{2.46}$$

If $\theta < a + 1 + r$, $\mu_{(r)}$ is infinite.

EXERCISE 2.15. (Irwin, 1975). Show that, for distribution (2.44),

$$\Pr[X \geq x] = \frac{a^{[x]}}{\theta^{[x]}}.$$

The *generalized* Waring distribution (Irwin, 1968, 1975) is constructed by introducing a further (integer-valued) parameter ϕ. It is defined by

$$\Pr[X = j] = \frac{(\theta - a)^{[\phi]}}{\theta^{[\phi]}} \frac{a^{[j]}\phi^{[j]}}{(\theta + \phi)^{[j]}}, \tag{2.47}$$

and its probability generating function is

$$\frac{(\theta - a)^{[\phi]}}{\theta^{[\phi]}} F(a, \phi; \theta + \phi; s) \tag{2.48}$$

[$\phi = 1$ gives the Waring distribution (2.44) and its pgf (2.45)].

The rth factorial moment is

$$\mu_{(r)}(X) = \frac{\phi^{[r]}a^{[r]}}{(\theta - a - 1)^{(r)}}. \tag{2.49}$$

Note that (2.47), (2.48), and (2.49) reduce to (2.44), (2.45), and (2.46), respectively, if $\phi = 1$.

EXERCISE 2.16. In what sense may the generalized Waring distributions be regarded as generalizations of hypergeometric distributions?

2.7. MULTIVARIATE DISTRIBUTIONS

Multivariate discrete distributions can be generated by methods analogous to those described in Sections 2.2 through 2.6, using urns containing balls of *more* than two different colors. We describe the derivations relatively briefly.

2.7.1. Multinomial Distributions

We have an urn containing m balls, of which m_1 have color $\mathscr{C}_j$ (for $j = 1, \ldots, k$), with $m_1 + m_2 + \cdots + m_k = m$. We now draw n balls at random from the urn one at a time, replacing the ball drawn each time after noting its color. Let $N_1, N_2, \ldots, N_k$ denote the numbers of balls of color $\mathscr{C}_1, \mathscr{C}_2, \ldots, \mathscr{C}_k$, respectively, among the n drawn balls. What is the probability of the event

$$\bigcap_{j=1}^{k} (N_j = n_j)?$$

For any one ordered sequence of drawings with this many balls of each color, the probability of occurrence is $\prod_{j=1}^{k} (m_j/m)^{n_j} = \prod p_j^{n_j}$, where $p_j = m_j/m$ denotes the probability of drawing a ball of color $\mathscr{C}_j$ from the urn at any one draw. The number of different such orders is the multiple combination

$$\binom{n}{n_1\ n_2 \cdots n_k} = \frac{n!}{\prod_{j=1}^{k} n_j!}.$$

Hence

$$\Pr\left[\bigcap_{j=1}^{k} (N_j = n_j)\right] = n! \prod_{j=1}^{k} \left(\frac{p_j^{n_j}}{n_j!}\right). \tag{2.50}$$

The joint distribution of $N_1, N_2, \ldots, N_k$ is called *multinomial* (with parameters $n, p_1, p_2, \ldots, p_k$). Moments can be evaluated from the formula

$$E\left[\prod_{j=1}^{k} N_j^{(\alpha_j)}\right] = n^{\left(\sum_{j=1}^{k} \alpha_j\right)} \prod_{j=1}^{k} p_j^{\alpha_j}. \tag{2.51}$$

In particular,

$$\begin{aligned} E[N_j] &= np_m, \qquad \operatorname{var}(N_j) = np_j(1 - p_j),\\ \operatorname{cov}(N_i, N_j) &= -np_i p_j,\\ \operatorname{corr}(N_i, N_j) &= -\{p_i p_j (1 - p_i)^{-1}(1 - p_j)^{-1}\}^{1/2}. \end{aligned} \tag{2.52}$$

The conditional distribution of a subset $N_{a_1}, N_{a_2}, \ldots, N_{a_t}$, given that the remaining $(k - t)$ variables have specified values, summing to S, is *also* multinomial with parameters $(n - S)$, $p_{a_1}p^{-1}, p_{a_2}p^{-1}, \ldots, p_{a_t}p^{-1}$, where $p = \sum_{i=1}^{t} p_{a_i}$. Also, the marginal distribution of N_j is binomial with parameters n, p_j. The moments of N_j, given in (2.52), are consistent with this fact.

It is interesting to note that the joint distribution of $N_1, N_2, \ldots, N_k$ is the same as that of the joint distribution of independent *Poisson variables* $Y_1, Y_2, \ldots, Y_k$ with expected values $np_1, np_2, \ldots, np_k$ and subject to the condition $Y_1 + Y_2 + \cdots + Y_k = n$. We use this result in the next chapter and also in Chapter 6.

If $X_1, X_2, \ldots, X_m$ have a symmetric joint multinomial distribution with parameters $n, m^{-1}, m^{-1}, \ldots, m^{-1}$, then

$$\Pr\left[\bigcap_{j=1}^{m} (X_j \geq k)\right] = I_{m^{-1}}^{(m)}(k, n). \tag{2.53}$$

[The quantity on the right-hand side is defined in (1.57).]

This is a generalization of (2.4). [Note that $I_p^{(1)}(k, n) = I_p(k, n - k + 1)$.] Tables of $I_{m^{-1}}^{(m)}(k, n)$ to 10 decimal places for $m = 2(1)10$, $k = 1(1)10$ are available in Sobel et al. (1975). Values of n $(\geq mk)$ increase by 5 until the first tabulated value to exceed 0.999 is encountered.

EXERCISE 2.17.

(i) An urn contains ng balls, n each of g different colors. In succession (and without replacing drawn balls from the urn) n people each take g balls from the urn. Show that the probability that each person will take g different colors from the urn is $(n!)^g(g!)^n/(gn)!$.

[*Hint*: The number of ways of assigning the ng balls to n groups of g is

$$\binom{gn}{g\ g \cdots g} = \frac{\{(gn)!\}}{(g!)^n}.$$

The first person can choose one out of n available balls in each of the g colors. He or she can do this in n ways. The next person has $(n - 1)$ balls of each color left. He can choose one of each color in $(n - 1)^g$ ways, and so on. The total number of ways in which each person obtains one ball of each color is thus $n^g(n - 1)^g \cdots 3^g \cdot 2^g \cdot 1^g = (n!)^g$.]

(ii) Find an approximate formula, valid for large n, and show that if $g = 2$ and n is large the probability is approximately $2^{-n}\sqrt{\pi n}$.

[*Hint*: Use Stirling's formula (1.50).]

2.7.2. Multivariate Hypergeometric Distributions

We have an urn containing balls of $(k + 1)$ different colors $\mathscr{C}_0, \mathscr{C}_1, \ldots, \mathscr{C}_k$. There are m_j balls of color $\mathscr{C}_j$ $(j = 0, 1, \ldots, k)$, and the total number of balls in the urn is $m = \sum_{j=0}^{k} m_j$. n balls are chosen at random from the urn. Then the numbers $X_0, X_1, \ldots, X_k$ of balls of color $\mathscr{C}_0, \mathscr{C}_1, \ldots, \mathscr{C}_k$, respectively, among the n chosen have the joint distribution

$$\Pr\left[\bigcap_{j=0}^{k} (X_j = x_j)\right] = \frac{\prod_{j=0}^{k} \binom{m_j}{x_j}}{\binom{m}{n}} \qquad \left(\sum_{j=0}^{k} x_j = n, 0 \leq x_j \leq m_j\right). \tag{2.54}$$

By methods similar to those used in the univariate case, it is easy to show that the joint descending $(r_0, r_1, \ldots, r_k)$ factorial moment is

$$\mu_{(r_0, r_1, \ldots, r_k)} = E\left[\prod_{j=0}^{k} X_j^{(r_j)}\right] = \frac{n^{(r)} \prod_{j=0}^{k} m_j^{(r_j)}}{m^{(r)}}, \tag{2.55}$$

where $r = \sum_{j=0}^{k} r_j$.

Although there are $(k + 1)$ random variables $X_0, \ldots, X_k$ involved, the distribution (2.54) is really only a k-variate distribution, since the variables must satisfy the condition

$$\sum_{j=0}^{k} X_j = n,$$

and so any one of them can be expressed in terms of the others.

The marginal distribution of X_j is hypergeometric with parameters n, m_j, m. From (2.55) we find that the covariance between X_i and X_j is

$$\begin{aligned} \operatorname{cov}(X_i, X_j) &= m_i m_j[n(n-1)m^{-1}(m-1)^{-1} - n^2 m^{-2}] \\ &= -n\left(\frac{m_i}{m}\right)\left(\frac{m_j}{m}\right)\left(\frac{m-n}{m-1}\right) \\ &= -np_i p_j(m-n)(m-1)^{-1}, \end{aligned} \tag{2.56}$$

where $p_j = m_j/m$ $(j = 0, 1, \ldots, k)$. Hence the correlation between X_i and X_j is

$$\rho(X_i, X_j) = -\sqrt{\frac{p_i p_j}{q_i q_j}} \tag{2.57}$$

where $q_j = 1 - p_j$ $(j = 0, 1, \ldots, k)$.

Note the similarity between (2.52) and (2.57).

2.7.3. Multivariate Waiting-Time Distributions

We sample (with replacement) from an urn containing proportions $p_0, p_1, \ldots, p_k$ of balls colored $\mathscr{C}_0, \mathscr{C}_1, \ldots, \mathscr{C}_k$. For given n, the number of times $\mathscr{C}_0, \mathscr{C}_1, \ldots, \mathscr{C}_k$ have appeared, $X_0, X_1, \ldots, X_k$, respectively, have a joint multinomial distribution with parameters $n, p_0, p_1, \ldots, p_k$ $(\sum_{j=0}^{k} p_j = 1)$. Now suppose that we increase n, by one at a time, until the value of X_0 reaches a prespecified number g_0. What is the joint distribution of the values of $X_1, X_2, \ldots, X_k$ by the time X_0 reaches g_0? Clearly,

$$\Pr\left[\bigcap_{j=1}^{k}(X_j = x_j)\right]$$

$$= \Pr\left[(X_0 = g_0 - 1)\bigcap_{j=1}^{k}(X_j = x_j) \quad \textit{with} \quad n = g_0 - 1 + \sum_{j=1}^{k} x_j\right] \times p_0$$

$$= \begin{pmatrix} g_0 - 1 + \sum_{j=1}^{k} x_j \\ g_0 - 1\ x_1 x_2 \cdots x_k \end{pmatrix} p_0^{g_0} \prod_{j=1}^{k} p_j^{x_j} \qquad (x_j \geq 0). \tag{2.58}$$

This represents a *negative multinomial distribution*—the probabilities are terms in the expansion of

$$p_0^{g_0}\left(1 - \sum_{j=1}^{k} p_j\right)^{-g_0}. \tag{2.59}$$

The joint distribution obtained when $X_0, X_1, \ldots, X_k$ have (for given n) the joint multivariate hypergeometric distribution (2.54) (and $g_0 < m_0$) is

$$\Pr\left[\bigcap_{j=1}^{k}(X_j = x_j)\right] = \binom{m_0}{g_0}\prod_{j=1}^{k}\binom{m_k}{x_j}\left[\binom{m}{\sum_{j=1}^{k} x_j + g_0 - 1}\right]^{-1} \tag{2.60}$$

$$\left(\sum_{j=1}^{k} x_j \leq m - g_0 + 1\right).$$

This is a *negative multivariate hypergeometric distribution.*

2.7.4. Multivariate Series Distributions

The distributions described in Section 2.6 have natural multivariate generalizations. Here we just list them:

Patil (1965) has studied *multivariate power series distributions* defined by

$$\Pr\left[\bigcap_{i=1}^{m}(X_i = j_i)\right] = \frac{a_{j_1 j_2 \cdots j_m} \prod_{i=1}^{m} \theta_i^{j_i}}{f(\theta_1, \theta_2, \ldots, \theta_m)}, \tag{2.61}$$

where $f(\theta_1, \theta_2, \ldots, \theta_m) = \sum_{j_1=0}^{\infty} \cdots \sum_{j_m=0}^{\infty} a_{j_1 j_2 \cdots j_m}[\prod_{i=1}^{m} \theta_i^{j_i}]$.

Berg (1976) has studied *multivariate factorial series distributions* defined by

$$\Pr\left[\bigcap_{i=1}^{m}(X_i = j_i)\right] = \left[\prod_{i=1}^{m}\frac{\theta_i^{(j_i)}}{j_i!}\right]\frac{a_{j_1 j_2 \cdots j_m}}{f(\theta_1, \ldots, \theta_m)}, \tag{2.62}$$

where $a_{j_1 j_2 \cdots j_m} = \Delta_1^{j_1}\Delta_2^{j_2}\cdots\Delta_m^{j_m}f(\mathbf{0}_1, \mathbf{0}_2, \ldots, \mathbf{0}_m)$ (suffices are used to denote the zero on which each Δ operates).

EXERCISE 2.18. Show that the probability generating function for the distribution (2.61) is

$$\frac{f(s_1\theta_1, s_2\theta_2, \ldots, s_m\theta_m)}{f(\theta_1, \theta_2, \ldots, \theta_m)},$$

hence obtain an expression for the joint factorial moment $\mu_{(r_1, \ldots, r_m)}$.

2.7.5. Multinormal Distributions

Just as the standardized binomial variable has a limiting distribution (the normal or Gaussian distribution—see Section 2.2.2) as $n \to \infty$, so the set of standardized variables corresponding to the joint multinomial distribution (2.50) has a joint limiting distribution as $n \to \infty$. This limiting distribution is a (standardized) *multinormal distribution.*

We first describe multinormal distributions in general, and then discuss the particular distributions obtained as limits of standardized multinormal distributions.

The m random variables $X_1, X_2, \ldots, X_m$ have a joint multinormal distribution if their joint density function is

$$K \exp[-(\text{quadratic form in } x_1, x_2, \ldots, x_m)],$$

where K is a constant. The quadratic form is *positive definite*; that is, it is never negative, and only zero for at most one set of values of $x_1, x_2, \ldots, x_m$.

We now use matrix notation for compactness and write the density as

$$(2\pi)^{-(1/2)m}|\mathbf{C}|^{-1/2}\exp[-\tfrac{1}{2}(\mathbf{x} - \boldsymbol{\xi})'\mathbf{C}(\mathbf{x} - \boldsymbol{\xi})], \tag{2.63}$$

where $\mathbf{C}$ is a symmetric positive definite $m \times m$ matrix, and $\mathbf{x}' = (x_1, \ldots, x_m)$ and $\boldsymbol{\xi}' = (\xi_1, \ldots, \xi_m)$ are $1 \times m$ vectors. The value $(2\pi)^{-(1/2)m}|\mathbf{C}|^{-1/2}$ for K ensures that the integral of the density function (2.63) over all values of $x_1, x_2, \ldots, x_m$ (from $-\infty$ to ∞) is equal to 1.

It can be shown that

1. $E[\mathbf{X}] = \boldsymbol{\xi}$.
2. The variance-covariance matrix of $X_1, X_2, \ldots, X_m$ is $\mathbf{C}^{-1}$ (the inverse of $\mathbf{C}$). (This means that the variance of X_i is the ith diagonal element of $\mathbf{C}^{-1}$, and the covariance of X_i and X_j is the element in the ith row and jth column

of $\mathbf{C}^{-1}$. This is also, of course, the element in the jth row and ith column of $\mathbf{C}^{-1}$, since $\mathbf{C}$, and so also $\mathbf{C}^{-1}$, is symmetric.)

3. The conditional distribution of any subset of the X's given any other (disjoint) subset is also multinormal. In particular, each X_j has a normal distribution.

The result referred to at the beginning of this section is that, if $N_1, \ldots, N_m$ have a joint multinomial distribution with parameters $n, p_1, \ldots, p_m$ [see (2.50)], then the limiting joint distribution of any subset of the standardized variables

$$X_j = \frac{N_j - np_j}{\sqrt{np_j(1 - p_j)}} \qquad (j = 1, \ldots, m)$$

is a standardized multinormal distribution.

For a *standardized* multinormal distribution we must have $E[X_j] = 0$, $\text{var}(X_j) = 1$ $(j = 1, 2, \ldots, m)$, so in (2.63) we must have $\boldsymbol{\xi} = \mathbf{O}$, and the diagonal elements of $\mathbf{C}^{-1}$ must each be 1. This means that any off-diagonal element is a *correlation* (as well as a *covariance*).

The notation $\Phi(a_1, a_2, \ldots, a_m : \mathbf{V})$ is used to denote $\Pr[\bigcap_{j=1}^{m} (X_j < a_j)]$ when $X_1, X_2, \ldots, X_m$ have a joint standardized multinormal distribution with the variance-covariance matrix $\mathbf{V}\,(=\mathbf{C}^{-1})$. Generally this quantity is not easily evaluated, though there are tables for special cases (e.g., when all a's are equal and when all off-diagonal elements of $\mathbf{V}$ are equal). For the case $m = 2$ (*bivariate normal* distributions), where there is only one parameter (ρ, the correlation between the two variables), there are quite extensive tables available from which $\Phi(a_1, a_2; \mathbf{V})$ with $\mathbf{V} = \begin{pmatrix} 1 & \rho \\ \rho & 1 \end{pmatrix}$ can be evaluated (cf. Johnson and Kotz, Chapter 35, 1972).

2.7.6. Dirichlet Distributions

From the formula for the Dirichlet integral (1.55) it can be seen that the function of $x_1, x_2, \ldots, x_m$

$$f(x_1, \ldots, x_m) = \frac{\Gamma\left(\sum_{j=0}^{m} \alpha_j\right)}{\prod_{j=0}^{m} \Gamma(\alpha_j)} \left(\prod_{j=1}^{m} x_j^{\alpha_j - 1}\right)\left(1 - \sum_{j=1}^{m} x_j\right)^{\alpha_0 - 1}$$

$$\left[0 < \alpha_j, x_j \text{ for all } j, \sum_{j=1}^{m} x_j \leq 1\right] \tag{2.64}$$

[and $f(x_1, \ldots, x_m) = 0$ for all other values of $x_1, x_2, \ldots, x_m$] is the density function of a proper distribution. This distribution is called a *Dirichlet distribution* with parameters $\alpha_1, \alpha_2, \ldots, \alpha_m; \alpha_0$. Using $X_1, X_2, \ldots, X_m$ to denote the random variables corresponding to $x_1, x_2, \ldots, x_m$, respectively, we see that $\Pr[\bigcap_{j=1}^m (X_j < x_j)]$ is equal to the incomplete Dirichlet integral (1.56), with p_j replaced by x_j $(j = 1, 2, \ldots, m)$. In particular, using the notation of (1.57), when $\alpha_1 = \alpha_2 = \cdots = \alpha_m = \alpha$,

$$\Pr\left[\bigcap_{j=1}^{m} (X_j < x)\right] = I_x^{(m)}(\alpha, m\alpha + \alpha_0 - 1).$$

Using formula (1.55) it is easy to show that, if $X_1, X_2, \ldots, X_m$ have the joint density function (2.64), then

$$\mu_{r_1, r_2, \ldots, r_m}(X_1, \ldots, X_m) = E\left[\prod_{j=1}^{m} X_j^{r_j}\right] = \frac{\prod_{j=1}^{m} \alpha_j^{[r_j]}}{\left(\sum_{j=0}^{m} \alpha_j\right)^{[r]}}, \tag{2.65}$$

where $r = \sum_{j=1}^m r_j$.

For np_j's not too small, the Dirichlet distribution with parameters $(n-1)p_0; (n-1)p_1, \ldots, (n-1)p_m$; approximates to the joint distribution of $X_j = N_j/n$ $(j = 1, \ldots, m)$, where $N_0, N_1, N_2, \ldots, N_m$ have a joint multinomial distribution with parameters $n, p_0, p_1, p_2, \ldots, p_m$ $(\sum_{i=0}^m p_i = 1)$ (Johnson, 1959). The marginal distribution of X_i is standard beta (see Section 2.2.4) with parameters $\alpha_i, \sum_{j=0}^m \alpha_j - \alpha_i$.

2.8. MIXTURE DISTRIBUTIONS

We consider a situation in which there are k groups of urns, each with their own composition, sampling, and operating rules. There is also a "control" urn which decides which of the regular groups of urns is to be used. This control urn contains a proportion p_j of balls numbered j $(j = 1, \ldots, k)$. If the selected ball has number i, this directs us to sample from the ith regular group of urns. Note that we have encountered a similar model in the derivation of Bayes' formula (Section 1.3.3).

Denoting the cumulative distribution function of the random variable X when it arises from the sampling and operating system of the jth group of urns by $F_j(x)$ $(j = 1, \ldots, k)$ we see that

$$\begin{aligned} F_X(x) = \Pr[X \le x] &= \sum_{i=1}^{k} \Pr[i\text{th group selected}] F_i(x) \\ &= \sum_{i=1}^{k} p_i F_i(x). \end{aligned} \tag{2.66}$$

The distribution of X is a *mixture* of the k-component distributions in proportions $p_1 : p_2 : \cdots : p_k$.

EXERCISE 2.19. Show that, if X has the distribution (2.66), then

$$E[X] = \sum_{i=1}^{k} p_i \xi_i = \bar{\xi}, \qquad \operatorname{var}(X) = \sum_{i=1}^{k} \sigma_i^2 + \sum_{i=1}^{k} p_i(\xi_i - \bar{\xi})^2, \tag{2.67}$$

where ξ_i, σ_i^2 are the expected value and variance, respectively, of the distribution $F_i(\cdot)$.

Among mixture distributions are included *contagious* distributions. These arise when the distributions arising from the k groups of urns belong to the same family, with a parameter depending on the group.

Another way in which mixtures can arise is in the form of "random sums" (Chatfield and Theobald, 1973). Suppose we have two urns, urn P containing balls numbered 0, 1, 2, ..., k in proportions $p_0, p_1, p_2, \ldots, p_k$ $(\sum_{i=0}^{k} p_i = 1)$, and urn A containing m balls with numbers $a_1, a_2, \ldots, a_m$ on them. We pick a ball at random from P and then select at random as many balls from A as is indicated by the number on the ball chosen from P. The distribution of the sum of the numbers on the balls chosen from A will then be a mixture of the distributions for a fixed number (say N) of chosen balls with N having the ("mixing") distribution

$$\Pr[N = n] = p_n \qquad (0 \le n \le k).$$

This result is valid whether the sampling from A is with or without replacement.

EXERCISE 2.20. In the situation described above suppose that $p_0 = p_1 = \cdots = p_k = (k+1)^{-1}$, $m = k + 1$, and $a_j = j - 1$ $(j = 1, \ldots, k + 1)$. The N numbers from A are obtained by sampling with replacement. Show that the generating function of $\sum_{j=1}^{N} A_j$ (where A_j is the number on the chosen ball at the jth drawing) is

$$\frac{(1-t)\left[1 - \left\{\dfrac{1 - t^{k+1}}{(k+1)(1-t)}\right\}^{k+1}\right]}{[(k+1)(1-t) - (1 - t^{k+1})]}.$$

2.9. EXCHANGEABLE VARIATES*

If we analyze the sampling methods leading to the binomial and hypergeometric distributions (Sections 2.2 and 2.3, respectively), we are led to an important concept of modern probability theory called *exchangeability*.

* This section was written by J. Galambos.

Suppose an urn contains m red and $(n - m)$ white balls. We draw balls successively, one at a time. Sampling may be with or without replacement. We denote by A_j the event that the jth draw results in a red ball. In order to emphasize the sampling scheme decided on, we denote by $P_0(A)$ the probability of an event A when sampling is with replacement, and by $P_1(A)$ the probability of an event A if we do not replace the balls previously drawn. It is quite evident that, for any $k \geq 1$ and for integers $1 \leq i_1 < i_2 < \cdots < i_k$,

$$P_0\left(\bigcap_{j=1}^{k} A_{i_j}\right) = \left(\frac{m}{n}\right)^k. \tag{2.68}$$

It is, however, somewhat surprising at first glance that $P_1(\bigcap_{j=1}^{k} A_{i_j})$ is also independent of the actual subscripts $1 \leq i_1 < i_2 < \cdots < i_k$. In fact, for $k \geq 1$ and for $1 \leq i_1 < i_2 < \cdots < i_k$,

$$P_1\left(\bigcap_{j=1}^{k} A_{i_j}\right) = \frac{m(m-1)\cdots(m-k+1)}{n(n-1)\cdots(n-k+3)} = \frac{m^{(k)}}{n^{(k)}}. \tag{2.69}$$

Since the i_1th ball chosen is equally likely to be any one of the original n balls, of which m are red, (2.69) is clearly valid for $k = 1$. For arbitrary k, fix the k positions $1 \leq i_1 < i_2 < \cdots < i_k = j$ and let $g_{j,k}(n, m)$ be the number of outcomes when, drawing j balls without replacement, there are red balls in positions $i_1, i_2, \ldots, i_k$. If $k > m$, then evidently

$$g_{j,k}(n, m) = 0. \tag{2.70}$$

If $k \leq m$, we note that we can choose the k red balls in positions $i_1, \ldots, i_k$ in $m^{(k)}$ ways (allowing for order) and the remaining $(j - k)$ positions in any order we wish from the remaining $(n - k)$ balls. Hence

$$g_{j,k}(n, m) = m^{(k)} f_{j-k}(n - k), \tag{2.71}$$

where $f_{j-k}(n - k)$ = number of ways of choosing $(j - k)$ balls in order (without replacement) from $(n - k)$ balls. Clearly,

$$f_{j-k}(n - k) = (n - k)^{(j-k)}.$$

Noticing that (2.70) is included formally in (2.71) and that

$$f_j(n) = n(n-1)\cdots(n-k+1) f_{j-k}(n-k),$$

we thus have

$$P_1(A_{i_1}, A_{i_2}, \ldots, A_{i_k}) = \frac{g_{j,k}(n, m)}{f_j(n)} = \frac{m(m-1)\cdots(m-k+1)}{n(n-1)\cdots(n-k+1)},$$

as claimed in (2.69).

Comparing formulas (2.68) and (2.69) once again, we recognize a common property. Namely, we have sequences of events (a chosen ball being red) in which the probability of the joint occurrence of the i_1th, i_2th, ..., i_kth, with $1 \le i_1 < i_2 < \cdots < i_k$, does not depend on the actual subscripts but on k only. We now express this property in a formal definition.

Definition 2.9.1. Let $A_1, A_2, \ldots$ be a finite or infinite (countable) sequence of events. If, for *any* possible choices $1 \le i_1 < i_2 < \cdots < i_k$ of k subscripts,

$$\Pr(A_{i_1} \cap A_{i_2} \cap \cdots \cap A_{i_k}) = p_k$$

depends on k only but not on the actual subscripts i_t $(1 \le t \le k)$, we call the events *exchangeable*.

If the events A_t are independent and have the same probability of occurring for each t, then the A_t are evidently exchangeable. However, (2.69) shows that the class of exchangeable events is larger than that of independent events with equal probabilities.

It is important to note the difference between the A_j's, defined as above, according to whether the balls are drawn with or without replacement. In the first case (with replacement) the number N of balls to be drawn can be increased indefinitely, while in the second case (without replacement) N is limited by n. Therefore, in the first case we can regard $A_1, A_2, \ldots, A_N$ as a segment of an infinite sequence of exchangeable events, but in the second case our sequence is from a finite set $A_1, A_2, \ldots, A_n$ and no further term can be added to it without destroying the exchangeability.

Here are some further examples of exchangeable events.

Example 2.4. We have two urns. In the first urn there are two red and three white balls, and in the second one, two red and two white balls. The experiment consists of choosing an urn at random with a probability $\frac{2}{3}$ of choosing the first urn, and then, from the urn chosen, we draw n balls with replacement. Let A_j be the event that the jth draw is a red ball. Then the events $A_1, A_2, \ldots, A_n$ are exchangeable.

If we denote by B or C, respectively, the event that the first or that the second urn was chosen in the first step, then, conditionally, given B or C, the events $A_1, A_2, \ldots, A_n$ are independent. Since $B \cap C$ is impossible and $B \cup C$ is certain to occur, we have from (1.89a), for $1 \le i_1 < i_2 < \cdots < i_k \le n$,

$$\Pr\left[\bigcap_{j=1}^{k} A_{i_j}\right] = \Pr\left[\bigcap_{j=1}^{k} A_{i_j} \middle| B\right]\Pr[B] + \Pr\left[\bigcap_{j=1}^{k} A_{i_j} \middle| C\right]\Pr[C]$$

$$= (\tfrac{2}{5})^k \tfrac{2}{3} + (\tfrac{1}{2})^k \tfrac{1}{3}.$$

Thus the criterion for exchangeability is satisfied.

The preceding example is a special case of the following one.

Example 2.5. Let X be a random variable with

$$\Pr[0 \le X \le 1] = 1.$$

Let the events $A_1, A_2, \ldots$ be conditionally independent, given X, and let the conditional probability

$$\Pr[A_j|X] = X \qquad (j \geq 1).$$

Then $A_1, A_2, \ldots$ are exchangeable.

As a matter of fact, for $1 \leq i_1 < i_2 < \cdots < i_k$,

$$\Pr\left[\bigcap_{j=1}^{k} A_{i_j} \middle| X\right] = X^k,$$

and thus

$$\Pr\left[\bigcap_{j=1}^{k} A_{i_j}\right] = E[X^k], \tag{2.72}$$

the kth moment of X, independently of i_t $(1 \leq t \leq k)$. [*Note*: This result is valid whatever the distribution of X, provided it is limited to $0 \leq x \leq 1$.]

Example 2.6. Let an urn contain a fixed number of balls, M red and $(n - M)$ white. The number M has been selected by a random procedure. We draw from the urn without replacement, and again A_j denotes the event that the jth draw results in a red ball. Here, $1 \leq j \leq n$. Then the events A_j $(1 \leq j \leq n)$, are exchangeable.

Our claim of exchangeability follows by another appeal to (1.89a). Indeed, given $\{M = m\}$ and if we set $P_1[A] = \Pr[A|M = m]$, then formula (2.69) is applicable. Therefore, putting

$$v_m = \Pr[M = m],$$

(1.89a) yields, for $1 \leq i_1 < i_2 < \cdots < i_k \leq n$,

$$\Pr\left[\bigcap_{j=1}^{k} A_{i_j}\right] = \sum_{s=k}^{n} \frac{s^{(k)}}{n^{(k)}} v_s. \tag{2.73}$$

Formula (2.73) evidently implies exchangeability.

Examples 2.5 and 2.6 are the most general examples for exchangeable events, in the sense of the following theorems of De Finetti (1930) and Kendall (1967).

Theorem 2.1 (De Finetti, 1930). Let $A_1, A_2, \ldots$ be an infinite sequence of exchangeable events. Then there is a random variable X such that, for any $k \geq 1$,

$$p_k = \Pr\left[\bigcap_{j=1}^{k} A_j\right] = E[X^k]. \tag{2.74}$$

Theorem 2.2 (Kendall, 1967). Let $A_1, A_2, \ldots, A_n$ be a finite set of exchangeable events. Then there is an integer-valued random variable M such that $0 \leq M \leq n$ and that (2.73) holds.

Proofs of these theorems are outlined in the appendix to this chapter. Theorems 2.1 and 2.2 have some interesting theoretical implications. By comparing Theorem 2.2 and Example 2.6 we see that the existence of M such that (2.73) holds is a sufficient *and necessary* condition for exchangeability.

We now introduce a new notation. For a sequence $B_1, B_2, \ldots, B_n$ of events, let $f_n(B)$ denote the number of B's that occur. Then by Theorem 2.1 and by the method of inclusion and exclusion (see Section 1.3.2) we have the following.

Corollary 2.1. Let $A_1, A_2, \ldots, A_n$ be a "segment" of an infinite sequence of exchangeable events. Then

$$\begin{aligned}\Pr[f_n(A) = s] &= E\left[\binom{n}{s}x^s(1-x)^{n-s}\right]\\ &= \int_0^1 \binom{n}{s}x^s(1-x)^{n-s}\,dF(x). \qquad (2.75)\end{aligned}$$

In other words, the distribution of the number of occurrences in a finite segment of an infinite sequence of exchangeable events is always a *mixture* of the binomial $B(n, p)$ distribution with some proper distribution for p over the interval $[0, 1]$.

Similarly, from Theorem 2.2, we have

Corollary 2.2. Let $A_1, A_2, \ldots, A_n$ be exchangeable events. Then the distribution of $f_n(A)$ is always of the form

$$\Pr[f_n(A) = s] = \int_0^N \frac{\binom{M}{s}\binom{N-M}{n-s}}{\binom{N}{n}}\,dF(M) \qquad (\text{for some } N \geq n), \quad (2.76)$$

where the distribution function $F(\cdot)$ is discrete with possible jumps on the nonnegative integers $M \leq N$.

However, Corollary 2.2 is *not* a characterization of exchangeable events. Galambos (1973) has shown that (2.76) is valid for *any* sequence of events.

Theorem 2.3 (Galambos, 1973). For an arbitrary sequence $B_1, B_2, \ldots, B_n$ of events, $f_n(B)$ has the distribution given in (2.76). (The proof of this theorem is not presented here.)

Therefore the number of occurrences in an arbitrary sequence of events always follows the same kind of law as applies to the number of red balls in Example 2.6.

So far we have dealt with exchangeable *events*. We now define exchangeable *random variables*.

Definition 2.9.2. The random variables $X_1, X_2, \ldots$ are called *exchangeable* if the events $(X_1 \leq x_1), (X_2 \leq x_2), \ldots$ are exchangeable for all $x_1, x_2, \ldots$.

For infinite sequences, Theorem 2.1 still applies. We quote here this theorem without proof; it can be proved in the same manner as we proved Theorem 2.1 by turning to the indicators of $\{X_j < x\}$. [For details, see Loève (1963, p. 365).]

Theorem 2.4. Let $X_1, X_2, \ldots$ be an infinite sequence of exchangeable random variables. Then there is a σ field $\mathscr{F}$ and a distribution function $G(\cdot)$ such that, given $\mathscr{F}$, the random variables $X_1, X_2, \ldots,$ are conditionally independent with distribution function $G(\cdot)$.

EXERCISE 2.21. Let A_1, A_2, and A_3 be three events with $\Pr(A_1) = .6$, $\Pr(A_1 \cap A_2) = .3$, and $\Pr(A_1 \cap A_2 \cap A_3) = .2$. Show that A_1, A_2, and A_3 are not exchangeable. [*Hint*: Assume that they are exchangeable and calculate the differences occurring in (2.80) (see APPENDIX to this chapter, opposite).]

EXERCISE 2.22. Let A_1, A_2, and A_3 be three exchangeable events with $\Pr(A_1) = .7$, $\Pr(A_1 \cap A_2) = .5$, and $\Pr(A_1 \cap A_2 \cap A_3) = .4$. Can A_1, A_2, and A_3 be enlarged into a set of four exchangeable events? [*Hint*: Analyze the differences occurring in (2.80) (see APPENDIX to this chapter, opposite).]

EXERCISE 2.23. Urn 1 contains two red and three white balls and urn 2 one red and one white. We first choose an urn, and then from the urn chosen we draw five balls with the following rule. If urn 1 is chosen, then the selection of balls is without replacement, while from urn 2 the drawing is with replacement. Let A_j $(1 \leq j \leq 5)$ be the event that the jth ball drawn is red. Show that these events are exchangeable.

EXERCISE 2.24. Let $A_1, A_2, \ldots$ be an infinite sequence of exchangeable events. Prove that

$$\Pr(A_1 \cap A_2) \geq [\Pr(A_1)]^2.$$

Show by an example that the above inequality does not hold in general for finite sequences of exchangeable events. [*Hint*: Let V_n be the variance of $\sum_{j=1}^{n} I(A_j)$, where $I(A_j)$ is the indicator of A_j. Calculate $\lim V_n n^{-2}$, as $n \to +\infty$. Or, alternatively, apply the Cauchy inequality in De Finetti's theorem 2.1.]

APPENDIX. Proofs of Theorems on Exchangeability

J. Galambos

Proof of Theorem 2.1. We first introduce indicator variables $X_1, X_2, \ldots$ for the events $A_1, A_2, \ldots$.

Let

$$X_j = \begin{cases} 1 & \text{if } A_j \text{ occurs,} \\ 0 & \text{otherwise.} \end{cases}$$

Since the A's are exchangeable, $E(X_j) = \Pr(A_j) = p_1$ for all j and $E[X_i X_j] = p_2$ for all $i \neq j$.

Now put $Y_n = n^{-1} \sum_{j=1}^{n} X_j$ and take $m > n$:

$$\begin{aligned} E[(Y_n - Y_m)^2] &= n^{-2} E\left[\left(\sum_{j=1}^{n} X_j\right)^2\right] + m^{-2} E\left[\left(\sum_{j=1}^{m} X_j\right)^2\right] \\ &\quad - 2(mn)^{-1} E\left[\left(\sum_{j=1}^{n} X_j\right)\left(\sum_{j=1}^{m} X_j\right)\right] \\ &= \frac{np_1 + n(n-1)p_2}{n^2} + \frac{mp_1 + m(m-1)p_2}{m^2} \\ &\quad - \frac{2np_1 + 2(mn - n)p_2}{mn} \\ &= \frac{m-n}{mn}(p_1 - p_2) < n^{-1}(p_1 - p_2), \end{aligned}$$

and so

$$\lim_{\substack{n \to \infty \\ (m > n)}} E[(Y_n - Y_m)^2] = 0. \tag{2.77}$$

It can be shown, using a theorem of Riesz (Loève, 1963, p. 161), that (2.77) implies that there *exists* a random variable X such that

$$\lim_{n \to \infty} E[(Y_n - X)^2] = 0. \tag{2.78}$$

Since $0 \leq Y_n \leq 1$, this implies that

$$\Pr[0 \leq X \leq 1] = 1.$$

(If $\Pr[X > 1 + \delta] = \varepsilon > 0$, then $E[(Y_n - X)^2] > \varepsilon\delta^2 > 0$ for all n.)

For any $k \geq 1$,

$$|Y_n^k - X^k| = |Y_n - X|(Y_n^{k-1} + Y_n^{k-2}X + \cdots + Y_n X^{k-2} + X^{k-1})$$
$$\leq k|Y_n - X|,$$

since $0 \leq X \leq 1, 0 \leq Y_n \leq 1$.

Hence

$$|E(Y_n^k - X^k)| \leq E[|Y_n^k - X^k|] \leq kE[|Y_n - X|] \leq k\{E[(Y_n - X)^2]\}^{1/2}$$

[since for any variable T, $\text{var}(T) = E[T^2] - \{E[|T|]\}^2 \geq 0$] and so, from (2.78),

$$\lim_{n\to\infty} |E(Y_n^k - X^k)| = 0. \tag{2.79}$$

This means that

$$p_k = \lim_{n\to\infty} E(Y_n^k) = E[X^k].$$

Note that this is an *existence* theorem. It assures us of the existence of a random variable X with the stated properties, but does not enable us to construct it.

Heath and Sudderth (1976) give an alternative simple proof of Corollary 2.1.

Proof of Theorem 2.2. We use the successive differences of the p's (with $p_0 = 1$) (see Section 1.1.2).

$$\Delta p_j = p_{j+1} - p_j \qquad = E[X^j(X-1)] < 0$$
$$\Delta^2 p_j = \Delta(\Delta p_j) = p_{j+2} - 2p_{j+1} + p_j \qquad = E[X^j(X-1)^2] > 0$$

Writing

$$v_m = (-1)^{n-m}\binom{n}{m}\Delta^{n-m}p_m = (-1)^{n-m}\binom{n}{m}\Delta^{n-m}E^m p_0, \tag{2.80}$$

where E is the displacement operator (1.12), we have

$$\sum_{m=0}^{n} v_m = (\Delta - E)^n p_0 = p_0 = 1,$$

and

$$\sum_{s=m}^{n} \frac{\binom{n-m}{s-m}}{\binom{n}{s}} v_s = \sum_{s=m}^{n} (-1)^{n-s}\binom{n-m}{s-m}\Delta^{n-s}p_s$$
$$= \sum_{t=0}^{n-m} (-1)^t \binom{n-m}{t}\Delta^{n-m-t}p_{m+t}$$
$$= (\Delta - E)^{n-m}p_m = p_m.$$

In this case we have thus actually *constructed* a distribution

$$\Pr[X = m] = v_m \qquad (m = 0, 1, \ldots, n),$$

which satisfies (2.73).

References

Baldessari, B. (1972) Una presentazione unificata degli schemi con estrazioni bernoulliane, *Metron*, **30**, 234–255.

Berg, S. (1974) Factorial series distributions, with applications to capture-recapture problems, *Scand. J. Stat.*, **1**, 145–152.

Berg, S. (1975). A note on the connection between factorial series distribution and zero-truncated power series distribution, *Scand. Actuarial J.*, 233–237.

Berg, S. (1976) Certain properties of the multivariate factorial series distributions, *Scand. J. Stat.*, 3.

Bernard, S. (1976) *An urn model study of variability within a compartment*, Manuscript, Health Physics Division, Oak Ridge National Laboratory, Oak Ridge, Tennessee.

Chatfield, C. and Theobald, C. M. (1973) Mixtures and random sums, *J. Appl. Stat.*, **22**, 281–287.

Consul, P. C. (1974) A simple urn model dependent upon predetermined strategy, *Sankhyā, Ser. B*, **36**, 391–399.

Consul, P. C. and Mittal, S. P. (1975) A new urn model with predetermined strategy, *Biom. Z.*, **17**(2), 67–75.

De Finetti, B. (1930) Funzione caratteristica di un fenomeno aleatorio, *Atti Accad. Naz. Lincei, Rend. Cl. Fis. Mat. Nat.*, (6), **4**, 86–133.

Galambos, J. (1973) A general Poisson limit theorem of probability theory, *Duke Math. J.*, **40**, 581–586.

Heath, D. and Sudderth, W. (1976) De Finetti's theorem on exchangeable variates, *Am. Stat.*, **30**, 188–189.

Irwin, J. O. (1955) A unified derivation of some well-known frequency distributions, *J. R. Stat. Soc. Ser. A*, **118**, 394–404.

Irwin, J. O. (1968) The generalized Waring distribution applied to accident theory, *J. R. Stat. Soc. Ser. A*, **126**, 1–44.

Irwin, J. O. (1975) The generalized Waring distribution I, II, III, *J. R. Stat. Soc. Ser. A*, **138**, 18–32, 204–228, 374–384.

Johnson, N. L. and Kotz, S. (1969, 1970, 1972) *Distributions in Statistics*, Vols. 1–4, New York: John Wiley and Sons.

Kendall, D. G. (1967) On finite and infinite sequences of exchangeable events, *Stud. Sci. Math. Hung.*, **2**, 319–327.

Lessing, R. (1973) An alternative expression for the hypergeometric moment generating function, *Am. Stat.*, **27**(3), 115.

Loève, M. (1963) *Probability Theory*, 3rd ed., New York: Van Nostrand.

Murty, K. G. (1975) A note on an estimation problem, College of Engineering, University of Michigan, Ann Arbor.

Noack, A. (1950) A class of random variables with discrete distributions, *Ann. Math. Stat.* **21**, 127–132.

Patil, G. P. (1962) Certain properties of the generalized power series distribution, *Ann. Inst. Stat. Math. Tokyo*, **14**, 179–182.

Patil, G. P. (1965) On multivariate generalized power series distribution and its application to the multinomial and negative multinomial, in *Classical and Contagious Discrete Distributions*, G. P. Patil ed.), Calcutta: Statistical Publishing Company, pp. 183–194.

Sobel, M. Uppuluri, V. R. R. and Frankowski, K. (1975) *Dirichlet Distributions, Type I*, Department of Statistics, University of Minnesota.

V'ndev, D. L. (1974) A generalization of an urn scheme in probability theory, *Bolg. Akad. Nauk.*, pp. 27–29 (in Bulgarian).

3

Occupancy and Related Problems

3.1. OCCUPANCY PROBLEMS

3.1.1. Classical Occupancy

Imagine that we have m urns and a supply of balls arriving one after the other. Each ball is assigned to an urn chosen randomly, so that the probability of assignment to the jth urn is p_j for each ball. Clearly, the p_j's must sum to unity, that is,

$$\sum_{j=1}^{m} p_j = 1.$$

Problems of *occupancy* arise when we are concerned with the distribution of urns containing at least one ball (*occupied* urns) after a specified number of balls has been assigned. Equivalently, we may consider the distribution of the number of empty (*unoccupied*) urns.

For a single urn, it is simple to calculate the probability of emptiness after n balls have been assigned. For the jth urn, this probability is $(1 - p_j)^n$. Similarly, the probability that the ith and jth urns are both unoccupied is $(1 - p_i - p_j)^n$, and so the probability that just one of the ith and jth urns is empty is

$$(1 - p_i)^n + (1 - p_j)^n - (1 - p_i - p_j)^n.$$

It is rather more difficult to establish a formula for the *number* of empty urns, say X. If $N_1, N_2, \ldots, N_m$ denote the number of balls assigned to the 1st, 2nd, ..., mth urn, respectively (with $\sum_{j=1}^{m} N_j = n$), we have

$$P(n_1, n_2, \ldots, n_m) = \Pr[N_1 = n_1, N_2 = n_2, \ldots, N_m = n_m]$$
$$= \binom{n}{n_1 \cdots n_m} \prod_{j=1}^{m} p_j^{n_j}. \tag{3.1}$$

The joint distribution of $N_1, N_2, \ldots, N_m$ is *multinomial* (cf. Section 2.6.1), and the value of $\Pr[X = k]$ can be obtained by summing over the appropriate sets of values of $n_1, n_2, \ldots, n_m$. (In this case over all sets for which exactly k of the n's are zero.) The following arguments, however, give more insight into the nature of the distribution of X.

It is easy to see that

$$\Pr[X = m - 1] = \sum_{j=1}^{m} p_j^n, \tag{3.2}$$

because in this case only one urn can contain any balls and all n balls must be assigned to it. The probability that all are assigned to the jth urn is p_j^n for $j = 1, 2, \ldots, m$, and the m possible choices of urn are mutually exclusive.

Also, using E_j to denote the event that the jth urn is empty, we have, using the *inclusion-exclusion* principle (Section 1.3.2),

$$\Pr[X = 0] = 1 - \Pr\left[\bigcup_{j=1}^{m} E_j\right]$$
$$= 1 - \sum_{j=1}^{m} \Pr[E_j] + \sum_{j<j'}^{m}\sum \Pr[E_j \cap E_{j'}] - \cdots$$
$$+ (-1)^{m-1} \sum_{j=1}^{m} \Pr\left[\bigcap_{i \neq j}^{m} E_i\right]$$
$$= 1 - \sum_{j=1}^{m} (1 - p_j)^n + \sum_{j<j'}^{m}\sum (1 - p_j - p_{j'})^n - \cdots$$
$$+ (-1)^{m-1} \sum_{j=1}^{m} p_j^n. \tag{3.3}$$

(Note that $\Pr[\bigcap_{j=1}^{m} E_j] = 0$ and, if $n < m$, $\Pr[X = 0] = 0$. Formula (3.3) will in fact give the value zero if $n < m$, as can be shown from Example 1.9 in Section 1.1.2.)

The probability $P_{12\cdots k}$ that the 1st, 2nd, ..., kth urns are empty and *none* of the other urns are empty is

$$P_{12\cdots k} = \left(1 - \sum_{i=1}^{k} p_i\right)^n \times \Pr[\text{no empty urns among } (k+1)\text{st}, \ldots, m\text{th urns} \mid n \text{ balls}].$$

The latter probability relates to a situation with $(m - k)$ urns with probabilities $p_j(1 - \sum_{i=1}^k p_i)^{-1}$ for $j = k + 1, \ldots, m$.

Using (3.3), with appropriate changes in the values of m and p_j, we find that

$$P_{12\cdots k} = \left(1 - \sum_{j=1}^k p_j\right)^n - \sum_{j'=k+1}^m \left(1 - p_{j'} - \sum_{j=1}^k p_j\right)^n + \sum\sum_{j' \neq j'' > k}^m \left(1 - p_{j'} - p_{j''} - \sum_{j=1}^k p_j\right)^n - \cdots. \tag{3.4}$$

Summing expressions similar to this for all $\binom{m}{k}$ possible sets of k events chosen from $E_1, E_2, \ldots, E_m$, we find

$$\begin{aligned}\Pr[X = k] &= \sum_{\mathbf{a}}^{(k)} P_{a_1 \cdots a_k} \\ &= \sum_{\mathbf{a}}^{(k)}\left(1 - \sum_{j=1}^k p_{a_j}\right)^n - \binom{k+1}{k}\sum_{\mathbf{a}}^{(k+1)}\left(1 - \sum_{j=1}^{k+1} p_{a_j}\right)^n \\ &\quad + \binom{k+2}{k}\sum_{\mathbf{a}}^{(k+2)}\left(1 - \sum_{j=1}^{k+2} p_{a_j}\right)^n - \cdots + (-1)^{m-k}\binom{m}{k}\sum_{j=1}^m p_j^n,\end{aligned} \tag{3.5}$$

where $\sum_{\mathbf{a}}^{(k)}$ denotes summation over all subsets $\mathbf{a} = (a_1, a_2, \ldots, a_k)$ of k integers from the m integers $(1, 2, \ldots, m)$. $\left[\text{There are } \binom{m}{k} \text{ terms in the sum } \sum_{\mathbf{a}}^{(k)}.\right]$

The value of $\Pr[X \geq k]$ is obtained by changing k to $(k - 1)$ in each of the combinatorial symbols in (3.5)—for example, $\binom{k+1}{k}$ becomes $\binom{k}{k-1}$. (Cf. (1.85) and (1.87)—see also Price (1946).)

EXERCISE 3.1. Show that the expected number of empty urns after a specified number of trials is a minimum when each ball is equally likely to be assigned to any one of the m urns (i.e., $p_1 = \cdots = p_m = m^{-1}$). [*Hint*: This problem may be solved by noting that $(1 - x)^n$ is a convex function of x for $0 < x < 1$ and applying Jensen's inequality. It is not essential, however, to use this result.]

In the special case $p_1 = p_2 = \cdots = p_m = m^{-1}$ $\left[\text{noting that } \binom{m}{k+r}\binom{k+r}{k} = \binom{m}{k}\binom{m-k}{r}\right]$, we have

$$\Pr[X = k] = \binom{m}{k}\Big[(1 - km^{-1})^n - \binom{m-k}{1}\{1 - (k+1)m^{-1}\}^n + \cdots$$

$$- (-1)^{m-k}\binom{m-k}{m-k-1}\{1 - (m-1)m^{-1}\}^n\Big]$$

$$= \binom{m}{k} m^{-n}\Delta^{m-k}\mathbf{0}^n \qquad (k = 0, 1, \ldots, m). \qquad (3.6)$$

[See (1.15).]

EXERCISE 3.2. Show that the sum of the coefficients of all terms in the multinomial expansion of $(\sum_{i=1}^{m} x_i)^n$, which include every one of $x_1, x_2, \ldots, x_n$, is equal to $\Delta^m\mathbf{0}^n$. [*Hint*: Put $k = 0$ in (3.6).]

The result from Exercise 3.2 is used in Section 4.7.3.

From this formula it is clear that, for $n < m - k$, $\Pr[X = k] = 0$, as must be the case in general. [In fact there must be *at least* $(m - n)$ empty urns.]

Formula (3.6) represents the *classical occupancy* distribution. The quantities m and n are parameters of this distribution. (The distribution of the number of *occupied* urns $Y = m - X$ is sometimes also called the *classical occupancy distribution.*) We have

$$\Pr[Y = k] = \binom{m}{k} m^{-n}\Delta^{k}\mathbf{0}^n \qquad (k = 0, 1, \ldots, m). \qquad (3.6')$$

A particular form of occupancy problem, which has been widely discussed, is the birthday problem which we have already encountered in Exercise 1.22. This in turn has many variants, but the essential feature is coincidence of birthdays among a group of individuals. For the purpose of the problem it is usually assumed that (1) there are 365 days in a year, and (2) it is equally likely that any birthday will fall on any day of the year. We then determine the probability that, among a group of n persons, at least r will have a birthday in common.

This corresponds to the classical occupancy problem with the number of urns m equal to 365 and the number of balls n equal to the number of persons.

EXERCISE 3.3. An urn contains N white balls ($N = 1, 2, \ldots$). We repeatedly draw a ball at random, observe its color, and replace it by a black ball. Eventually, all N white balls will have been drawn and the urn will contain only black balls. Find the probability that this will occur *at or before* the nth draw.

Example 3.1. An interesting application of occupancy distributions in computer theory is implicit in the work of Denning and Schwartz (1970; 1972). They use the concept of a current *working set* $W(t, \tau)$ at time t when using a sequence of *references*. If each reference is defined as a page number, then it is supposed that $W(t, \tau)$ includes only page numbers referenced on at least one of the last τ *occasions*. The distribution of the number of distinct pages in the working set $W(t, \tau)$ is then an occupancy distribution, with τ trials, each possible "page" being an "urn." If it is supposed that the probability of referencing a specific urn is constant (and independent of other references), then the appropriate distribution is that of the classical occupancy problem.

The relevant distribution is (3.5). Vantilborgh (1974) has examined the accuracy of a normal approximation in the equiprobable case $p_i = m^{-1}$ [for which (3.6) is appropriate] and finds it to be quite good. Additional generalizations of this example are discussed in Section 5.5.4.

Example 3.2 (Mertz and Davies, 1968). A simple model relating numbers of predators and prey (pupae) is based on the assumptions:

(i) All predators are equally predaceous and act independently of each other.
(ii) A prey once attacked is regarded as a corpse.
(iii) Predators attack prey and prey corpses with equal probabilities, and a corpse remains liable to attack for the duration of the experiment.

We denote the number of surviving prey at time t by $X(t)$ and the number of attacks generated by the predators *up* to time t by $K(t)$. For the moment generating functions we use the notation

$$G(u; t) = E[u^{K(t)}]$$
$$H(v; t) = E[v^{X(t)}].$$

If the attacks are delivered randomly on the prey (so that some may have more than one attack, that is, the corpses are attacked), then $X(t)$ will have the classical occupancy distribution

$$\Pr[X(t) = x \mid K(t) = k] = \binom{X(0)}{x}\{X(0)\}^{-k}\Delta^{X(0)-x}\mathbf{0}^k.$$

Averaging over the distribution of $K(t)$, we obtain

$$\Pr[X(t) = x] = \binom{X(0)}{x}\Delta^{X(0)-x} \sum_k \Pr[K(t) = k]\mathbf{0}^k\{X(0)\}^{-k}$$

$$= \binom{X(0)}{x}\Delta^{X(0)-x}G\left(\frac{\mathbf{0}}{X(0)}; t\right).$$

[Note that Δ does not operate on the 0 in $X(0)$.]

Finally,

$$H(v; t) = \sum_x \binom{X(0)}{x} v^x \Delta^{X(0)-x}G\left(\frac{\mathbf{0}}{X(0)}; t\right) = (v + \Delta)^{X(0)}G\left(\frac{\mathbf{0}}{X(0)}; t\right). \tag{3.7}$$

Example 3.3. Physicists are interested in the expected number of cells (urns in our terminology) containing a specified number r of particles (balls in our terminology). (This is the expected value of M_r, as defined in Example 2.2.)

For M–B statistics, the probability that just r out of μ particles will be in one particular cell, out of γ cells, is simply the binomial probability

$$\binom{\mu}{r}\gamma^{-r}(1 - \gamma^{-1})^{\mu-r} = \binom{\mu}{r}\gamma^{-\mu}(\gamma - 1)^{\mu-r} \qquad (r = 0, 1, \ldots, \mu). \tag{3.8}$$

This is the expected proportion of cells with r particles ($E[M_r]/\gamma$).

For B–E statistics, however, we note that there are $\binom{\mu + \gamma - 1}{\gamma - 1}$ ways of assigning μ particles to γ cells. If we first assign r particles to one particular cell, the remaining $(\mu - r)$ particles can be assigned to the remaining $(\gamma - 1)$ cells in $\binom{\mu + \gamma - r - 2}{\gamma - 2}$ ways.

The required probability (=expected proportion of urns with r particles) is the ratio

$$\frac{\binom{\mu + \gamma - r - 2}{\gamma - 2}}{\binom{\mu + \gamma - 1}{\gamma - 1}} = (\gamma - 1)\frac{\mu!(\mu + \gamma - r - 2)!}{(\mu - r)!(\mu + \gamma - 1)!}. \tag{3.9}$$

(See, for example, Decomps and Kastler, 1963.)

A further discussion of B–E statistics can be found in Section 3.2.2. The Fermi–Dirac (F–D) system is discussed in Exercise 3.18.

The occupancy distribution also applies in what may, on first consideration, appear to be quite different problems. For example, suppose we have a single urn containing m white balls. We then draw balls one at a time from the urn, replacing the drawn ball by a black ball on each occasion. The number of black balls remaining in the urn after the rth draw (and subsequent replacement by a black ball) has the same distribution as the number of occupied cells, out of m originally empty cells, after random assignment of n balls, that is, the classical occupancy distribution [see (3.6)].

This can be seen by regarding the m balls as "urns" and regarding their being drawn as corresponding to the assignment of a "ball" to an "urn" in the classic occupancy model. (It may help to think of a ball as being painted black each time it is drawn—after the first occasion this makes no difference to the status of the ball.)

To obtain the moments of the distribution (3.5) we write

$$X = X_1 + X_2 + \cdots + X_m,$$

where

$$\begin{aligned} X_j &= 0 \qquad \text{if the } j\text{th urn is occupied,} \\ &= 1 \qquad \text{if the } j\text{th urn is empty.} \end{aligned}$$

As we have already seen,

$$\Pr[X_j = 1] = (1 - p_j)^n$$

$$\Pr[(X_j = 1) \cap (X_{j'} = 1)] = (1 - p_j - p_{j'})^n.$$

Hence

$$E[X] = \sum_{j=1}^{m} E[X_j] = \sum_{j=1}^{m} (1 - p_j)^n \tag{3.10}$$

$$\begin{aligned} \text{var}(X) &= \sum \text{var}(X_j) + 2 \sum_{j<j'}\sum \text{cov}(X_j, X_{j'}) \\ &= \sum_{j=1}^{m} (1 - p_j)^n\{1 - (1 - p_j)^n\} \\ &\quad + 2 \sum_{j<j'}\sum \{(1 - p_j - p_{j'})^n - (1 - p_j)^n(1 - p_{j'})^n\} \\ &= \sum_{j=1}^{m} (1 - p_j)^n\left\{1 - \sum_{j=1}^{m} (1 - p_j)^n\right\} + 2 \sum_{j<j'}\sum (1 - p_j - p_{j'})^n. \end{aligned} \tag{3.11}$$

For the distribution (3.6), (descending) factorial moments are obtained by noting that

$$\begin{aligned}\mu_{(r)}(X) &= \sum_{k=0}^{m} k^{(r)}\binom{m}{k} m^{-n}\Delta^{m-k}\mathbf{0}^n \\ &= m^{(r)}\sum_{k=r}^{m}\binom{m-r}{k-r} m^{-n}\Delta^{m-k}\mathbf{0}^n \\ &= m^{-n}m^{(r)}\sum_{k'=0}^{m-r}\binom{m-r}{k'}\Delta^{m-r-k'}\mathbf{0}^n = \left(\frac{m-r}{m}\right)^n m^{(r)} \qquad (3.12)\end{aligned}$$

$$(r = 0, 1, \ldots, m-1).$$

[Of course, $\mu_{(r)}(X) = 0$ for $r \geq m$.]

From (3.12), we derive

$$\mu_1'(X) = m(1-m^{-1})^n,$$

$$\operatorname{var}(X) = m(m-1)(1-2m^{-1})^n + m(1-m^{-1})^n - m^2(1-m^{-1})^{2n}. \quad (3.13)$$

EXERCISE 3.4. Suppose that k indistinguishable balls are randomly placed in m urns, each urn having the same probability m^{-1} of assignment. If there are r balls in a particular urn, then r is the *occupancy number* of the urn. Show that the expected number of urns with occupancy number r is:

$$a_r = m\binom{k}{r}\left(\frac{1}{m}\right)^r\left(1-\frac{1}{m}\right)^{k-r} \qquad (r = 0, 1, \ldots, k).$$

This is $m[\Pr(X = r)]$, when X has a binomial distribution with parameters k, m^{-1}.

EXERCISE 3.5. In Exercise 3.4, letting $m = 365$, $k = 60$, verify that

$$\begin{aligned} a_0 &= 309.6 \\ a_1 &= 51.03 \\ a_2 &= 4.14 \\ a_3 &= 0.22 \end{aligned}$$

and interpret these results in terms of the coincidences of birthdays. (For $m = 365$, $k = 20$ we find $a_2 \doteqdot 1$. This is the well-known result that among 20 persons we may on the average find about one double birthday.)

In many applied problems, the occupancy of the urns alone is not the main interest. We are interested, rather, in the numbers of urns containing specified numbers of balls.

Let M_r be the number of urns containing exactly r balls. The joint distribution of $M_0, M_1, \ldots, M_n$ is

$$\Pr\left[\bigcap_{j=0}^{n}(M_j = m_j)\right] = \frac{m!n!}{m^n \prod_{j=0}^{n}[(j!)^{m_j}m_j!]} \quad \left(\sum_{j=0}^{n} m_j = m, \sum_{j=0}^{n} jm_j = n\right). \tag{3.14}$$

[For any set of values $m_0, m_1, m_2, \ldots, m_n$ there are $m!(\prod_{j=0}^{n} m_j!)^{-1}$ ways of assigning the urns to the $(m + 1)$ classes. Given that j balls have to be assigned to each of a specified m_j urns ($j = 0, 1, \ldots, n$) this can be done in $n!\{\prod_{j=0}^{n}(j!)^{m_j}\}^{-1}$ ways.]

This result was obtained by von Mises (1939).

Using formula (1.87), with

$$\begin{aligned} S_k &= \frac{n!}{(r!)^k(m - kr)!}\left(\frac{1}{m}\right)^{kr}\left(1 - \frac{k}{m}\right)^{m-kr} \\ &= \frac{n!}{(r!)^k(m - kr)!} m^{-m}(m - k)^{m-kr}, \end{aligned}$$

we find

$$\begin{aligned} \Pr[M_r = m_r] &= S_{m_r} - \binom{m_r + 1}{m_r} S_{m_r+1} + \binom{m_r + 2}{m_r} S_{m_r+2} - \cdots \\ &\quad + (-1)^{m-m_r}\binom{m}{m_r} S_m \\ &= n!m^{-m}\sum_{k=0}^{m-m_r}(-1)^k\binom{m_r + k}{m_r}\frac{(m - k)^{m-kr}}{(r!)^k(m - kr)!}. \end{aligned} \tag{3.15}$$

The moments of this distribution are determined using epgf's in Section 3.1.2.

EXERCISE 3.6. For distribution (3.15) show that

$$E[M_r] = \sum_{j=1}^{m}\binom{n}{j} p_j^r(1 - p_j)^{n-r}.$$

Note that the variable X, as defined at the beginning of this section, is in fact M_0. If we need to draw attention to the values of n (number of balls) and m (number of urns), we use the notation $M_0(n, m)$. Using this notation, we now establish two recurrence formulas:

$$\begin{aligned} \Pr[M_0(n + 1, m) = k] &= (1 - km^{-1})\Pr[M_0(n, m) = k] \\ &\quad + (k + 1)m^{-1}\Pr[M_0(n, m) = k + 1] \end{aligned} \tag{3.16a}$$

and

$$\Pr[M_0(n, m) = k] = (1 - m^{-1})^n \Pr[M_0(n, m-1) = k-1] + \sum_{j=1}^{n} \binom{n}{j} m^{-j}(1 - m^{-1})^{n-j} \Pr[M_0(n-j, m-1) = k]. \tag{3.16b}$$

Formula (3.16a) can be obtained by considering what happens if n balls are thrown into m urns and then one further ball is added. If k of the m urns are unoccupied after the first n balls have been thrown, then the conditional probability that the $(n + 1)$th ball will fall into an already occupied urn is $1 - km^{-1}$. There can be k unoccupied urns after $(n + 1)$ balls have been thrown, either because

1. k urns are unoccupied after n balls have been thrown and the $(n + 1)$th ball falls into an already occupied urn, or
2. $(k + 1)$ urns are unoccupied after n balls have been thrown and the $(n + 1)$th ball falls into an unoccupied urn.

Formula (3.16b) can be obtained by labeling the urns $1, 2, \ldots, m$. If no balls fall into the first urn [a probability of $(1 - m^{-1})^n$], then the n balls will be distributed among the remaining $(m - 1)$ urns in such a way that $(k - 1)$ of these urns are empty. This gives the first term on the right-hand side of (3.16b).

The probability that there will be g (≥ 1) balls in the first urn after n throws is $\binom{n}{g} m^{-g}(1 - m^{-1})^{n-g}$. The conditional probability that the remaining $(n - g)$ balls will be distributed among the other $(m - 1)$ urns in such a way that exactly k urns remain empty is $\Pr[M_0(n - g, m - 1) = k]$. This yields the second term on the right-hand side of (3.16b).

3.1.2. Generating Functions of Occupancy Distributions

In Section 3.1.1, M_j was defined as the number of urns containing exactly j balls. The epgf of M_j [cf. (1.166b)] which we use is

$$H_j^{(m)}(z, x) = \sum_{n=0}^{\infty} \sum_{g=0}^{\infty} \frac{m^n z^n}{n!} x^g \Pr[M_j = g \mid n]. \tag{3.17a}$$

We now show that

$$H_j^{(m)}(z, x) = \left\{ e^z + \frac{z^j}{j!}(x - 1) \right\}^m. \tag{3.17b}$$

To establish this, we first divide our collection of urns into two groups, containing m' and m'' $(=m - m')$ urns, respectively. Then, in an obvious notation,

$$\Pr[M_j = m_j|n] = \sum_{n'+n''=n} \sum_{g'+g''=g} \binom{n}{n'} \left(\frac{m'}{m}\right)^{n'} \left(\frac{m''}{m}\right)^{n''}$$
$$\times \Pr[M_j' = g'|n'] \Pr[M_j'' = g''|n''].$$

[This is a special case of the total probability formula (1.89a).]

We now multiply both sides of the equation by $(mz)^n x^g/n!$ and sum over both n and g, obtaining

$$H_j^{(m)}(z, x) = \sum_{n=0}^{\infty} \sum_{g=0}^{\infty} \sum_{n'+n''=n} \sum_{g'+g''=g} \left\{\frac{(m'z)^{n'}}{n'!} x^{g'} \Pr[M_j' = g'|n']\right\}$$
$$\times \left\{\frac{(m''z)^{n''}}{n''!} x^{g''} \Pr[M_j'' = g''|n'']\right\}$$
$$\left[\text{note that } \frac{(mz)^{n'}}{n'!} \left(\frac{m'}{m}\right)^{n'} = \frac{(m'z)^{n'}}{n'!}\right]$$
$$= H_j^{(m')}(z, x) H_j^{(m'')}(z, x).$$

Proceeding by induction we obtain

$$H_j^{(m)}(z, x) = [H_j^{(1)}(z, x)]^m, \tag{3.18}$$

where

$$H_j^{(1)}(z, x) = \sum_{n=0}^{\infty} \sum_{g=0}^{\infty} \frac{(mz)^n}{n!} x^g \Pr[M_j = g|n; 1].$$

Where there is only one urn, then necessarily it contains all the available balls, and so

$$\Pr[M_j = g|n; 1] = \begin{cases} 1 & \text{if } g = 1, n = j, \quad \text{or} \quad g = 0, n \neq j, \\ 0 & \text{otherwise.} \end{cases}$$

Hence

$$H_j^{(1)}(z, x) = \frac{z^j}{j!} x + \left(e^z - \frac{z^j}{j!}\right)$$
$$= e^z + (x - 1) \frac{z^j}{j!},$$

and so, from (3.18),

$$H_j^{(m)}(z, x) = \left[e^z + (x-1)\frac{z^j}{j!}\right]^m.$$

The (descending) factorial moments of M_j are

$$\begin{aligned} E[M_j^{(r)} | n] &= \left(\frac{n!}{m^n}\right) \times \left(\text{coefficient of } z^n \text{ in } \frac{d^r}{dx^r}\left[e^z + (x-1)\frac{z^j}{j!}\right]^m \Bigg|_{x=1}\right) \\ &= \left(\frac{n!}{m^n}\right) \times \left(\text{coefficient of } z^n \text{ in } m^{(r)} e^{(m-r)z}\left(\frac{z^j}{j!}\right)^r\right) \\ &= \frac{n!}{m^n} m^{(r)} \frac{(m-r)^{n-rj}}{(n-rj)!} \frac{1}{(j!)^r} \\ &= m^{(r)} \frac{n!}{(n-rj)!(j!)^r}\left(\frac{1}{m}\right)^{rj}\left(1 - \frac{r}{m}\right)^{n-rj}. \end{aligned} \quad (3.19)$$

[If we put $j = 0$, we obtain $m^{(r)}(1 - rm^{-1})^n$, agreeing with (3.12).]

We now return to the case (discussed at the beginning of this chapter) where the probabilities of assignment to each of the m urns are not all equal. We can use formula (1.87) to evaluate probabilities in this case, as indicated above. Here we describe how a generating function relevant to the distribution of the maximum frequency (i.e., the largest number of balls contained in a single urn) can be evaluated. We define [cf. (1.166b)]

$$H(z) = \sum_{j=0}^{\infty} P_x(m, j)\frac{z^j}{j!}, \quad (3.20)$$

where $P_x(m, j)$ = probability that the maximum frequency exceeds x = probability that at least one frequency exceeds x. The probability that, when j balls are thrown, the number of balls falling in the ith urn is x_i for $i = 1, \ldots, m$ is

$$\Pr\left[\bigcap_{i=1}^{k}(X_i = x_i)\right] = \frac{j!}{\prod_{i=1}^{m} x_i!} \prod_{i=1}^{m} p_i^{x_i} \quad \left(\sum_{i=1}^{k} x_i = j\right), \quad (3.21)$$

where p_i = Pr[ball falls in the ith urn], and of course $\sum_{i=1}^{k} p_i = 1$.

Hence

$$P_x(m, j) = j! \sum_x^{(j)} \left\{\prod_{i=1}^{m} \frac{p_i^{x_i}}{x_i!}\right\},$$

summation being over $\max(x_1, x_2, \ldots, x_m) > x$, and $x_1 + x_2 + \cdots + x_m = j$.

It follows that

$$H(z) = \sum_j z^j \sum_x^{(j)} \left\{ \prod_{i=1}^m \frac{p_i^{x_i}}{x_i!} \right\}$$

$$= \sum_j \sum_x^{(j)} \left[\prod_{i=1}^m \frac{(p_i z)^{x_i}}{x_i!} \right] \tag{3.22}$$

(remembering that $\sum_{i=1}^m x_i = j$).

If no terms were excluded from $\sum_x^{(j)}$ the right-hand side of (3.22) would equal

$$\prod_{i=1}^m \left[\sum_{x_i=0}^{\infty} \frac{(p_i z)^{x_i}}{x_i!} \right] = \prod_{i=1}^m e^{p_i z} = e^z.$$

In fact, all sets of values $x_1, \ldots, x_m$ for which $\max(x_1, x_2, \ldots, x_m) \le x$ are excluded, so we must subtract

$$\prod_{i=1}^m \left\{ \sum_{h=0}^x \frac{(p_i z)^h}{h!} \right\},$$

giving

$$H(z) = e^z - \prod_{i=1}^m \left\{ \sum_{h=0}^x \frac{(p_i z)^h}{h!} \right\}. \tag{3.23}$$

(See Richards, 1968.)

3.1.3. Location and Occupancy Vectors

Occupancy problems occur in many kinds of applications. Suppose a population is divided into m strata and we wish to draw a sample containing v_j individuals from the jth stratum ($j = 1, \ldots, m$). If individuals are chosen at random and the population is large, we have a situation like that described in the beginning of this chapter (Section 3.1.1), with p_j = proportion of individuals in the population belonging to the jth stratum.

We are not here concerned with simple occupancy of each urn (stratum) but with "sufficiently multiple" occupancy, as specified by the numbers $v_1, v_2, \ldots, v_m$. The notions of *location* and *occupancy* vectors are useful in this connection. [These notions are discussed in more detail by von Mises (1964) and Thomasian (1969).]

Consider the distribution of n balls among k urns. A total of $n \ge 1$ balls labeled $1, 2, \ldots, n$ is distributed among $k > 1$ urns (boxes or cells) labeled $1, 2, \ldots, k$. Which ball is in which urn can be described by the n-dimensional vector $\mathbf{x} = (x_1, \ldots, x_n)$ called the *location* vector, where x_i is the *number of the urn* in which the ith ball is located. x_i is called the *location* number. There are k^n location vectors.

Example 3.5. $n = 10$, $k = 3$, and the location vector $\mathbf{x} = (2, 2, 3, 1, 1, 2, 3, 3, 2, 3)$ is given by

$$\begin{matrix} \text{Urn 1} & \text{Urn 2} & \text{Urn 3} \\ \begin{pmatrix} 4 \\ 5 \end{pmatrix} & \begin{pmatrix} 1 & 6 \\ 2 & 9 \end{pmatrix} & \begin{pmatrix} 8 & 3 \\ 7 & 10 \end{pmatrix}. \end{matrix}$$

One can also define a k-dimensional occupancy vector $\mathbf{r} = (r_1, r_2, \ldots, r_k)$, where r_j is the number of balls in urn j. The r_j's are called the *occupancy numbers*. In this example $\mathbf{r} = (2, 4, 4)$.

Clearly, $\sum_{j=1}^{k} r_j = n$. Moreover, there are exactly $\binom{n + k - 1}{k - 1}$ different occupancy vectors. This can be established by the following argument.

To an occupancy vector $\mathbf{r} = (r_1, r_2, \ldots, r_k)$ we correspond a $\mathbf{r}'$-tuple of 0's and 1's, constructed as follows: the first $(r_1 + 1)$ coordinates of $\mathbf{r}'$ are r_1 0's followed by 1; the next $(r_2 + 1)$ coordinates of $\mathbf{r}'$ are r_2 0's followed by 1, and so on through $r_{k-1} + 1$. The last r_k coordinates of $\mathbf{r}'$ are r_k0's. [For example, if $r = (3, 2, 0, 4, \ldots, 3, 1)$, then

$$\mathbf{r}' = (\underbrace{0, 0, 0}_{r_1}, 1, \underbrace{0, 0}_{r_2}, 1, \underbrace{}_{r_3} 1, \underbrace{0, 0, 0, 0}_{r_4}, 1, \ldots, 1, \underbrace{0, 0, 0}_{r_{k-1}}, 1, \underbrace{0}_{r_k}).]$$

Each $\mathbf{r}'$ has a corresponding $\mathbf{r} = (r_1, r_2, \ldots, r_k)$, where r_j is the number of zeros between the $(j - 1)$st and jth 1 in $\mathbf{r}'$.

Each $\mathbf{r}'$ is a $(n + k - 1)$-tuple with $(k - 1)$ coordinates equal to 1 and n coordinates equal to 0. Each $\mathbf{r}'$ is obtained from one (and only one) $\mathbf{r}$. Hence the number of occupancy vectors is equal to the number of $\mathbf{r}$'s. But there are $\binom{n + k - 1}{n} = \binom{n + k - 1}{k - 1}$ different $(n + k - 1)$-tuples with $(k - 1)$ coordinates equal to 1 and n coordinates equal to 0.

Note that we could have used this result to derive the number of equally likely B–E arrangements in Table 1.2 in Section 1.3.4.

EXERCISE 3.7. For $k = 3$, $n = 4$, list all $\binom{4 + 3 - 1}{2} = 15$ occupancy vectors.

EXERCISE 3.8. Using elementary combinatorial arguments, prove that the number of *location vectors* which have the same corresponding occupancy vector $(r_1, r_2, \ldots, r_k)$ where $r_1 + r_2 + \cdots + r_k = n$, is $n!/r_1!r_2!r_3! \cdots r_k!$.

EXERCISE 3.9. (Thomasian (1969)) Verify the following recursion formula for multinomial coefficients using the notation of an occupancy vector. If $r_1 + r_2 + \cdots + r_k = n + 1$, where $r_i \geq 1$, then

$$\frac{(n+1)!}{r_1!r_2!\cdots r_k!} = \sum_{j=1}^{k} \frac{n!}{r_1!r_2!\cdots r_{j-1}!(r_j - 1)!r_{j+1}!\cdots r_k!}. \tag{3.24}$$

[*Hint*: Use $(n + 1)! = n!(r_1 + \cdots + r_k)$ on the left-hand side of the recursion formula.]

3.2. RELATED OCCUPANCY DISTRIBUTIONS

3.2.1. Some Modifications of Classical Occupancy

There are many variants of the classical occupancy problem arising both from practical requirements and mathematical interest. In this section we describe a few of the more important variations. Later sections deal with further modifications of special interest.

Example 3.5. There are n men in a room; n women enter the room, and each selects a man at random and holds one of his hands if not already held by another woman (so that no man can be selected by more than two women).

Find the joint distribution of the numbers of men with (i) both hands, (ii) one hand, (iii) neither hand held by a woman.

(This problem, which arose in a chemical industry enquiry, was communicated to us by P. Hagis, Jr. and C. Schmidt.)

This is an occupancy problem with n urns (men) and a maximum of two balls (hands held) per man.

Let N_j denote the number of men with j hands held ($j = 0, 1, 2$). Then

$$\sum_{j=0}^{2} jN_j = n = \sum_{j=0}^{2} N_j,$$

so that $N_2 = N_0$, $N_1 = n - 2N_0$, and we need only find the distribution of N_0.

The event $N_0 = r$ will occur if no woman takes any of the corresponding $2r$ hands out of a total of $2n$ available hands. Hence

$$\Pr[N_0 = r] = \frac{\binom{2(n-r)}{n}}{\binom{2n}{n}} = \frac{\{2(n-r)\}^{(n)}}{(2n)^{(n)}} \qquad (0 \leq r \leq \tfrac{1}{2}n).$$

EXERCISE 3.10. Suppose that, in the situation described in Example 3.5, there are m men and w women. Without finding the distribution of N_0, N_1, or N_2, show that

$$\frac{E[N_2]}{E[N_1]} = \frac{1}{2}\frac{(w-1)}{(2m-w)}.$$

[*Hint*: For any one man, the probability that neither of his hands will be held is $\binom{2m-2}{w}\Big/\binom{2m}{w}$.]

Note: $\text{var}(N_2) = \dfrac{w(w-1)(2m-w)(2m-w+1)}{2(2m-1)^2(2m-3)}$. This can be established by noting that

(i) $N_2 = \sum_{i=1}^{m} X_i$,

where

$$X_i = \begin{cases} 1 & \text{if } i\text{th man has both hands held,} \\ 0 & \text{otherwise.} \end{cases}$$

(ii) $E[X_i X_j]$ = probability that both the ith man and the jth man have both hands held $= w^{(4)}/(2m)^{(4)}$.

In the classical occupancy situation, if we consider a subset of t out of the m available urns, the number of balls T dropped into any one of this subset is distributed binomially with parameters n, tm^{-1}.

For a given value of T the probability that exactly k out of the subset of t urns are occupied is

$$\Pr[Y = k \mid T] = \binom{t}{k}\Delta^k\left(\frac{\mathbf{0}}{t}\right)^T.$$

Averaging this over the distribution of T, we obtain

$$\begin{aligned}\Pr[Y = k] &= \binom{t}{k}\Delta^k \sum_{T=0}^{n}\binom{n}{T}(1 - tm^{-1})^{n-T}(m^{-1}\mathbf{0})^T \\ &= \binom{t}{k}\Delta^k(1 - tm^{-1} + m^{-1}\mathbf{0})^n. \qquad (3.25)\end{aligned}$$

This result was used by Jones (1959) and Glasser (1963) in connection with random number generation.

Note that this distribution is a *mixture* of classical occupancy distributions with a binomial mixing distribution for the number of balls assigned.

More general mixing distributions for the number of balls assigned have been considered by Ivchenko and Medvedev (1965). They obtain some limiting distributions which are discussed in Chapter 6.

A slightly different case is now discussed. Suppose there are m urns. Each ball is equally likely (with probability p) to be assigned to any one of b urns ($bp \leq 1$) of the total population. We call these class I urns. The remaining $(m - b)$ urns we call class II urns. The probability of assignment to some one of the class II urns is $(1 - bp)$.

We first evaluate the probability that after n balls have been assigned exactly j of the b class I urns will be occupied.

We denote by V_n the number of different class I urns *occupied* and by R_1 the number of balls assigned to a class I urn (*any one* among the b in this class). Then R_1 has a binomial distribution with parameters n, bp.

The distribution of V_n, given R_1, is the same as that of the number of different urns occupied when there are just b equally likely urns (i.e., each in proportion b^{-1}) and R_1 *balls*. This is the classical occupancy distribution just discussed. Hence, from equation (3.6), putting $m = b$, $m - k = j$, $n = R_1$,

$$\Pr[V_r = j \mid R_1] = \binom{b}{j} \frac{\Delta^j \mathbf{0}^{R_1}}{b^{R_1}}. \tag{3.26}$$

Averaging over the distribution of R_1,

$$\Pr[V_n = j] = \binom{b}{j} E\left[\frac{\Delta^j \mathbf{0}^{R_1}}{b^{R_1}}\right] = \binom{b}{j} E\left[\Delta^j \left(\frac{\mathbf{0}}{b}\right)^{R_1}\right]. \tag{3.27}$$

Using the formula

$$E[A^{R_1}] = (1 - bp + bpA)^n, \tag{3.28}$$

where A is an arbitrary constant, we obtain (putting $A = \mathbf{0}/b$),

$$\Pr[V_n = j] = \binom{b}{j} \Delta^j (1 - bp + p\mathbf{0})^n. \tag{3.29}$$

Since

$$\Delta^j f(x) = \sum_{h=0}^{j} (-1)^{j-h} \binom{j}{h} f(x + h) = \sum_{h=0}^{j} (-1)^h \binom{j}{h} f(x + j - h)$$

[cf. (1.15)], we can express (3.29) as

$$\Pr[V_n = j] = \binom{b}{j} \sum_{h=0}^{j} (-1)^{j-h} \binom{j}{h} \{1 - (b - h)p\}^n \tag{3.29'}$$

or

$$\Pr[V_n = j] = \binom{b}{j} \sum_{h=0}^{j} (-1)^h \binom{j}{h} \{1 - (b - j + h)p\}^n. \tag{3.29''}$$

This result can be used to obtain conditional distributions of the number of occupied urns, given that a certain number, say g, out of m are already occupied, in the *classical* occupancy case.

We regard these g as class II urns, and the remaining $b = m - g$ unoccupied urns as class I urns.

So if assignment of the first s, out of n, balls occupies g urns, then the probability that a further $(k - g)$ urns will be occupied (making a total of k occupied urns), after assigning the remaining $(n - s)$ balls is obtained by putting $b = m - g$, $p = m^{-1}$, and $j = k - g$, and replacing n by $n - s$ in (3.29), (3.29′), or (3.29″). Using (3.29), we obtain

$$\binom{m-g}{k-g}\Delta^{k-g}\{1 - (m-g)m^{-1} + m^{-1}\mathbf{0}\}^{n-s} = \binom{m-g}{k-g}\Delta^{k-g}\left(\frac{g+\mathbf{0}}{m}\right)^{n-s}. \tag{3.30}$$

EXERCISE 3.11. Show that the probability of hitting j targets (not already hit) with s shots, having hit g targets previously, where there are m targets altogether (each equally likely to be hit) and it is assumed that each shot hits a target, is

$$\binom{m-g}{j}\Delta^{j}\{gm^{-1} + (1 - gm^{-1})\mathbf{0}\}^{s}.$$

Give two other equivalent formulas. (For additional military applications see Section 5.5.1.)

EXERCISE 3.12. Suppose we have b_1 class-1 urns, b_2 class-2 urns, and b_3 class-3 urns. The probability of assignment to any specified class-j urn is p_j, with $b_1p_1 + b_2p_2 + b_2p_3 = 1$. Find the joint distribution of the numbers of occupied class-1 and class-2 urns after the assignment of n balls.

EXERCISE 3.13. For the distribution (3.29) we can represent V_n as the sum of b independent indicator variables $Z_1, Z_2, \ldots, Z_b$, where

$$Z_t = \begin{cases} 1 & \text{if the } t\text{th urn in class 1 is occupied,} \\ 0 & \text{otherwise.} \end{cases}$$

Using this fact, show that

$$E[V_n] = b\{1 - (1-p)^n\}$$
$$\text{var}(V_n) = b\{(1-p)^n - (1-2p)^n\} - b^2\{(1-p)^{2n} - (1-2p)^n\}. \tag{3.31}$$

Among further variations on problems of occupancy we may note the following.

1. We may be concerned with the distribution of the minimum number of balls in any one urn, or with the maximum. As we have already seen

[equation (3.1)], the joint distribution of the number of balls in the urns is multinomial, so it is effectively the distributions of minimum and maximum of a set of multinomial variables that we require. (See Young, 1961; Johnson and Young, 1960.)

2. The probabilities p_j may change according to the number of balls already in the corresponding urn, or simply with the successive assignments of balls. (These cases are discussed in Chapter 4.)

3. The balls may be of different colors (in a specified proportion), and we are interested in obtaining certain *minimum* numbers of each in each urn.

We consider a few specific problems in detail. Suppose that we draw (with replacement) balls from an urn in which the proportion of balls with color $\mathscr{C}_j$ is p_j $(j = 1, 2, \ldots, k; \sum_{j=1}^{k} p_j = 1)$.

We wish to evaluate the probability that a specified color $\mathscr{C}_a$ $(a > r)$ is not observed until after all of a specified set of colors $\mathscr{C}_1, \ldots, \mathscr{C}_r$ $(r < k)$ have been observed.

The required probability is clearly the sum, with respect to h, of

$$\begin{aligned}&\Pr[\mathscr{C}_a \text{ observed for the first time on the } (h+1)\text{th draw}]\\ &\quad\times \Pr[\text{each of } \mathscr{C}_1, \mathscr{C}_2, \ldots, \mathscr{C}_r \text{ observed at least once in the first}\\ &\qquad h \text{ draws} \mid \mathscr{C}_a \text{ not observed in any of these draws}].\end{aligned}$$

This is

$$\sum_{h=r}^{\infty} (1 - p_a)^h p_a \left\{1 - \sum_{j=1}^{r} \bar{p}_j^h + \sum_{1 \le j < j'}^{r} \sum \bar{p}_{j,j'}^h - \cdots\right\}, \tag{3.32a}$$

where $\bar{p}_j = 1 - (1 - p_a)^{-1} p_j$, $\bar{p}_{j,j'} = 1 - (1 - p_a)^{-1}(p_j + p_{j'})$, and so on.

More generally, we can write

$$\begin{aligned}\bar{p}_j &= 1 - \Pr[\mathscr{C}_j | \overline{\mathscr{C}}_a]\\ \bar{p}_{j,j'} &= 1 - \Pr[\mathscr{C}_j \cup \mathscr{C}_{j'} | \overline{\mathscr{C}}_a], \ldots .\end{aligned} \tag{3.32b}$$

This is a special case of the following problem, due to Thorp (1964). "Consider a series of repeated independent trials with the outcomes of each trial being events in a given (fixed) sample space. Let $E_a, E_1, \ldots, E_r$ be $(r + 1)$ events, with E_a disjoint from each of the E_i. What is the probability that all the events E_i $(i \ge 1)$ will occur before the event E_a occurs?" If the event E_i is "ball has color $\mathscr{C}_i$," we have the problem we have been considering. In the present case it is not specified that $E_1, E_2, \ldots, E_r$ are mutually exclusive, though E_a must be disjoint from each of $E_1, E_2, \ldots, E_r$.

The solution of Thorp's problem is obtained by using formulas (3.32) for $\bar{p}_j, \bar{p}_{j,j'}, \ldots$.

Note that, although the problem was not stated directly in terms of the occupancy models used earlier in the chapter, it can be so stated by regarding the *colors* as different *urns* and the sampling of a ball as the assignment of a ball to the urn corresponding to its color.

If the maximum allowed number of balls in any one urn is k, the total number of ways of distributing the n balls among the m urns is the coefficient of x^n in

$$\begin{aligned}(1 + x + x^2 + \cdots + x^k)^m &= (1 - x^{k+1})^m(1 - x)^{-m} \\ &= (1 - x^{k+1})^m \sum_{j=0}^{\infty} \binom{m + j - 1}{m - 1} x^j.\end{aligned}$$

This coefficient is

$$N(n, m, k) = \sum_{j=0}^{m} (-1)^j \binom{m}{j} \binom{m + n - j(k + 1) - 1}{m - 1}.$$

(Of course, if $n \geq km$ this takes the value zero.) Dividing $N(n, m, k)$ by m^n gives the probability that no urn contains more than k balls.

Freund and Pozner (1956) point out that the number of ways of distributing n balls among m urns with no more than k balls per urn, subject to the condition that *exactly* h urns have *exactly* g balls each is

(number of ways of choosing the urns) $\times$ [number of ways of distributing $(n - gh)$ balls among $(m - h)$ urns so that no urn contains more than k balls, or exactly g balls]

$$\begin{aligned}&= \binom{m}{h} \times \left[\text{coefficient of } x^{n-gh} \text{ in } \left(\sum_{u=0}^{m} x^u - x^g\right)^{m-h}\right] \\ &= N(n, m, k \mid g, h) \text{ say.}\end{aligned}$$

Hence the conditional distribution of the number of urns containing g balls, given that no urn contains more than k balls, is

$$\Pr[\text{number of urns with } g \text{ balls} = h] = \frac{N(n, m, k \mid g, h)}{N(n, m, k)}.$$

EXERCISE 3.14. For the preceding distribution:

(i) Show that the tth factorial moment is

$$\frac{m^{(t)} N(n - tg, m - t, k)}{N(n, m, k)}.$$

(ii) Show that the $(t_1, t_2, \ldots, t_r)$th factorial moment of the joint conditional distribution of the numbers $(H_1, H_2, \ldots, H_r)$ of urns containing exactly $g_1, g_2, \ldots, g_r$ balls respectively is

$$E\left[\prod_{j=1}^{r} H_j^{(t_j)}\right] = \frac{m^{\left(\sum_{j=1}^{r} t_j\right)} N\left(n - \sum_{j=1}^{r} t_j g_j, m - \sum_{j=1}^{r} g_j, k\right)}{N(n, m, k)}$$

(see Freund and Pozner, 1956.)

Similar kinds of distributions are discussed by Baticle (1933, 1935, 1946), who suggests applications to commercial problems.

The classical occupancy problem and the modifications so far discussed do not take into account the positions of occupied urns relative to each other.

A simple kind of situation arises when the m urns are set out along a line, that is, one-dimensionally. Suppose we have n balls of which n_i are of color $\mathscr{C}_i$ $(i = 1, \ldots, k)$, and we assign these balls to urns so that

1. No urn contains more than one ball, and
2. All n_i balls of color $\mathscr{C}_i$ are in consecutive urns $(i = 1, \ldots, k)$. [Clearly, for condition 1 to be satisfied $\sum_{i=1}^{k} n_i = n \leq m$.]

We wish to find the number of different arrangements (balls of the same color being supposed indistinguishable). To find this we temporarily regard each arrangement as one of k groups of urns (each urn of a group containing balls of the same color) and $(m - n)$ empty urns (Pease, 1975).

These $(m - n + k)$ entities of two kinds (groups and empty urns) can be arranged in

$$\binom{m - n + k}{k}$$

different orders, this being the number of ways of choosing the positions of the k groups of nonempty urns. In each case there are $k!$ different ways of ordering the latter (the $k!$ ways of ordering $\mathscr{C}_1, \mathscr{C}_2, \ldots, \mathscr{C}_k$), and so the total number of different arrangements is

$$k!\binom{m - n + k}{k}.$$

If the order of the colors of the groups is specified then of course the number of arrangements is just

$$\binom{m - n + k}{k}.$$

Note that these results do not depend on the numbers $n_1, n_2, \ldots, n_k$ of the balls of each of the colors, but only on their total n and on the number k of different colors, in addition to the number m of urns. (Section 1.3.4 is relevant to these problems.)

EXERCISE 3.15. How many arrangements are there if there must be at least one empty urn between any two urns containing balls of different colors?

3.2.2. Occupancy Distributions with Bose–Einstein Statistics

If n (indistinguishable) balls are assigned to m (distinguishable) urns subject to the condition that each urn must contain at least one ball, there are $\binom{n-1}{m-1}$ ways of making this assignment (see Table 1.2 in Section 1.3.4). If it is supposed that each such arrangement is equally likely (B–E statistics, see Section 1.3.4), then each arrangement has a probability of $\binom{n-1}{m-1}^{-1}$.

Let M_s denote the number of urns containing s balls, and denote the *number* of balls in the ith urn by N'_i $(i = 1, \ldots, m)$, with the corresponding *order statistics* (i.e., the N''s arranged in increasing magnitude) $N_1 \leq N_2 \leq \cdots \leq N_m$. [We note that the random variables $N'_1, N'_2, \ldots, N'_m$ are exchangeable (see Section 2.9).]

EXERCISE 3.16. Show that

$$\Pr[M_0 = m_0] = \frac{\binom{m}{m_0}\binom{n-1}{m-m_0-1}}{\binom{n+m-1}{n}} \tag{3.33a}$$

and

$$E[M_0^{(k)}] = \frac{m^{(k)}\binom{n+m-k-1}{n}}{\binom{n+m-1}{n}}. \tag{3.33b}$$

Hill (1974) points out that (for a an integer)

$$\Pr[N_r \leq a] = \Pr\left[\sum_{j=1}^{a} M_j \geq r\right] \tag{3.33c}$$

(since each side is the probability that at least r urns contain no more than a balls).

Now

$$\Pr\left[\sum_{j=1}^{a} M_j = r|m, n\right] = \binom{m}{r} \sum_{j=0}^{r} (-1)^j \binom{r}{j} \Pr\left[\bigcap_{i=1}^{m-r+j} (N'_i > a)|m, n\right] \tag{3.34}$$

[from (1.87)], because $\sum_{j=1}^{a} M_j = r$ if exactly $(m - r)$ of the events $(N'_j > a)$ occur.

Inserting (3.34) in (3.33c), we obtain

$$\Pr[N_r \leq a|m, n] = \sum_{i=0}^{m-r} \binom{m}{r+i} \sum_{j=0}^{r+i} (-1)^j \binom{r+i}{j} \times \Pr\left[\bigcap_{h=1}^{m-r-i+j} (N'_h > a)|m, n\right]. \tag{3.35}$$

For $r = m$, we have, for the distribution of the *maximum* number of balls in any one urn,

$$\Pr[N_m \leq a|m, n] = \sum_{j=0}^{m} (-1)^j \binom{m}{j} \Pr\left[\bigcap_{i=1}^{j} (N'_i > a)|m, n\right]. \tag{3.36}$$

If, as we have supposed, B–E statistics apply, then

$$\Pr\left[\bigcap_{i=1}^{j} (N'_i > a)|m, n\right] = \binom{n-1}{m-1}^{-1} \binom{n - ja - 1}{m-1}. \tag{3.37}$$

(See Exercise 1.26.)

Replacing a by na in (3.35) and using (3.37), we obtain

$$\Pr[n^{-1}N_r \leq a|m, n] = \sum_{i=0}^{m-r} \binom{m}{r+i} \sum_{j=0}^{r+i} (-1)^j \binom{r+i}{j} \binom{n-1}{m-1}^{-1} \times \binom{n - (m - r - i + j)na - 1}{m-1}$$

and

$$\lim_{n\to\infty} \Pr[n^{-1}N_r \leq a|m, n] = \sum_{i=0}^{m-r} \binom{m}{r+i} \sum_{j=0}^{r+i} (-1)^j \binom{r+i}{j} \times [1 - (m - r + j)a]^{m-1}. \tag{3.38}$$

Note that

$$\binom{n-1}{m-1}^{-1} \binom{n - n\alpha a - 1}{m-1} = \frac{(1 - \alpha a - n^{-1})(1 - \alpha a - 2n^{-1}) \cdots (1 - \alpha a - (m-1)n^{-1})}{(1 - n^{-1})(1 - 2n^{-1}) \cdots (1 - (m-1)n^{-1})}. \tag{3.39}$$

In particular, if $r = m$, we have

$$\lim_{n\to\infty} \Pr[n^{-1}N_m \le a|m, n] \sum_{j=0}^{m} (-1)^j \binom{m}{j} (1 - ja)^{m-1} = (-1)^m \Delta^m (1 - a\mathbf{0})^{m-1}. \tag{3.40}$$

Limiting distributions as $m \to \infty$ are discussed in Chapter 6.

EXERCISE 3.17. Show that, for M–B statistics (see Section 1.3.4),

$$\Pr[N_m \le a|m, n] = \sum_{j=0}^{m} (-1)^j \binom{m}{j} \Pr\left[\bigcap_{i=1}^{j} (N_i' > a)\right], \tag{3.41}$$

where $N_1', \ldots, N_m'$ have a joint multinomial distribution with parameters $m^{-1}, \ldots, m^{-1}; m$. [Compare this with (3.40).]

EXERCISE 3.18. Consider the F–D occupancy model of distributing r indistinguishable balls into m distinguishable urns (cells), subject to the conditions

(i) No cell can contain more than one ball (i.e., $r \le m$).

(ii) All $\binom{m}{r}$ distinguishable outcomes subject to (i) are equally probable.

Let $\mathbf{r}^{(j)} = (r_1^{(j)}, r_2^{(j)}, \ldots, r_m^{(j)})$ be the occupancy vector after j balls have been thrown. r_i denotes the number of balls in the ith urn. For $i = 1, \ldots, m$, define $p_i^{(j)} = \Pr[\mathbf{r}^{(j)} = (r_1, \ldots, r_i + 1, \ldots, r_m)|\mathbf{r}^{(j-1)} = (r_1, \ldots, r_i, \ldots, r_m)]$. This is the probability that the jth ball is thrown into the ith urn, given the occupancy vector $\mathbf{r}^{(j-1)} = (r_1, \ldots, r_m)$.

Show that

$$p_i^{(j)} = \begin{cases} \dfrac{1}{n - j + 1} & \text{if } r_i = 0, \\ 0 & \text{if } r_i = 1. \end{cases}$$

[*Hint*: Observe that to generate a F–D model one can choose an empty urn uniformly at random at each toss (Bolmarcich and Belkin, 1974).]

3.2.3. Waiting-Time Problems

Waiting-time problems arise when we wish to consider the number of balls needed to satisfy specified occupancy conditions for the urns. We might, for example, consider the number of assignments (one ball at a time) needed for

1. All urns to be occupied.
2. At least g urns to have at least h balls each.
3. g specified urns to have at least h balls each, and so on.

Sometimes waiting-time calculations are simplified by using the (conditional) distribution of numbers of balls in the urns, given N, the number of balls thrown. In particular, if the probabilities of falling into the m cells are each equal to $1/m$, then the joint distribution of $N_1, N_2, \ldots, N_m$, the numbers of balls in the m cells, given that $N_1 + N_2 + \cdots + N_m = N$ is multinomial, with

$$\Pr\left[\bigcap_{j=1}^{m} (N_j = n_j) \middle| \sum_{j=1}^{m} N_j = N\right] = m^{-N} N! \left\{\prod_{j=1}^{m} n_j!\right\}^{-1}.$$

This is also the joint distribution of $N_1, N_2, \ldots, N_m$ conditional on $N_1 + N_2 + \cdots + N_m = N$, if $N_1, N_2, \ldots, N_m$ are independent random variables with a common Poisson distribution. (This is a special case of a result stated in Section 2.7.1.)

This fact has been exploited by Klamkin and Newman (1967) and Dwass (1969) to obtain simple asymptotic formulas for certain sequential occupancy probabilities.

In particular, let W denote the waiting time until one urn has k balls in it. If $N_1, N_2, \ldots, N_m$ are independent Poisson variables, each with expected value $m^{-1}t$ (so that N is Poisson with expected value t), then the probability that each urn contains fewer than k balls is

$$\left(\sum_{j=0}^{k-1} e^{-t/m} \frac{(t/m)^j}{j!}\right)^m. \tag{3.42}$$

This probability can also be expressed as

$$\sum_{n=0}^{\infty} \Pr[W > n] \Pr[N = n] = \sum_{n=0}^{\infty} \Pr[W > n] \left\{e^{-t} \frac{t^n}{n!}\right\}. \tag{3.43}$$

From the identity

$$\left(\sum_{j=0}^{k-1} e^{-t/m} \frac{(t/m)^j}{j!}\right)^m = \sum_{n=0}^{\infty} \Pr[W > n] e^{-t} \frac{t^n}{n!}, \tag{3.44}$$

we wish to determine $\Pr[W > n]$.

If both sides of (3.44) are multiplied by $t^r e^t$ and then integrated with respect to t between 0 and ∞, we obtain

$$\begin{aligned} \int_0^{\infty} \left[T_k\left(\frac{t}{m}\right)\right]^m t^r e^{-t}\, dt &= \sum_{n=0}^{\infty} \Pr[W > n] \int_0^{\infty} e^{-t} \frac{t^{n+r}}{n!}\, dt \\ &= \sum_{n=0}^{\infty} \Pr[W > n] (n + r)^{(r)} \\ &= \sum_{n=r}^{\infty} \Pr[W + r > n] n^{(r)} \\ &= \frac{1}{r + 1} \mu_{(r+1)}(W + r), \end{aligned} \tag{3.45}$$

where

$$T_k\left(\frac{t}{m}\right) = \sum_{j=0}^{k-1} \frac{(t/m)^j}{j!}.$$

Replacing r by $(r - 1)$ in (3.45), we have

$$\mu_{(r)}(W + r - 1) = r \int_0^\infty \left[T_k\left(\frac{t}{m}\right)\right]^m t^{r-1} e^{-t}\, dt. \tag{3.46}$$

In particular, for $r = 1$, we have

$$E[W] = \int_0^\infty \left[T_k\left(\frac{t}{m}\right)\right]^m e^{-t}\, dt.$$

Putting $t = m^{1-k^{-1}} u$, we have

$$E[W] = m^{1-k^{-1}} \int_0^\infty \left[T_k\left(\frac{u}{m^{k^{-1}}}\right)\exp\left(\frac{-u}{m^{k^{-1}}}\right)\right]^m du. \tag{3.47}$$

We now note that

$$\begin{aligned} T_k(y)e^{-y} &= \left[1 + \left\{\frac{y^k}{k!} + \frac{y^{k+1}}{(k+1)!} + \cdots\right\}\left\{1 + \frac{y}{1!} + \cdots + \frac{y^{k-1}}{(k-1)!}\right\}^{-1}\right]^{-1} \\ &= 1 - \frac{y^k}{k!} + (\text{terms in } y^{k+1}, y^{k+2}, \ldots), \end{aligned}$$

and so

$$\lim_{m\to\infty}\left[T_k\left(\frac{u}{m^{k^{-1}}}\right)\exp\left(\frac{-u}{m^{k^{-1}}}\right)\right]^m = \lim_{m\to\infty}\left[\left(1 - \frac{u^k}{k!}\frac{1}{m}\right)^m\right] = \exp\left(\frac{-u^k}{k!}\right).$$

We also have

$$T_k(y)e^{-y} \le \left(1 + \frac{y^k}{k!}\right)^{-1},$$

and so

$$\left\{T_k\left(\frac{u}{m^{k^{-1}}}\right)\exp\left(\frac{-u}{m^{k^{-1}}}\right)\right\}^m \le \left(1 + \frac{u^k}{k!}\right)^{-1}.$$

The integral on the right-hand side of (3.47) is dominantly convergent, and so we can take limits through the integral sign, giving

$$\lim_{m\to\infty} \frac{E[W]}{m^{1-k^{-1}}} = \int_0^\infty \exp\left(-\frac{u^k}{k!}\right) du = (k!)^{k^{-1}}\Gamma(k^{-1} + 1). \tag{3.48}$$

This is a limiting result on occupancy distributions. Many similar (and more complex) results can be found in Chapter 6.

Further results are given in the following exercises.

EXERCISE 3.19. Show that

$$\lim_{m \to \infty} \frac{E[(W + r - 1)^{(r)}]}{m^{r(1-k^{-1})}} = \frac{r}{k}(k!)^{r/k}\Gamma\left(\frac{r}{k}\right). \tag{3.49}$$

EXERCISE 3.20. In one form of the classical birthday problem (cf. Exercise 1.22) we seek the expected number of persons needed to find two with a common birthday (assuming 365 days to the year and each day independently equally likely to be the birthday of a given person). The exact value has been found to be 24. Obtain the approximation given by using formula (3.48).

(Note that this problem is *different* from that discussed in Exercises 3.4 and 3.5. In those exercises we sought to make the expected number of *urns* with $r\,(=2)$ balls equal to 1; here we seek the expected number of *balls* needed so that at least one urn contains two balls. Yet another possibility is to seek the number of balls needed so that the probability of there being an urn with two or more balls in it exceeds $\frac{1}{2}$.)

EXERCISE 3.21. Give an approximate solution to an extended birthday problem when four persons with a common birthday are sought.

EXERCISE 3.22 (Nymann, 1975). In a table of random numbers, each of the numbers 000 through 999 appears independently with probability .001. A page of this table contains 200 numbers. We read through the page and stop when we reach a number that also appears (for a second time) later on the same page. Denoting the number of random numbers read by N, find the least value of n for which $\Pr[N \le n] > .90$. (*Hint*: $\Pr[N > n] = (1000)^{-200} \times [1000 \times 999 \times \cdots \times (1000 - n + 1)](1000 - n)^{200-n}$.)

An early modification of the waiting-time problem was put forward by von Schelling (1951). He introduced the idea of time units with one draw taking place at the conclusion of each time unit, but with the proviso that when a ball of a certain color (e.g., white) was drawn, drawings were suspended for k time units. Von Schelling (1951) was specifically interested in evaluating the conditional distributions of the number N of white balls already drawn, given that a white ball was drawn at the tth time unit. In fact it can easily be seen that the important conditioning fact is that a ball *was* drawn at the tth time unit, rather than the color of the selected ball.

We first note that

$$\Pr[N = 1] = q^{t-1}p,$$

where p = proportion of white balls in the urn and $q = 1 - p$.

If the ball chosen at the tth time unit were the second white one, then the previous white one could have appeared at the 1st, 2nd, ..., $(t - k - 1)$th time units. There would be only $(t - k)$ drawings, of which two would be white and the remaining $(t - k - 2)$ not white. So

$$\Pr[N = 2] = (t - k - 1)q^{t-k-2}p^2.$$

If $N = 3$, then there were two white balls drawn previously. There would be only $(t - 2k)$ drawings, of which three would be white and the remaining $(t - 2k - 3)$ nonwhite. The time units at which the first two white balls are drawn can be selected in

$$\begin{aligned}&[t - 2(k + 1)] + [t - 2(k + 1) - 1] + \cdots + 2 + 1\\ &= \tfrac{1}{2}[t - 2(k + 1)][t - 2(k + 1) + 1]\\ &= \binom{t - 2(k + 1) + 1}{2} = \binom{t - 1 - 2k}{2}\end{aligned}$$

ways, and so

$$\Pr[N = 3] = \binom{t - 1 - 2k}{2} q^{t-2k-3}p^3.$$

For $N = 4$, there are $(t - 3k)$ drawings, four white and $t - 3k - 4$ nonwhite. The time units at which the first three white balls are chosen can be selected in

$$\binom{t - k - 1 - 1 - 2k}{2} + \binom{t - k - 2 - 1 - 2k}{2} + \cdots + \binom{2k + 3 - 1 - 2k}{2} = \binom{t - 3k - 1}{3}$$

ways.

$$\left[\text{We use the identity } \sum_{j=r}^{c} \binom{j}{r} = \binom{c + 1}{r + 1}.\right]$$

So

$$\Pr[N = 4] = \binom{t - 1 - 3k}{3} q^{t-3k-4}p^4$$

and, generally,

$$\begin{aligned}\Pr[N = n] &= \binom{t - 1 - k(n - 1)}{n - 1} q^{t-k(n-1)-n}p^n\\ &= pq^{t-1}\binom{t - 1 - k(n - 1)}{n - 1}\left(\frac{p}{q^{k+1}}\right)^{n-1}.\end{aligned} \tag{3.50}$$

Another kind of waiting-time distribution is obtained by continuing to assign balls to urns only until one urn contains two balls, that is, until a ball is assigned to an urn that is already occupied.

Clearly, if after $(r - 1)$ assignments no urn contains more than one ball, there must be $(r - 1)$ occupied and $(n - r + 1)$ unoccupied urns.

The probability of this occurring is

$$P_{r-1} = \frac{m(m-1)\cdots(m-r+2)}{m^{r-1}} = \frac{m^{(r-1)}}{m^{r-1}}.$$

If X denotes the number of balls ("waiting-time") needed to assign more than one ball to some urn, then

$$\Pr[X = r] = P_{r-1} - P_r = \frac{m^{(r-1)}}{m^{r-1}} - \frac{m^{(r)}}{m^r} = m^{-r}m^{(r-1)}(r-1). \quad (3.51)$$

The jth ascending factorial moment of X is

$$\sum_{r=1}^{m+1} r^{[j]}(r-1)m^{-r}m^{(r-1)} = \sum_{r=1}^{m+1} (r-1)^{[j+1]}m^{-r}m^{(r-1)}. \quad (3.52)$$

This problem has been tackled by other means by McCabe (1970) and Banjevič (1974), among others.

The problem can also be expressed in terms of a single urn containing balls of different colors, sampling with replacement, and continuing until we obtain for the first time the repetition of a color already drawn. The following example, which is couched in these terms, also includes an interesting additional condition.

Example 3.6. An urn contains n balls, each of a different color. Balls are drawn with replacement until a ball of some color is drawn twice within at most $(k + 1)$ successive drawings. Find the distribution of the number of drawings. [This problem was solved by Arnold (1972). It corresponds to a situation in which records are kept only of the last k drawings together with the current drawing.]

Denote the required number by $N_{(k)}$. Since there must be at least one color repeated in $(n + 1)$ successive drawings, we need consider only the cases $k \le n$.

We also note that $N_{(k)} \ge 2$. If $k = n$, then $N_{(n)} \le n + 1$. For $j \le k + 1$,

$$\Pr[N_{(k)} > j] = \Pr[\text{different colors in first } j \text{ drawings}]$$
$$= \frac{n^{(j)}}{n^j} \qquad (j = 0, 1, \ldots, k + 1)$$

and so

$$\Pr[N_{(k)} = j] = \frac{n^{(j-1)}}{n^{j-1}} - \frac{n^{(j)}}{n^j} \qquad (j = 0, 1, \ldots, k + 1).$$

If $j > k + 1$, then the conditional probability

$$\Pr[N_{(k)} = j \mid N_{(k)} > j - 1]$$

is equal to

Pr[jth ball drawn has one of k different colors (those of the preceding k balls)],

and so

$$\Pr[N_{(k)} = j \mid N_{(k)} > j - 1] = \frac{k}{n}$$

and

$$\Pr[N_{(k)} > j \mid N_{(k)} > j - 1] = 1 - kn^{-1}.$$

By repeated application of this result we have, for $j > k + 1$,

$$\begin{aligned}\Pr[N_{(k)} > j] &= (1 - kn^{-1})^{j-k-1}\Pr[N_{(k)} > k + 1] \\ &= \frac{(1 - kn^{-1})^{j-k-1}n^{(k+1)}}{n^{k+1}}. \end{aligned} \tag{3.53}$$

Hence

$$\Pr[N_{(k)} = j] = \frac{(1 - kn^{-1})^{j-k-2}kn^{(k+1)}}{n^{k+2}} \qquad (\text{for } j > k + 1).$$

Also,

$$E[N_{(k)}] = \sum_{j=0}^{\infty} \Pr[N_{(k)} > j] = \sum_{j=0}^{k} \frac{n^{(j)}}{n^j} + n^{(k+1)}(kn^k)^{-1}. \tag{3.54}$$

[Note that $n^{(j)}/n^j = 1$ for $j = 0, 1$, and (3.53) is also valid for $j = k + 1$.] An application of this result can be found in Section 5.4.6.

3.2.4. Estimating the Number of Urns

Suppose there are equal numbers of balls of k different colors in an urn, and that sampling with replacement is continued until n balls have been drawn. The probability that n_1 balls of color 1, n_2 of color 2, ..., n_k of color k are drawn, is

$$\frac{n!}{n_1!n_2!\cdots n_k!}k^{-n}. \tag{3.55}$$

The probability (*likelihood*) that only j different colors are observed is

$$\binom{k}{j} k^{-n} \sum \cdots \sum \frac{n!}{n_{a_1}! \cdots n_{a_j}!}, \tag{3.56}$$

the summation being over all $n_{a_i} > 0$ $(i = 1, \ldots, k)$ satisfying $\sum_{i=1}^{j} n_{a_i} = n$. If k is not known, we may wish to estimate it.

As in a similar case in Section 3.2.1, although this model is not stated directly in terms of the occupancy models used earlier in this chapter, it can be so stated, by regarding the *colors* as k different *urns* and the sampling of a ball as an assignment of a ball to the urn corresponding to its color. Our problem is then to estimate the total number of urns k, given the number of balls in the occupied urns. The likelihood (3.56) is a multiple (not depending on k) of

$$L_k = \binom{k}{j} k^{-n}.$$

Note that the likelihood depends only on j, the number of different colors observed, and not on the number of repetitions $n_{a_1}, n_{a_2}, \ldots, n_{a_j}$.

The ratio of likelihoods is

$$R_k = \frac{L_{k+1}}{L_k} = \frac{k+1}{k+1-j}\left(\frac{k}{k+1}\right)^n.$$

We have

1. $R_k \to 1$ as $k \to \infty$.
2. R_k decreases with k increasing for $k < n(j-1)/(n-j)$ and increases with k increasing for $k > n(j-1)/(n-j)$.

Clearly, we need consider only values of k not less than j. We have

$$R_j = (j+1)\left(\frac{j}{j+1}\right)^n.$$

If $R_j < 1$, then $R_k < 1$ for all k $(\geq j)$. $L_{\hat{k}}$ is then a maximum for $\hat{k} = j$. If $R_j > 1$, then there is a value $\hat{k}$ such that $R_{\hat{k}-1} > 1 \geq R_{\hat{k}}$. In this case either $\hat{k}$ is the maximum likelihood estimator of k, or $\hat{k}$ and $\hat{k}+1$ produce equal maximum values for the likelihood.

So in any event, $\hat{k}$ is a maximum likelihood estimator of k if

$$\left(1 - \frac{j}{\hat{k}}\right)^{-1}\left(1 - \frac{1}{\hat{k}}\right)^n > 1 \geq \left(1 - \frac{j}{\hat{k}+1}\right)^{-1}\left(1 - \frac{1}{\hat{k}+1}\right)^n. \tag{3.57}$$

An alternative form of these inequalities is

$$\frac{\log(1 - j\hat{k}^{-1})}{\log(1 - \hat{k}^{-1})} < n \leq \frac{\log\{1 - j(\hat{k}+1)^{-1}\}}{\log\{1 - (\hat{k}+1)^{-1}\}}. \tag{3.58}$$

Driml and Ullrich (1967) provide a table of the integral part of

$$\frac{\log\{1 - j(k + 1)^{-1}\}}{\log\{1 - (k + 1)^{-1}\}}$$

as a function of j and $i = k - j$, which assists in obtaining $\hat{k}$ from (3.58). We reproduce part of this table here as Table 3.1.

TABLE 3.1

Integral Part of $[\log\{1 - j(k + 1)^{-1}\}]/[\log\{1 - (k + 1)^{-1}\}]$

j \ $k - j = i$	0	1	2	3	4	5	6	7	8	9
2	3									
3	5		4							
4	8	7	6		5					
5	10	9	8	7					6	
6	13	11	10	9		8				
7	16	13	12	11		10			9	
8	19	16	14	13	12			11		
9	22	18	16	15	14		13			12
10	26	21	19	17	16		15		14	

Example 3.7. Suppose that after 11 balls have been chosen, 6 different colors have been observed. Entering the table with $j = 6$, we find that the tabulated value 11 corresponds to $k - j = 1$. Hence the maximum likelihood estimator of the number of different colors is $6 + 1 = 7$. (If seven different colors had been observed, the value of the maximum likelihood estimator would have been $7 + 3 = 10$.)

It can be shown that $\hat{k}$ is not an unbiased estimator of k. It is, however, possible to find a function of J, the number of different colors in the sample, that has expected value equal to k (for $k \leq n$, n being the sample size). We first note that $(k - J)$ has the classical occupancy distribution and

$$\Pr[J = j] = \binom{k}{j} k^{-n} \Delta^j \mathbf{0}^n \qquad [j = 0, 1, \ldots, \min(n, k)]. \tag{3.6$'$}$$

Noting that $\sum_{j=0}^{g} \binom{g}{j} \Delta^j \mathbf{0}^h = g^h$ (and $\Delta^0 \mathbf{0} = 0$), we see that for any h the expected value of

$$T_h = \frac{\Delta^J \mathbf{0}^h}{\Delta^J \mathbf{0}^n}$$

is, for $k \leq n$,

$$E[T_h] = k^{-n} \sum_{j=0}^{k} \binom{k}{j} \Delta^j \mathbf{0}^h = k^{-n} k^h = k^{h-n}. \tag{3.59}$$

Hence, provided $k \leq n$,

$$E[T_{n+1}] = k;$$

so

$$\frac{\Delta^J \mathbf{0}^{n+1}}{\Delta^J \mathbf{0}^n}$$

is an unbiased estimator of k whenever k belongs to a set of no more than $\{1, 2, \ldots, n\}$. (Harris, 1968.)

By a similar argument, if $k \leq n$ and $f(\cdot)$ is any function,

$$E\left[\frac{\Delta^J f(\mathbf{0})}{\Delta^J \mathbf{0}^n}\right] = f(k) \qquad (\text{for } k \leq n). \tag{3.60}$$

3.3. RANDOMIZED OCCUPANCY MODELS

3.3.1. Introduction

In many statistical applications the underlying probabilistic problem can be described in terms of the following urn model: n balls are randomly allocated to k urns [so that each ball has probability $1/k$ of falling into the ith urn ($i = 1, 2, \ldots, k$)]. Each ball has probability p of staying in its urn and probability $1 - p$ of "falling through" or "leaking" (hence not being available to "fill" an urn). The random variable of interest here is the number X of occupied urns or, equivalently, the number $(k - X)$ of empty urns. Several examples of situations leading to this urn model in biology, nuclear physics, and other fields are given in Feller (1968) and Harkness (1969). The case $p = 1$ is the classical occupancy distribution (see Section 3.1). We are concerned here with the case $p < 1$, giving the *extended occupancy distribution.*

Johnson et al. (1974) tabulated critical points in the general case for various values of n, k, and p ($\geq .75$). We discuss some simple approximations to the distribution of occupied urns, which shed additional light on the structure and possible interpretations of this model.

3.3.2. A Randomized Occupancy Distribution

The classical occupancy distribution (corresponding to $p = 1$) gives the probability that exactly x urns are *occupied*, when n balls are used, as

$$H_1(x; k, n) = k^{-n}\binom{k}{x}\sum_{i=0}^{x}(-1)^i\binom{x}{i}(x-i)^n = k^{-n}\binom{k}{x}\Delta^x\mathbf{0}^n$$

$$(x = 0, 1, \ldots, k). \qquad (3.61)$$

[See (3.6′).]

The number of balls not falling through N is distributed binomially with parameters n, p, and so the extended occupancy distribution is the corresponding mixture of classical occupancy distributions. The overall probability $H_p(x; k, n)$ of exactly x occupied urns is then the expected value of $H_1(x; k, N)$, with N distributed binomially with parameters n, p. Hence

$$\begin{aligned} H_p(x; k, n) &= \sum_{j=0}^{n}\binom{n}{j}p^j(1-p)^{n-j}H_1(x; k, j) \\ &= \binom{k}{x}\sum_{i=0}^{x}(-1)^i\binom{x}{i}\sum_{j=0}^{n}\binom{n}{j}\left\{\frac{(x-i)p}{k}\right\}^j(1-p)^{n-j} \\ &= \binom{k}{x}\sum_{i=0}^{x}(-1)^i\binom{x}{i}\left\{1 - p + \frac{p}{k}(x-i)\right\}^n \\ &= k^{-n}\binom{k}{x}\Delta^x\{(1-p)k + p\mathbf{0}\}^n \qquad (x = 0, 1, \ldots, k). \quad (3.62) \end{aligned}$$

(using (1.15).)

These probabilities satisfy the recursion relation

$$\begin{aligned} kH_p(x; k, n) = {} & p(k + 1 - x)H_p(x - 1; k, n - 1) \\ & + \{px + (1-p)k\}H_p(x; k, n - 1). \quad (3.63) \end{aligned}$$

This can be established directly from formula (3.62), or by the following probabilistic argument.

If there are x occupied urns after n balls have been used then, after $(n - 1)$ were used, there must have been either (1) $(x - 1)$ occupied urns (probability $H_p(x - 1; n - 1)$) and the nth ball fell in an unoccupied urn and did not fall through (probability $p[1 - (x - 1)/k]$), or (2) x occupied urns (probability $H_p(x; k, n - 1)$) and the nth ball fell either in one of these x urns (probability x/k) or in one of the remaining (empty) $(k - x)$ urns and fell through [probability $(1 - p)(1 - x/k)$].

The rth factorial moment is

$$\begin{aligned}\mu'_{(r)} &= k^{-n}\sum_{x=r}^{k} x^{(r)}\binom{k}{x}\Delta^x\{(1-p)k + p\mathbf{0}\}^n \\ &= k^{-n}k^{(r)}\Delta^r\sum_{x=r}^{k}\binom{k-r}{x-r}\Delta^{x-r}\{(1-p)k+p\mathbf{0}\}^n \\ &= k^{-n}k^{(r)}\Delta^r(k-pr+p\mathbf{0})^n \qquad \text{[by using (1.21)]} \\ &= k^{(r)}\Delta^r\left\{1-\frac{p}{k}(r-\mathbf{0})\right\}^n. \end{aligned} \tag{3.64}$$

From (3.64), we obtain expressions for the expected value

$$k\left\{1-\left(1-\frac{p}{k}\right)^n\right\} \tag{3.65a}$$

and the variance

$$\begin{aligned} & k(k-1)\left\{1-2\left(1-\frac{p}{k}\right)^n+\left(1-\frac{2p}{k}\right)^n\right\}+k\left\{1-\left(1-\frac{p}{k}\right)^n\right\} \\ &\quad -k^2\left\{1-\left(1-\frac{p}{k}\right)^n\right\} \\ &= k\left\{1-\left(1-\frac{p}{k}\right)^n\right\}\left(1-\frac{p}{k}\right)^n - k(k-1)\left\{\left(1-\frac{p}{k}\right)^{2n}-\left(1-\frac{2p}{k}\right)^n\right\}. \end{aligned} \tag{3.65b}$$

The rth factorial moment of the number of *empty* urns has the simpler form

$$k^{(r)}\left(1-\frac{pr}{k}\right)^n \tag{3.66}$$

[as indicated by Harkness (1969)].

3.3.3. Tables and Approximations

Johnson et al. (1974) give tables which present, for specified values of k, x, p, α, the smallest n, say n_1, such that $\Pr[X \le x \mid k, n_1, p] = \sum_{y=0}^{x} H_p(y, k, n_1) \le \alpha$. Values of the corresponding upper-tail critical values, the greatest n_2 such that $\Pr[X \ge x \mid k, n_2, p] = \sum_{y=x}^{k} H_p(y, k, n_2) \le \alpha$ are also tabulated. The recursive relation (3.63) was utilized in these computations. Table 3.2 is based on these tables: the columns $p = 1$ are taken from Nicholson (1961).

Harkness (1969) suggested that a binomial distribution with parameters k, $1-(1-p/k)^n$ would give a good approximation to the distribution (3.62).

This approximation gives the correct mean [see (3.65a)] and is exact for $k = 1$. It would be expected to give good results for p small, since the mixing distribution of N (see Section 2.7) may then be approximated by a Poisson distribution (with parameter Np) and, as Harkness shows, a mixture of *classical* occupancy distributions with such a Poisson mixing distribution is in fact a binomial distribution. (See Exercise 3.27.)

If the binomial approximation is good, then Molenaar's (1970) normal approximation to the binomial may be used to obtain approximate expressions for n_1 and n_2. Writing

$$\omega(n) = 1 - \left(1 - \frac{p}{k}\right)^n, \tag{3.67}$$

we have from Molenaar's (1970, p. 110) "quick work" approximation for $\alpha_j \leq .05$ the following equations in n (for $n = n_1$ and n_2, respectively):

$$2(x+1)^{1/2}\{1 - \omega(n)\}^{1/2} - 2(k-x)^{1/2}\{\omega(n)\}^{1/2} \doteqdot -V_{\alpha_1} \tag{3.68a}$$

$$2x^{1/2}\{1 - \omega(n)\}^{1/2} - 2(k-x-1)^{1/2}\{\omega(n)\}^{1/2} \doteqdot V_{\alpha_2} \tag{3.68b}$$

where $\Phi(V_\alpha) = 1 - \alpha$. The equations for $\alpha_j > .05$ are of similar form.

In order to solve equations of form

$$A\sqrt{1 - \omega(n)} - B\sqrt{\omega(n)} = V \tag{3.69}$$

we write $\omega(n) = \sin^2\theta$ and $A/\sqrt{A^2 + B^2} = \sin\phi$ and obtain

$$\sin(\phi - \theta) = \frac{V}{\sqrt{A^2 + B^2}},$$

whence

$$\begin{aligned} \omega(n) &\doteqdot \sin^2\left[\sin^{-1}\frac{A}{\sqrt{(A^2 + B^2)}} - \sin^{-1}\frac{V}{\sqrt{(A^2 + B^2)}}\right] \\ &\doteqdot \left\{\frac{A}{\sqrt{(A^2 + B^2)}}\sqrt{1 - \frac{V^2}{A^2 + B^2}} - \frac{BV}{\sqrt{A^2 + B^2}}\right\}^2 \\ &\doteqdot (A^2 + B^2)^{-2}\left(A\sqrt{(A^2 + B^2 - V^2)} - BV\right)^2. \end{aligned} \tag{3.70}$$

(When $A^2 + B^2 < V^2$, this equation is not applicable but, since $A^2 + B^2$ is about $4k$, while the largest V^2 we need is $V_{0.01}^2 \doteqdot 5.4$, this imposes little restriction.)

Once $\omega(n_j)$ has been found, we can calculate the corresponding n_j from (3.67).

Some trial calculations with $k = 20$ indicated that, when $p \geq .75$, this approximation gives values of n_1 which are too big by about 2 to 4 units and of

TABLE 3.2

Percentage Points of Classical and Extended Occupancy Distributions

1. LOWER-TAIL CRITICAL POINTS

		$\alpha = .025$								$\alpha = .050$							
		$p = .75$		$p = .90$		$p = .95$		$p = 1$		$p = .75$		$p = .90$		$p = .95$		$p = 1$	
k	x	n_1	α_1	n_1	α_1	n_1	α_1	n_1	α_1	n_1	α_1	n_1	α_1	n_1	α_1	n_1	α_1
2	1	10	.018	8	.017	7	.022	7	.016	8	.047	7	.030	6	.042	6	.031
3	1	7	.023	6	.012	5	.020	5	.012	6	.046	5	.031	5	.020	4	.037
	2	17	.023	14	.020	13	.021	12	.023	15	.040	12	.041	11	.045	11	.035
4	1	7	.012	5	.014	5	.008	4	.016	6	.027	4	.044	4	.027	4	.016
	2	12	.021	10	.015	9	.018	8	.023	11	.033	8	.049	8	.034	7	.046
	3	25	.022	20	.024	19	.023	18	.023	22	.041	18	.041	17	.040	16	.040
5	1	6	.020	5	.009	4	.017	4	.008	5	.047	4	.030	4	.017	3	.040
	2	10	.024	8	.019	8	.011	7	.016	9	.042	7	.042	7	.026	6	.040
	3	17	.022	14	.019	13	.020	12	.021	15	.045	12	.045	11	.050	11	.035
	4	33	.023	27	.023	26	.021	24	.024	29	.045	24	.042	22	.048	21	.050
6	1	6	.015	4	.023	4	.011	4	.005	5	.040	4	.023	4	.011	3	.028
	2	10	.013	7	.023	7	.013	6	.020	8	.049	7	.023	6	.034	6	.020
	3	14	.025	12	.015	11	.016	10	.019	13	.039	10	.046	10	.030	9	.037
	4	22	.025	18	.024	17	.022	16	.022	20	.044	16	.047	15	.047	14	.049
	5	41	.025	34	.023	32	.024	31	.021	36	.049	30	.045	28	.048	27	.043
7	1	6	.013	4	.019	4	.008	4	.003	5	.035	4	.019	3	.044	3	.020
	2	9	.018	7	.014	6	.022	6	.011	8	.036	6	.039	6	.022	5	.038
	3	13	.020	10	.023	10	.013	9	.016	12	.034	9	.045	9	.027	8	.036
	4	19	.020	15	.021	14	.021	13	.023	17	.041	14	.034	13	.035	12	.038
	5	28	.023	23	.022	22	.019	20	.024	25	.046	21	.038	19	.048	18	.046
	6	50	.024	41	.025	39	.024	37	.023	44	.047	36	.049	34	.049	33	.043

TABLE 3.2 *(Continued)*

		$\alpha = .025$								$\alpha = .050$							
		$p = .75$		$p = .90$		$p = .95$		$p = 1$		$p = .75$		$p = .90$		$p = .95$		$p = 1$	
k	x	n_1	α_1	n_1	α_1	n_1	α_1	n_1	α_1	n_1	α_1	n_1	α_1	n_1	α_1	n_1	α_1
8	1	6	.011	4	.016	4	.006	3	.016	5	.032	4	.016	3	.038	3	.016
	2	9	.013	7	.010	6	.015	6	.007	8	.029	6	.029	5	.048	5	.026
	3	12	.022	10	.013	9	.015	8	.020	11	.038	9	.027	8	.035	7	.050
	4	17	.019	13	.025	13	.014	12	.015	15	.045	12	.043	11	.048	11	.030
	5	23	.024	19	.020	18	.018	17	.017	21	.044	17	.043	16	.042	15	.042
	6	34	.023	29	.021	26	.023	25	.020	30	.050	25	.044	23	.050	22	.046
	7	59	.024	49	.023	46	.024	44	.022	52	.047	43	.047	40	.050	38	.050
9	1	6	.010	4	.014	4	.005	3	.012	5	.029	4	.014	3	.033	3	.012
	2	8	.024	6	.022	6	.011	5	.018	8	.024	6	.022	5	.038	5	.018
	3	12	.015	9	.018	8	.023	8	.012	11	.028	8	.041	8	.023	7	.033
	4	16	.017	12	.024	12	.013	11	.015	14	.043	11	.045	11	.026	10	.031
	5	21	.020	17	.018	16	.016	15	.016	19	.040	15	.045	14	.045	13	.048
	6	28	.022	23	.020	21	.025	20	.022	25	.049	21	.039	19	.050	18	.048
	7	40	.023	33	.022	31	.022	29	.023	36	.046	29	.050	28	.043	26	.048
	8	>60	—	56	.025	53	.024	50	.025	60	.048	50	.046	47	.047	44	.050
10	1	6	.010	4	.012	4	.004	3	.010	5	.027	4	.012	3	.029	3	.010
	2	8	.020	6	.018	6	.008	5	.014	7	.045	6	.018	5	.031	5	.014
	3	11	.021	9	.012	8	.016	7	.022	10	.041	8	.031	7	.043	7	.022
	4	15	.018	12	.014	11	.015	10	.018	14	.030	12	.014	10	.033	9	.041
	5	19	.022	16	.014	14	.023	13	.024	17	.050	14	.040	13	.041	12	.045
	6	25	.021	20	.022	19	.019	18	.018	23	.039	18	.049	17	.046	16	.045
	7	33	.022	27	.021	25	.023	24	.020	30	.044	24	.050	23	.043	22	.039
	8	46	.024	38	.022	36	.021	34	.022	42	.044	34	.048	32	.048	31	.041
	9	>60	—	>60	—	60	.025	57	.025	>60	—	56	.050	53	.050	51	.046

		$\alpha = .025$								$\alpha = .050$							
		$p = .75$		$p = .90$		$p = .95$		$p = 1$		$p = .75$		$p = .90$		$p = .95$		$p = 1$	
k	x	n_2	α_2	n_2	α_2	n_2	α_2	n_2	α_2	n_2	α_2	n_2	α_2	n_2	α_2	n_2	α_2
4	4									4	.030						
5	5	5	.009	5	.023					6	.034	5	.023	5	.030	5	.038
6	5	5	.022							5	.022						
	6	7	.012	6	.008	6	.011	6	.015	8	.031	7	.032	7	.042	6	.015
7	5									5	.036						
	6	6	.008	6	.023					7	.031	6	.023	6	.031	6	.043
	7	10	.025	8	.013	8	.018	8	.024	11	.045	9	.033	9	.044	8	.024
8	5									5	.049						
	6	6	.014							6	.014	6	.041				
	7	8	.012	7	.009	7	.013	7	.019	9	.033	8	.037	7	.013	7	.019
	8	12	.020	10	.014	10	.020	9	.011	13	.034	11	.031	11	.042	10	.028
9	6	6	.020							6	.020						
	7	8	.023	7	.018					8	.023	7	.018	7	.026	7	.038
	8	10	.014	9	.017	9	.024	8	.008	11	.030	10	.044	9	.024	9	.035
	9	15	.025	12	.014	12	.021	11	.013	16	.040	14	.046	13	.038	12	.029
10	6									6	.027						
	7	7	.008							8	.035	7	.029	7	.042		
	8	9	.009	8	.008	8	.012	8	.018	10	.027	9	.034	9	.049	8	.018
	9	12	.013	11	.021	10	.011	10	.017	14	.046	12	.045	11	.031	11	.044
	10	17	.012	15	.024	14	.019	13	.014	19	.041	16	.038	15	.033	15	.046

n_2 which are too small by about 1 to 2 units, at any rate for $\alpha = .025$ and .05. It is evident from the discussion above that for smaller values of p the approximation should be more satisfactory. [Moreover, for p close to $\frac{1}{2}$ the alternative quick formula suggested by Molenaar (1970, p. 110) may yield even more accurate results.]

Limiting distributions of (3.62) and further asymptotic formulas are presented in Park (1972) and Samuel-Cahn (1974). These are discussed in Chapter 6.

EXERCISE 3.23.

(i) Prove the recursion relation (3.63) for the randomized model [using formula (3.62)].

(ii) Derive the probability and moment generating functions of the distribution of the number of empty urns in the randomized model and verify that the mean and variance of number of empty urns are given by $k(1 - p/k)^n$ and $k(k - 1)(1 - 2p/k)^n + k(1 - p/k)^n - k^2(1 - p/k)^{2n}$, respectively.

EXERCISE 3.24. Write the explicit expression for the factorial moments given in (3.64) (i.e., without the $\mathbf{0}$).

EXERCISE 3.25. Derive (3.66) directly from the definition of factorial moments.

EXERCISE 3.26. Consider a binomial distribution with parameters k and $(1 - p/k)^n$ and compare the factorial moments of this distribution with the factorial moments of the distribution of the number of empty urns. [*Answer*: $k^{(r)}(1 - p/k)^{Nr}$ versus $k^{(r)}(1 - rp/k)^N$.]

EXERCISE 3.27. Show that a mixture of classical occupancy distributions with a Poisson mixing distribution yields a binomial distribution.

EXERCISE 3.28. Use Molenaar's approximation for $k = 15$, $X = 10$, $p = 0.8$, and upper-tail probability $\alpha_2 = .05$ to determine an approximate value for n_2. (The exact value is $n_2 = 12$.)

3.4. MULTIVARIATE OCCUPANCY DISTRIBUTIONS

3.4.1. Multivariate Occupancy Distributions

In the preceding sections we have considered mainly univariate distributions. We have, however, already met some multivariate distributions connected with even the simple classical occupancy distribution. We refer in particular to the joint distribution of $M_0, M_1, \ldots, M_n$, where [as in (3.14)] M_j is the number of urns containing exactly j balls. The calculation of a joint epgf of the M's is given later in this section.

Another multivariate distribution connected with simple occupancy is introduced in the following exercise.

EXERCISE 3.29. Suppose that m urns are divided into s groups of $m_1, m_2, \ldots, m_s$ urns, respectively ($m_1 + m_2 + \cdots + m_s = m$). Show that the probability that, after n balls have been assigned, there are just $k_1, k_2, \ldots, k_s$ empty urns in the s groups is

$$\left[\prod_{i=1}^{s}\binom{m_i}{k_i}\right] m^{-n}\Delta^{m-\sum_1^s k_i}\mathbf{0}^n. \tag{3.71}$$

More genuinely multivariate problems arise when we have balls of different colors, and we are concerned with the joint distributions of the numbers of these balls in the urns.

We first consider a case in which there are balls of two types (e.g., two colors). The analysis is rather more complicated than in Section 3.1.2 though straightforward provided we keep careful track of the algebra.

Suppose that n_1 balls of type 1 and n_2 balls of type 2 are placed independently in m urns. There are altogether ($n_3 = n_1 + n_2$) balls. A ball of type j falls into the ith urn with probability $p_i^{(j)}$ ($j = 1, 2; i = 1, \ldots, m$) (*multinomial model*).

We now introduce some new notation. Note that M_1, M_2, and M_3 below are quite distinct from the M's in the first paragraph of this section.

Let M_j denote the number of urns *not* containing balls of type j ($j = 1, 2$) and M_3 denote the number of urns *not* containing any balls. We write

$$P(\mathbf{m} \mid \{p_i^{(j)}\}) = \Pr\left[\bigcap_{j=1}^{3}(M_j = m_j) \mid \{p_i^{(j)}\}\right].$$

Then the joint pgf of M_1, M_2, and M_3 (given n_1 and n_2) is

$$\sum_{m_1}\sum_{m_2}\sum_{m_3}\left\{\prod_{t=1}^{3}x_t^{m_t}\right\}P(\mathbf{m} \mid \{p_i^{(j)}\}) = G(x_1, x_2, x_3; n_1, n_2 \mid \{p_i^{(j)}\}).$$

The joint epgf of M_1, M_2, and M_3 is

$$\begin{aligned} &G(z_1, z_2; x_1, x_2, x_3 \mid \{p_i^{(j)}\}) \\ &\quad = \sum_{n_1=1}^{\infty}\sum_{n_2=1}^{\infty}\frac{(z_1 m)^{n_1}}{n_1!}\cdot\frac{(z_2 m)^{n_2}}{n_2!}G(x_1, x_2, x_3; n_1, n_2 \mid \{p_i^{(j)}\}) \\ &\quad = \prod_{h=1}^{m}[x_1x_2x_3 + x_1\{\exp(z_2 p_h^{(2)}m) - 1\} \\ &\qquad + x_2\{\exp(z_1 p_h^{(1)}m) - 1\} \\ &\qquad + \{\exp(z_1 p_h^{(1)}m) - 1\}\{\exp(z_2 p_h^{(2)}m) - 1\}]. \end{aligned} \tag{3.72}$$

Proof. We first divide the m urns into two groups, one containing a single urn (say the hth) and the other the remaining $(m-1)$ urns. Let n'_g be the number of balls of type g in the single urn, and $n''_g = n_g - n'_g$ the number in the remaining $(m-1)$ urns. Then,

$$P(\mathbf{m} \mid \{p_i^{(j)}\}) = \sum_{\substack{n'_g + n''_g = n_g \\ (g=1,2)}}\sum \prod_{g=1}^{2}\left[\binom{n_g}{n'_g}(p_h^{(g)})^{n'_g}(1-p_h^{(g)})^{n''_g}\right]$$

$$\times \sum_{\substack{m'_g + m''_g = m_g \\ (g=1,2)}}\sum \delta(m'_1, m'_2; n'_1, n'_2)P(\mathbf{m}'' \mid \{p_i^{(j)}(1-p_h^{(j)})^{-1}\}, i \neq h), \quad (3.73)$$

where

$$\delta(m'_1, m'_2; n'_1, n'_2) = \begin{cases} 1 & \text{if } m'_1 n'_1 = m'_2 n'_2 = 0 \\ 0 & \text{otherwise} \end{cases} \quad (3.74)$$

and m'_g, m''_g are the numbers of urns in the two groups that do not contain any balls of type g. Note that m'_g must be 0 or 1, since there is only one urn in the first group. (The function $\delta(\cdot)$ is introduced because, if $n'_g > 0$ then $m'_g = 0$, while if $n'_g = 0$ then $m'_g = 1$.)

Multiplying both sides of (3.73) by

$$\frac{(mz_1)^{n_1}}{n_1!}\frac{(mz_2)^{n_2}}{n_2!}x_1^{m_1}x_2^{m_2}x_3^{m_3}$$

and summing with respect to n_1, n_2, m_1, m_2, m_3 we obtain, on the left-hand side,

$$G(z_1, z_2; x_1, x_2, x_3 \mid \{p_i^{(j)}\})$$

and, on the right-hand side,

$$\sum_{n_1=1}\sum_{n_2=1}\sum_{m_1=0}\sum_{m_2=0}\sum_{n'_g+n''_g=n_g}\sum\prod_{g=1}^{2}\left[\frac{(mz_g p_h^{(g)})^{n'_g}}{n'_g!}\frac{\{mz_g(1-p_h^{(g)})\}^{n''_g}}{n''_g!}\right]$$

$$\times \sum_{\substack{m'_g+m''_g=m_g \\ (g=1,2)}}\sum x_1^{m'_1}x_2^{m'_2}x_3^{m'_3}\delta(m'_1, m'_2; n'_1, n'_2)x_1^{m''_1}x_2^{m''_2}x_3^{m''_3}$$

$$\times P(\mathbf{m}'' \mid \{p_i^{(j)}(1-p_h^{(j)})^{-1}\}, i \neq h)$$

$$= \sum_{n_1=1}\sum_{n_2=1}\sum_{m_1=0}\sum_{m_2=0}\sum_{\substack{n'_g+n''_g=n_g \\ (g=1,2)}}\sum \frac{\{mz_g p_h^{(g)}\}^{n''_g}}{n''_g!}x_1^{m''_1}x_2^{m''_2}x_3^{m''_3}$$

$$\times P(\mathbf{m}'' \mid \{p_i^{(j)}(1-p_h^{(j)})^{-1}\}, i \neq h)$$

$$\times \frac{\{mz_g p_h^{(g)}\}^{n'_g}}{n'_g!}x_1^{m'_1}x_2^{m'_2}x_3^{m'_3}\delta(m'_1, m'_2; n'_1, n'_2)$$

$$= G(z_1(1-p_h^{(1)}), z_2(1-p_h^{(2)}); x_1, x_2, x_3 \mid \{p_i^{(j)}(1-p_h^{(j)})^{-1}\}, i \neq h)$$

$$\times G(z_1 p_h^{(1)}, z_2 p_h^{(2)}; x_1, x_2, x_3 \mid 1, 1).$$

Hence

$$G(z_1, z_2; x_1, x_2, x_3 \mid p_i^{(j)})$$
$$= G(z_1(1 - p_h^{(1)}), z_2(1 - p_h^{(2)}); x_1, x_2, x_3 \mid \{p_i^{(j)}(1 - p_h^{(j)})^{-1}\}, i \neq h)$$
$$\times\ G(z_1 p_h^{(1)}, z_2 p_h^{(2)}; x_1, x_2, x_3 \mid 1, 1). \tag{3.75}$$

We now proceed inductively, expressing the first $G(\cdot)$ function on the right-hand side of (3.75) as a product of two $G(\cdot)$ functions, and so on. Eventually, we obtain

$$G(z_1, z_2; x_1, x_2, x_3 \mid \{p_i^{(j)}\}) = \prod_{h=1}^{m} G(z_1 p_h^{(1)}, z_2 p_h^{(2)}; x_1, x_2, x_3 \mid 1, 1). \tag{3.76}$$

Finally, we observe that

$$P(m_1, m_2, m_3; n_1, n_2 \mid 1, 1) = \begin{cases} 1 & \text{if } m_1 n_1 = m_2 n_2 = 0, \\ 0 & \text{otherwise.} \end{cases} \tag{3.77}$$

Hence $G(z_1 p_h^{(1)}, z_2 p_h^{(2)}, x_1, x_2, x_3 \mid 1, 1)$ is the sum of terms:

(i) With $n_1 > 0, n_2 > 0$: $m_1 = 0, m_2 = 0, m_3 = 0$—total

$$\{\exp(z_1 p_h^{(1)} m) - 1\}\{\exp(z_2 p_h^{(2)} m) - 1\}.$$

(ii) With $n_1 > 0, n_2 = 0$: $m_1 = 0, m_3 = 0$—total $x_2\{\exp(z_1 p_h^{(1)} m) - 1\}$.

(iii) With $n_1 = 0, n_2 > 0$: $m_1 = 1, m_2 = 0, m_3 = 0$ – total

$$x_1\{\exp(z_2 p_g^{(1)} m) - 1\}.$$

(iv) With $n_1 = n_2 = 0$: $m_1 = m_2 = m_3 = 1$—total $x_1 x_2 x_3$.

These terms sum to give

$$\begin{aligned} G(z_1 p_h^{(1)}, z_2 p_h^{(2)}; x_1, x_2, x_3 \mid 1, 1) = {} & x_1 x_2 x_3 + x_1\{\exp(z_2 p_h^{(2)} m) - 1\} \\ & + x_2\{\exp(z_1 p_h^{(1)} m) - 1\} \\ & + \{\exp(z_1 p_h^{(1)} m) - 1\}\{\exp(z_2 p_h^{(2)} m) - 1\}. \end{aligned} \tag{3.78}$$

From (3.76) we then obtain (3.72).

A related formula has been obtained by Bolotnikov (1968). This applies to a situation where the assignment probabilities change at certain points. For the first n_1 balls, they are $p_1^{(1)}, \ldots, p_m^{(1)}$; for the next n_2, they are $p_1^{(2)}, \ldots, p_m^{(2)}$; and so on, concluding with the last n_t balls for which they are $p_1^{(t)}, \ldots, p_m^{(t)}$.

Bolotnikov considered the random variables

K_j = number of empty urns after the $(n_1 + n_2 + \cdots + n_j)$th ball has been assigned $(j = 1, \ldots, t)$

and evaluated the generating function

$$H(z_1, z_2, \ldots, z_t; x_1, x_2, \ldots, x_t; \{p_i^{(j)}\}) = \sum_{n_1=0}^{\infty} \cdots \sum_{n_t=0}^{\infty} G_{\mathbf{n}}(\mathbf{x}; \{p_i^{(j)}\}) \prod_{j=1}^{t} \frac{(mz_j)^{n_j}}{n_j!},$$

where

$$G_{\mathbf{n}}(\mathbf{x}; \{p_i^{(j)}\}) = \sum_{k_1 \geq k_2 \geq \ldots \geq k_t} \cdots \sum \Pr\left[\bigcap_{j=1}^{t} (K_j = k_j)\right] \prod_{j=1}^{t} x_j^{k_j}.$$

He showed that

$$H(\mathbf{z}; \mathbf{x}; \{p_i^{(j)}\}) = \prod_{h=1}^{m} f_h(\mathbf{z}; \mathbf{x}; \{p_i^{(j)}\}), \tag{3.79}$$

where

$$f_h(\mathbf{z}; \mathbf{x}; \{p_i^{(j)}\}) = \sum_{s=1}^{t-1} \left(\prod_{g=1}^{s-1} x_g\right)\left\{\exp\left(\sum_{g=s+1}^{t} mp_h^{(g)} z_g\right)\right\}\{\exp(mp_h^{(s)} z_s) - 1\}$$

$$+ \left(\prod_{g=1}^{t-1} x_g\right)\{\exp(mp_h^{(t)} z_t) - 1\} + \prod_{g=1}^{t} x_g. \tag{3.80}$$

EXERCISE 3.30. Suppose that we have three urns, I, II, and III. Twenty balls are assigned to the three urns. The probability of assignment to urn II remains at $\frac{1}{3}$ for every ball. For 10 of the balls the probability of assignment to urn I is $\frac{1}{6}$ and to urn II it is $\frac{1}{2}$; for the other 10 balls the probabilities are reversed. What is the probability that the numbers of balls allotted to urns I, II, and III are 6, 8, and 6, respectively? (Note that the two sets of 10 balls can be assumed to be chosen consecutively without affecting the answer.)

Yet another case is that described in Exercise 3.29. The joint distribution of the variables $K_1, \ldots, K_s$ is (cf. 3.71)

$$\Pr\left[\bigcap_{j=1}^{s} (K_i = k_i)\right] = m^{-n}\left[\prod_{i=1}^{s} \binom{m_i}{k_i}\right]\Delta^{m-\sum k_i} \mathbf{0}^n.$$

We introduce the epgf

$$\begin{aligned}
H(\mathbf{z};\mathbf{x}) &= m^{-n}\sum_{n=0}^{\infty}\sum_{k_1}\cdots\sum_{k_s}\left[\prod_{i=1}^{s}x_i^{k_i}\right]\Pr\left[\bigcap_{i=1}^{s}(K_i=k_i)\right]\frac{m^n z^n}{n!}\Delta^{m-\Sigma k_i}\mathbf{0}^n\\
&= \sum_{k_1}\cdots\sum_{k_s}\left[\prod_{i=1}^{s}\binom{m_i}{k_i}x_i^{k_i}\right]\Delta^{m-\Sigma k_i}e^{z\mathbf{0}}\\
&= \sum_{k_1}\cdots\sum_{k_s}\left[\prod_{i=1}^{s}\binom{m_i}{k_i}x_i^{k_i}\Delta^{m_i-k_i}\right]e^{z\mathbf{0}}\\
&= \prod_{i=1}^{s}(\Delta+x_i)^{m_i}e^{z\mathbf{0}}\\
&= \prod_{i=1}^{s}(E-1+x_i)^{m_i}e^{z\mathbf{0}}\\
&= \prod_{i=1}^{s}(e^z-1+x_i)^{m_i},
\end{aligned}\qquad(3.81)$$

since (polynomial in $E)e^{z\mathbf{0}}$ = (polynomial in e^z). [See the definition of E in (1.12).]

EXERCISE 3.31. In the notation of Section 3.1.2 show that the epgf of $M_{r_1}, M_{r_2}, \ldots, M_{r_s}$ is

$$H_{\mathbf{r}}^{(m)}(z;\mathbf{x}) = \left\{e^z + \sum_{i=1}^{s}\frac{z^{r_i}}{r_i!}(x_i-1)\right\}^m. \qquad (3.82)$$

Hence show that

$$E\left[\prod_{i=1}^{s}M_{r_i}^{(\alpha_i)}\right] = m^{(\Sigma\alpha_i)}\frac{n!}{(n-\sum r_i\alpha_i)!\prod_{i=1}^{s}(r_i!)^{\alpha_i}}\left(\frac{1}{m}\right)^{\Sigma r_i\alpha_i}\left(1-\frac{\sum\alpha_i}{m}\right)^{n-\Sigma r_i\alpha_i} \qquad (3.83)$$

and, in particular,

$$E[M_{r_j}(M_{r_j}-1)] = m(m-1)\frac{n!}{(n-2r_j)!(r_j!)^2}\left(\frac{1}{m}\right)^{2r_j}\left(1-\frac{2}{m}\right)^{n-2r_j} \qquad (3.84a)$$

$$E[M_{r_g}M_{r_j}] = m(m-1)\frac{n!}{(n-r_g-r_j)!r_g!r_j!}\left(\frac{1}{m}\right)^{r_g+r_j}\left(1-\frac{2}{m}\right)^{n-r_g-r_j}. \qquad (3.84b)$$

3.4.2. Classical Multivariate Occupancy Distributions

We now turn our attention to a multivariate generalization of the distribution (3.6) from Section 3.1.1.

The simple techniques employed in Section 3.2.1 can also be applied to obtain results for this generalization.

Suppose we have b_g urns of class I(g) for $g = 1, 2, \ldots, k$, and for each urn of class I(g) the probability of assignment of a ball is p_g. Class II, as in Section 3.2.1, contains all the remaining urns and has probability of assignment $1 - \sum_{g=1}^{k} b_g p_g = p_0$.

Suppose, now, that in $n = \sum_{g=0}^{k} N_g$ independent trials N_g trials yield balls in class I(g) (for $g = 1, 2, \ldots, k$) and N_0 yield class II balls. Then the conditional probability, given $N_1, N_2, \ldots, N_k$, that *exactly* j_g of the b_g class I(g) urns have been occupied ($g = 1, 2, \ldots, k$) is

$$\Pr[\mathbf{V}_n = \mathbf{j}|\mathbf{N}] = \prod_{g=1}^{k} \Pr[V_{n,g} = j_g | N_g] = \prod_{g=1}^{k} \frac{\binom{b_g}{j_g}\Delta^{j_g}\mathbf{0}^{N_g}}{b_g^{N_g}}. \tag{3.85}$$

Here $\mathbf{V}_n' = (V_{n,1}, \ldots, V_{n,k})$ with $V_{n,g}$ being the number of class I(g) urns that are occupied, and $\mathbf{j}' = (j_1, j_2, \ldots, j_k)$. (Note that *conditionally* on $\mathbf{N}$, the $V_{n,g}$'s are mutually independent.)

Now $N_0, N_1, \ldots, N_k$ have a joint multinomial distribution, with parameters n, $(1 - \sum_{g=1}^{k} b_g p_g)$, $b_1 p_1, \ldots, b_k p_k$. Proceeding as in the univariate case and noting that, for arbitrary constants $\alpha_1, \alpha_2, \ldots, \alpha_k$,

$$\begin{aligned} E\left[\sum_{g=1}^{k} \alpha_g^{N_g}\right] &= \left(1 - \sum_{g=1}^{k} b_g p_g + \sum_{g=1}^{k} b_g p_g \alpha_g\right)^n \\ &= \left(p_0 + \sum_{g=1}^{k} b_g p_g \alpha_g\right)^n, \end{aligned}$$

we obtain

$$\begin{aligned} \Pr[\mathbf{V}_n = \mathbf{j}] &= E[\Pr[\mathbf{V}_n = \mathbf{j}|\mathbf{N}]] \\ &= \prod_{g=1}^{k} \binom{b_g}{j_g} \Delta_g^{j_g} \left(p_0 + \sum_{g=1}^{k} p_g \mathbf{0}_g\right)^n, \end{aligned} \tag{3.86}$$

where the difference operator Δ_g operates only on $\mathbf{0}_g$.

Noting that

$$\Delta_1^{j_1}\Delta_2^{j_2}\cdots\Delta_k^{j_k}\mathbf{0}_1^{\alpha_1}\mathbf{0}_2^{\alpha_2}\cdots\mathbf{0}_k^{\alpha_k} = \prod_{g=1}^{k}(\Delta^{j_g}\mathbf{0}^{\alpha_g}), \tag{3.87}$$

we find that (3.86) can be put in the form

$$\Pr[\mathbf{V}_n = \mathbf{j}] = \prod_{g=1}^{k} \binom{b_g}{j_g} \sum_{\mathbf{h}=\mathbf{0}}^{\mathbf{j}} (-1)^{\Sigma_1^k (j_g - h_g)} \left\{ \prod_{g=1}^{k} \binom{j_g}{h_g} \right\} \left(p_0 + \sum_{g=1}^{k} h_g p_g \right)^n. \quad (3.86')$$

Noting further that

$$\binom{b_g}{j_g}\binom{j_g}{h_g} = \frac{b_g!}{(b_g - j_g)!\,h_g!\,(j_g - h_g)!} = \binom{b_g}{b_g - j_g, h_g, j_g - h_g},$$

we see that (3.86) can also be expressed as

$$\Pr[\mathbf{V}_n = \mathbf{j}] = \sum_{\mathbf{h}=\mathbf{0}}^{\mathbf{j}} (-1)^{\Sigma_1^k (j_g - h_g)} \prod_{g=1}^{k} \binom{b_g}{b_g - j_g, h_g, j_g - h_g} \left(p_0 + \sum_{g=1}^{k} h_g p_g \right)^n. \quad (3.86'')$$

Yet another form is

$$\Pr[\mathbf{V}_n = \mathbf{j}] = \sum_{\mathbf{h}=\mathbf{0}}^{\mathbf{j}} (-1)^{\Sigma_1^k h_g} \prod_{g=1}^{k} \binom{b_g}{b_g - j_g, h_g, j_g - h_g} \left\{ p_0 + \sum_{g=1}^{k} (j_g - h_g) p_g \right\}^n. \quad (3.86''')$$

If we remember that $p_0 = 1 - \sum_{g=1}^{k} b_g p_g$, the parallelism between (3.86′) and (3.29′) and between (3.86‴) and (3.29″) is apparent.

If we suppose that $i_1, i_2, \ldots, i_k$ urns of classes I(1), I(2), ..., I(k), respectively, have already been occupied, then, as in Section 3.2.1, we obtain the conditional probability that $j_1, j_2, \ldots, j_k$ urns *in all*, of classes I(1), I(2), ..., I(k), respectively, will have been observed after r further independent trials, by replacing b_g by $b_g - i_g$ and j_g by $j_g - i_g$ in (3.86), remembering the "hidden" b_g's in p_0. We obtain from (3.86″)

$$p_{\mathbf{ij}}^{(n)} = \sum_{\mathbf{h}=\mathbf{0}}^{\mathbf{j}-\mathbf{i}} (-1)^{\Sigma_1^k (j_g - i_g - h_g)} \prod_{g=1}^{k} \binom{b_g - i_g}{b_g - j_g, h_g, j_g - i_g - h_g} \times \left(p_0 + \sum_{g=1}^{k} (h_g - i_g) p_g \right)^n. \quad (3.88)$$

Example 3.8. As an example of the use of this formula we calculate the probability that, if there are three class I urns each with $p_1 = .2$, and three class II urns each with $p_2 = .1$, then, after assignment of five balls the numbers of occupied urns in classes I and II are equal.

We have $k = 2$, $n = 5$, $b_1 = b_2 = 3$, $p_0 = .1$, $p_1 = .2$, $p_2 = .1$. Inserting these values in the formula [which follows from (3.86′)]

$$\Pr[V_{n1} = V_{n2}] = \sum_{j=0}^{[n/2]} \binom{b_1}{j}\binom{b_2}{j} \sum_{h_1=0}^{j} \sum_{h_2=0}^{j} (-1)^{h_1+h_2} \binom{j}{h_1}\binom{j}{h_2} \times (p_0 + h_1 p_1 + h_2 p_2)^n,$$

we obtain

$$\begin{aligned}1\cdot(.1)^5 &+ 3^2[(.1)^5 - (.3)^5 - (.2)^5 + (.4)^5]\\ &+ 3^2[(.1)^5 - 2(.3)^5 - 2(.2)^5 + 4(.4)^5 + (.3)^5\\ &+ -2(.5)^5 + (.7)^5 + (.5)^5 - 2(.6)^5]\\ &= .24031\end{aligned}$$

We now find the regression of $V_{n,g'}$ on $V_{n,g}$. Since

$$\Pr[V_{n,g} = j_g \mid N_g] = \frac{\binom{b_g}{j_g}\Delta^{j_g}\mathbf{0}^{N_g}}{b_g^{N_g}},$$

we have, using Bayes' theorem (Section 1.3.3),

$$\Pr[N_g = n_g \mid V_{n,g} = j_g] = \frac{(\Delta^{j_g}\mathbf{0}^{n_g})\binom{n}{n_g}p_g^{n_g}(1 - b_gp_g)^{n-n_g}}{\Delta^{j_g}\{1 - b_gp_g + p_g\mathbf{0}\}}. \tag{3.89}$$

Given $N_g = n_g$, $V_{n,g'}$ will be distributed as $V_{n-n_g,g'}$, with $p_{g'}$ replaced by $p_{g'}(1 - b_gp_g)^{-1}$. Hence

$$E[V_{n,g'} \mid N_g = n_g] = b_{g'}[1 - \{1 - p_{g'}(1 - b_gp_g)^{-1}\}^{n-n_g}], \tag{3.90}$$

and $E[V_{n,g'} \mid V_{n,g} = j_g]$ is the expected value of (3.90) taken over the distribution (3.89).

This is

$$\begin{aligned}&b_{g'}\left[1 - [\Delta^{j_g}\{1 - b_gp_g + p_g\mathbf{0}\}^n]^{-1}\Delta^{j_g}\sum_{n_g=0}^{n}\binom{n}{n_g}(p_g\mathbf{0})^{n_g}(1 - b_gp_g - p_{g'})^{n-n_g}\right]\\ &\quad = b_{g'}\left[1 - \frac{\Delta^{j_g}(1 - b_gp_g - p_{g'} + p_g\mathbf{0})^n}{\Delta^{j_g}(1 - b_gp_g + p_g\mathbf{0})^n}\right].\end{aligned}$$

Thus the regression of $V_{n,g}$ on $V_{n,g'}$ is

$$E[V_{n,g'} \mid V_{n,g}] = b_{g'}\left[1 - \frac{\Delta^{V_{n,g}}(1 - b_gp_g - p_{g'} + p_g\mathbf{0})^n}{\Delta^{V_{n,g}}(1 - b_gp_g + p_g\mathbf{0})^n}\right]. \tag{3.91}$$

EXERCISE 3.32. Using methods of the kind described in Exercise 3.13, show that

$$\operatorname{cov}(V_{n,g}, V_{n,g'}) = b_gb_{g'}\{(1 - p_g - p_{g'})^n - (1 - p_g)^n(1 - p_{g'})^n\}$$

[cf. (3.31a)].

3.5. SEQUENTIAL OCCUPANCY PROBLEMS

In Section 2.5 we studied several waiting-time problems and described a few others.

In this section we consider a special class of waiting-time problem. We fix a required number of empty urns and count the number of balls necessary to achieve this required number. Suppose m is the number of urns and we wish to end up with k empty urns, then the required number of balls N can range from $m - k$ to ∞. The generic name for problems of this type is *sequential occupancy problems*. They are also often referred to as coupon collector problems or "dixie cup" problems. The following situation inspired these names. A certain ice cream company offers a picture of a famous ball player on the top of each of its dixie cups. If there are N different pictures all together, how many dixie cup tops must you collect in order to have at least one each of k different pictures?

Before we begin a formal discussion of problems of this type, observe the following relationship. If $N(k)$ represents the number of balls required to end up with k empty urns, X_n is the number of empty urns after n balls are assigned, and m is the total number of urns, then

$$\Pr\{N(k) \le n\} = \Pr\{X_n \le k\}. \tag{3.92}$$

The classical *sequential* occupancy problem corresponds to the classical occupancy problem. That is, it is the problem of finding the distribution of $N(k)$ when there are m urns and each ball is equally likely to fall into any particular urn with probability $1/m$. If $N(k) = n$, then the nth ball dropped must have fallen into an empty urn. Otherwise, $N(k) < n$. Therefore

$$\Pr\{N(k) = n\} = P\{X_{n-1} = k + 1\}\frac{k+1}{m},$$

because $(k + 1)/m = \Pr[\text{ball falls into an empty urn} \mid (k + 1) \text{ empty urns}]$. From (3.6), we know that

$$\Pr[X_{n-1} = k + 1] = m^{-n+1}\binom{m}{k+1}\Delta^{m-k-1}\mathbf{0}^{n-1}.$$

Hence

$$\Pr\{N(k) = n\} = \frac{1}{m^{n-1}}\binom{m-1}{k}\Delta^{m-k-1}\mathbf{0}^{n-1}. \tag{3.93}$$

Against the background of the coupon collector's problem, $N(k)$ can be regarded as the sum of $(m - k)$ variables $Y_1, Y_2, \ldots, Y_{m-k}$, say, where Y_j is equal to the number of coupons needed to obtain the jth new coupon once $(j - 1)$

have already been obtained. The Y's are mutually independent; $Y_1 = 1$; and, for $j \geq 2$, Y_j has the geometric distribution

$$\Pr[Y_j = y] = \left(\frac{m - j + 1}{m}\right)\left(\frac{j - 1}{m}\right)^{y-1} \qquad (y = 1, 2, \ldots).$$

Hence

$$E[Y_j] = \frac{m}{m - j + 1}, \qquad \operatorname{var}(Y_j) = \frac{m(j - 1)}{(m - j + 1)^2},$$

and

$$E[N(k)] = m \sum_{j=1}^{m-k} (m - j + 1)^{-1}, \tag{3.94a}$$

and

$$\operatorname{var}(N(k)) = m \sum_{j=2}^{m-k} \frac{j - 1}{(m - j + 1)^2}. \tag{3.94b}$$

The rth cumulant of Y_j is

$$\sum_{i=1}^{\infty} i^{r-1}\left(\frac{j - 1}{m}\right)^i,$$

and the rth cumulant of $N(k)$ is (cf. 1.153):

$$\kappa_r(N(k)) = \sum_{h=2}^{m-k} \sum_{i=1}^{\infty} i^{r-1}\left(\frac{h - 1}{m}\right)^i = \sum_{i=1}^{\infty} \frac{i^{r-1}}{m^i} \sum_{h=1}^{m-k-1} h^i. \tag{3.95}$$

What happens when each color has a different probability of appearing? This corresponds to a situation in which the probability p_i that a ball will fall into the ith urn is different from the probability p_j that it will land in the jth urn if $i \neq j$. Nath (1974) used formula (1.87) to give a solution to this problem.

EXERCISE 3.33. Show that, in this case, the quantities S_j in (1.86) are

$$\begin{aligned} S_j &= \sum_{1 \leq i_1 < i_2 < \cdots < i_j \leq m} (1 - p_{i_1} - p_{i_2} - \cdots - p_{i_j})^n \\ &= \sum_{1 \leq i_1 < \cdots < i_{m-j} \leq m} \cdots \sum (p_{i_1} + \cdots + p_{i_{m-j}})^n. \end{aligned}$$

The results of Exercises 1.20 and 3.33 give

$$\begin{aligned}\Pr[N(m-k) \geq n] &= S_{m-k+1} - \binom{m-k+1}{m-k} S_{m-k+2} \\ &\quad + \binom{m-k+2}{m-k} S_{m-k+3} - \cdots + (-1)^k \binom{m-2}{m-k} S_{m-1} \\ &= \sum_{j=0}^{k-2} (-1)^j \binom{m-k+j}{m-k} \\ &\quad \times \sum_{1 \leq i_1 < \cdots < i_{k-1+j} \leq m} \cdots \sum (p_{i_1} + \cdots + p_{i_{k-1+j}})^n.\end{aligned}$$

[Note that this is the probability that after n balls there are still at least $(m - k + 1)$ empty urns.]

A generating function for the survival distribution function (see Section 1.4.3) of $N(m - k)$ is

$$\begin{aligned}Q(t) &= \sum_{n=0}^{\infty} t^n \Pr[N(m-k) > n] \\ &= 1 + \sum_{n=1}^{\infty} \left[\sum_{j=0}^{k-2} (-1)^j \binom{m-k+j}{m-k} \right. \\ &\quad \left. \times \sum_{1 \leq i_1 < \cdots < i_{k-1+j} \leq m} \cdots \sum \{(p_{i_1} + \cdots + p_{i_{k-1+j}})t\}^n \right].\end{aligned}$$

Interchanging the order of summation, we have

$$\begin{aligned}Q(t) &= 1 + \sum_{j=0}^{k-2} (-1)^j \binom{m-k+j}{m-k} \\ &\quad \times \sum_{1 \leq i_1 < \cdots < i_{k-1+j} \leq m} \cdots \sum \sum_{n=1}^{\infty} \{(p_{i_1} + \cdots + p_{i_{k-1+j}})t\}^n \\ &= 1 + \sum_{j=0}^{k-2} (-1)^j \binom{m-k+j}{m-k} \\ &\quad \times \sum_{1 \leq i_1 < \cdots < i_{k-1+j} \leq m} \cdots \sum \frac{(p_{i_1} + \cdots + p_{i_{k-1+j}})t}{1 - (p_{i_1} + \cdots + p_{i_{k-1+j}})t}.\end{aligned} \tag{3.96}$$

Recalling Exercise 1.43,

$$\begin{aligned}E[N(m-k)] = Q(1) &= 1 + \sum_{j=0}^{k-2} (-1)^j \binom{m-k+j}{m-k} \\ &\quad \times \sum_{1 \leq i_1 < \cdots < i_{k-1+j} \leq m} \cdots \sum \left(\frac{p_{i_1} + \cdots + p_{i_{k-1+j}}}{1 - p_{i_1} - \cdots - p_{i_{k-1+j}}} \right).\end{aligned} \tag{3.97}$$

Noting that

$$Q^{(r)}(1) = r! \sum_{j=0}^{k-2} (-1)^j \binom{m-k+j}{m-k}$$

$$\times \sum_{1 \le i_1 < \cdots < i_{k-1+j} \le m} \cdots \sum \left(\frac{p_{i_1} + \cdots + p_{i_{k-1+j}}}{1 - p_{i_1} - \cdots - p_{i_{k-1+j}}} \right)^{r+1}, \tag{3.98}$$

we can evaluate higher moments of $N(m-k)$.

EXERCISE 3.34. Verify that, when $p_i = 1/m$ $(i = 1, 2, \ldots, m)$, (3.97) reduces to

$$E[N(m-k)] = m\left(\frac{1}{m} + \frac{1}{m-1} + \cdots + \frac{1}{m-k+1}\right)$$

[cf. (3.94a)].

EXERCISE 3.35. Pólya (1930) has considered the problem when g *different* coupons are obtained at each stage. Show that in this case the distribution of the necessary number of stages N to complete a set of n_0 different coupons is given by

$$\Pr[N \le n] = 1 - \binom{n_0}{1} S_1'^n + \binom{n_0}{2} S_2'^n - \cdots, \tag{3.99}$$

where

$$S_i' = \frac{(n_0 - i)^{(g)}}{n_0^{(g)}}.$$

Hence calculate the *expected value* of N. $\left[Ans\colon \sum_{i=1}^{n_0} (-1)^{i-1} \binom{n_0}{i} (1 - S_i')^{-1}. \right]$

As a variation on this model, consider the case in which the total number of coupons in existence is finite. This is analogous to random sampling from an urn containing c balls of m different colors, with c_j balls of color $\mathscr{C}_j$ and $c = c_1 + c_2 + \cdots + c_m$. We wish to consider the distribution of the number of drawings (*without replacement*) needed to obtain k different colors. This is a generalization of problem 12, p. 102, of Feller (1957) which reads:

"A pack of cards consists of s identical series, each containing n cards numbered $1, 2, \ldots, n$. A random sample of $r \ge n$ cards is drawn from the pack without replacement. What is the probability that it will take exactly r drawings to get a sample containing all the numbers?"

We first consider the case $m = 3$. That is, there are three different colors of coupons, say b blues, g greens, and w whites, such that $b + g + w = n$, the total number of coupons. We wish to collect until we have at least one blue and one green coupon. Let R be the number of coupons it is necessary to collect to reach this goal. Let $q_r = \Pr\{R \geq r + 1\}$, B_r = the event that no blue coupon has been collected among the first r coupons, G_r = the event that no green coupon has been collected among the first r coupons. Then,

$$\begin{aligned} q_r &= \Pr(B_r) + \Pr(G_r) - \Pr(B_r \cap G_r) \\ &= \frac{(g + w)^{(r)}}{n^{(r)}} + \frac{(b + w)^{(r)}}{n^{(r)}} - \frac{w^{(r)}}{n^{(r)}}. \end{aligned} \tag{3.100}$$

Hence

$$\Pr[R = r] = q_{r-1} - q_r = \frac{(g + w)^{(r-1)}b + (b + w)^{(r-1)}g - w^{(r-1)}(b + g)}{n^{(r)}}. \tag{3.101}$$

The generating function of the q_r's is

$$\begin{aligned} Q(s) &= \sum_{r=0}^{\infty} q_r s^r \\ &= \sum_{r=0}^{g+w} \frac{(g + w)^{(r)}}{n^{(r)}} s^r + \sum_{r=0}^{b+w} \frac{(b + w)^{(r)}}{n^{(r)}} s^r - \sum_{r=0}^{w} \frac{w^{(r)}}{n^{(r)}} s^r. \end{aligned} \tag{3.102}$$

Consider now a series of the form $\sum_{r=0}^{k} (k^{(r)}/n^{(r)})s^r$. This series can be written as

$$\sum_{r=0}^{k} \frac{\binom{k}{r}}{\binom{n}{r}} s^r = \frac{1}{\binom{n}{k}} \sum_{r=0}^{k} \binom{n - r}{n - k} s^r.$$

We also have

$$\begin{aligned} \sum_{r=0}^{k} \binom{n - r}{n - k}(1 + x)^r &= \sum_{r=0}^{k} \binom{n - r}{n - k} \sum_{t=0}^{r} \binom{r}{t} x^t = \sum_{t=0}^{k} \left[\sum_{r=t}^{k} \binom{n - r}{k - r}\binom{r}{r - t} \right] x^t \\ &= \sum_{t=0}^{k} \binom{n + 1}{k - t} x^t. \end{aligned} \tag{3.103}$$

We can now write

$$\frac{1}{\binom{n}{k}} \sum_{r=0}^{k} \binom{n - r}{n - k} s^r = \frac{1}{\binom{n}{k}} \sum_{r=0}^{k} \binom{n + 1}{k - r} (s - 1)^r.$$

Applying this expression to the three terms on the right-hand side of (3.102), we have

$$Q(s) = \frac{1}{\binom{n}{g+w}} \sum_{r=0}^{g+w} \binom{n+1}{g+w-r}(s-1)^r + \frac{1}{\binom{n}{b+w}} \sum_{r=0}^{b+w} \binom{n+1}{b+w-r}(s-1)^r$$
$$- \frac{1}{\binom{n}{w}} \sum_{r=0}^{w} \binom{n+1}{w-r}(s-1)^r. \tag{3.104}$$

As in the last problem,

$$E[R] = Q(1) = (n+1)\left[\frac{1}{b+1} + \frac{1}{g+1} - \frac{1}{b+g+1}\right]. \tag{3.105}$$

Notice that this expectation does not involve w.

Again,

$$\text{var}(R) = 2Q'(1) + Q(1) - Q^2(1) \qquad \text{(see Exercise 1.43)}$$
$$= (n+1)\left[\frac{b(g+w)}{(b+1)^2(b+2)} + \frac{g(b+w)}{(g+1)^2(g+2)} - \frac{w}{(b+g+1)^2}\right.$$
$$\left. - \frac{2(n+2)}{(b+g+1)(b+g+2)} + \frac{2(n+1)}{(b+1)(g+1)(b+g+1)}\right]. \tag{3.106}$$

Nath (1973) claims that empirical and graphical study make it apparent that the distribution of R in this problem closely resembles the distribution of the waiting time R' until two balls of the same color, say red, are obtained in sampling without replacement from a finite population consisting of $(b + g)$ red balls and $(n - b - g)$ nonred balls. This distribution is a negative hypergeometric distribution and is derived in Section 2.5.

In our present notation, the appropriate negative hypergeometric distribution is

$$\Pr[R' = r] = \frac{(r-1)(b+g)(n-b-g)^{(r-2)}}{n^{(r)}}. \tag{3.107}$$

Example 3.9. As an example of the similarity of the two distributions, (3.101) and (3.107), consider the cases in which $n = 5$, $b = g = 1$, $w = 3$ and, for the negative hypergeometric, $n = 5$, $b + g = 2$. By (3.101),

$$\Pr[R = 2] = \tfrac{1}{10}, \quad \Pr[R = 3] = \tfrac{1}{5}, \quad \Pr[R = 4] = \tfrac{3}{10}, \quad \Pr[R = 5] = \tfrac{2}{5},$$

while by (3.107),

$$\Pr[R' = 2] = \tfrac{1}{10}, \quad \Pr[R' = 3] = \tfrac{1}{5}, \quad \Pr[R' = 4] = \tfrac{3}{10}, \quad \Pr[R' = 5] = \tfrac{2}{5}.$$

EXERCISE 3.36. Find conditions under which the two distributions are identical, as in Example 3.9.

To generalize to the case of m colors, $\mathscr{C}_1, \mathscr{C}_2, \ldots, \mathscr{C}_m$, with c_1 $\mathscr{C}_1$ coupons, c_2 $\mathscr{C}_2$ coupons, $\ldots$, c_m $\mathscr{C}_m$ coupons, such that $c_1 + c_2 + \cdots + c_m = n$, the same type of reasoning can be used. Let R be the number of coupons it is necessary to collect in order to have at least one each of the first $(m - 1)$ colors, $\mathscr{C}_1, \mathscr{C}_2, \ldots, \mathscr{C}_{m-1}$. Let $q_r = P\{R \geq r + 1\}$.

EXERCISE 3.37. Show that

$$q_r = \frac{1}{n^{(r)}}\left[\sum_{i=1}^{m-1}(n - c_i)^{(r)} - \sum_{1 \leq i < j \leq m-1}(n - (c_i + c_j))^{(r)} + \sum_{1 \leq i < j < k \leq m-1}[n - (c_i + c_j + c_k)]^{(r)} - \cdots + (-1)^{m-2}\sum_{i=1}^{m-1} c_i^{(r)}\right] \quad (r \geq 0). \tag{3.108}$$

EXERCISE 3.38. Using (3.108), show that $Q(s) = \sum q_r s^r$ can be written

$$Q(s) = (n + 1)\left[\sum_{i=1}^{m-1}\frac{1}{c_i + 1}F(-(n - c_i), 1; c_{i+2}; -(s - 1)) - \sum_{1 \leq i < j \leq m-1}\frac{1}{c_i + c_j + 1}F(-(n - c_i - c_j), 1; c_i + c_j + 2; -(s - 1)) + \cdots + (-1)^{m-2}\left(\sum_{i=1}^{m}(n - c_i + 1)^{-1}F(-c_i, 1; n - c_i + 2; -(s - 1))\right)\right],$$

where $F(a, b; c; z) = 1 + \sum_{n=1}^{\infty}(a^{[n]}b^{[n]}z^n/c^{[n]}n!)$ is the hypergeometric function [see (1.58)].

Hence show that

$$E[R] = Q(1) = (n + 1)\left[\sum_{i=1}^{m-1}\frac{1}{c_i + 1} - \sum_{1 \leq i < j \leq m-1}(c_i + c_j + 1)^{-1} + \sum_{1 < i \leq j < k \leq m-1}(c_i + c_j + c_k + 1)^{-1} - \cdots + (-1)^{m-2}\sum_{i=1}^{m}(n - c_i + 1)^{-1}\right]. \tag{3.109}$$

Example 3.9a. (El-Neweihi et al. (1976)). Suppose an urn contains c_i balls of color $\mathscr{C}_i$ $(i = 1, \ldots, k)$ and no others. Balls are drawn from the urn one at a time, without replacement. What is the probability that the first color to be exhausted (i.e. all balls of that color drawn) is $\mathscr{C}_1$, the second is $\mathscr{C}_2, \ldots$ and the last (which must be the color of the last ball drawn) is $\mathscr{C}_k$?

The probability that the color of the last ball drawn is $\mathscr{C}_k$ is $c_k(c_1 + c_2 + \cdots + c_k)^{-1}$. Given that this is so, the conditional probability that $\mathscr{C}_{k-1}$ is the penultimate color to be exhausted is the probability that the last ball drawn, *after removing all balls of color* $\mathscr{C}_k$ *from the sequence of draws*, has color $\mathscr{C}_{k-1}$. This conditional probability is $c_{k-1}(c_1 + c_2 + \cdots + c_{k-1})^{-1}$. We then remove all balls of color $\mathscr{C}_{k-1}$ from the sequence and find that the conditional probability that a ball of color $\mathscr{C}_{k-2}$ occurs in last position is $c_{k-2}(c_1 + c_2 + \cdots + c_{k-2})^{-1}$. Proceeding in this way we find that the required probability is

$$\left(\prod_{i=2}^{k} c_i\right)\left\{\prod_{i=2}^{k}\left(\sum_{j=1}^{i} c_j\right)\right\}^{-1}.$$

The probability that $\mathscr{C}_1$ is the first color to be exhausted (whatever the order in which the remaining colors are exhausted) is

$$\left(\prod_{i=2}^{k} c_i\right)\sum_{(a_2,\ldots,a_k)}\cdots\sum\left\{\prod_{i=2}^{k}\left(c_1 + \sum_{j=2}^{i} c_{a_j}\right)\right\}^{-1}$$

where the summation is over all $(k-1)!$ possible ordering $(a_2, a_3, \ldots, a_k)$ of the integers $(2, 3, \ldots, k)$.

El-Neweihi et al. (1976) mention applications of the model described in this example to structural reliability of engineering systems.

EXERCISE 3.38a. By using (1.88) show that an alternative expression for the probability that $\mathscr{C}_1$ is the first color to be exhausted is

$$c_1 \sum_{\mathscr{S}} (-1)^{|\mathscr{S}|}\left(c_1 + \sum_{i\in\mathscr{S}} c_i\right)^{-1}$$

where summation is over all $(2^{k-1} - 1)$ subsets $\mathscr{S}$ of $(2, \ldots, k)$, and $|\mathscr{S}|$ is the number of elements of $\mathscr{S}$.

Note that the problem of finding the probability that the *first* color drawn is $\mathscr{C}_k$, the second $\mathscr{C}_{k-1}\ldots$ and the last $\mathscr{C}_1$ is equivalent to that described in Example 3.9a. This can be seen by making each sequence of colors correspond to the same sequence in reverse order.

3.6. COMMITTEE PROBLEMS

3.6.1. Individual Membership

Committee problems is a generic term for a class of problems in which in the corresponding urn model we have distinguishable balls. An early problem of this form was called the *chromosome problem.* It is stated by Feller (1957, p. 102, Problem 16) as follows:

"A cell contains n chromosomes, between any two of which an interchange of parts may occur. If r interchanges occur (which can happen in $\binom{n}{2}^r$ distinct ways) find the probability that exactly m chromosomes will be involved." (The initial source of this problem is Catcheside et al. (1945–1946)). At present this problem is considered biologically unrealistic. Moreover, it is a special case of the more general "committee" problem (Mantel and Pasternack (1968)):

"A group contains n individuals, any w of whom can be selected at random to form a committee. If r committees, each of size w, are formed (which can be done in $\binom{n}{w}^r$ distinct ways), find the probability that exactly m individuals will be committee members."

If the committee size w is 1, then this problem reduces to the classical occupancy problem. The n individuals from which the committees are drawn correspond to n urns. The r committees, each having only one seat, correspond to r balls thrown at random into the urns. The probability that exactly m individuals will be committee members equals the probability that exactly m urns will be occupied, or the probability that exactly $(n - m)$ urns will be empty.

If the committee size w is equal to 2, then we have the chromosome problem. (The n individuals correspond to n chromosomes, and the r committees to r interchanges.)

This model can be further generalized by allowing the sizes of the committees to differ so that committee i is of size w_i. Also, we can ask the more general question: What is the distribution of the random vector $\mathbf{f} = (f_0, f_1, \ldots, f_r)$, where f_j is the number of individuals belonging to exactly j of the r committees?

In Section 3.6.3 we consider the changes imposed on the above distribution where there is probability $p > 0$ that a person chosen for a committee refuses to serve on it. Note that this last generalization can be paralleled with a *randomized occupancy model* (Section 3.3) in which there is probability $p > 0$ that a ball will "leak out" of an urn.

We first consider the distribution of the number of committee members when there are r committees, all of equal size w, each drawn randomly from n individuals.

We denote by y_n the number of ways $\binom{n}{w}^r$ of forming r committees, each with w members. Then, $\Delta y_k = y_{k+1} - y_k$ can be interpreted as the increase in the number of ways of selecting r committees, each of size w, when a specified individual is added to a set of k individuals. Clearly, the specified individual must be a member of at least one of each of the additional possible sets of r committees. Hence Δy_k is the number of ways of selecting the committees from $(k + 1)$ individuals when one *specified* individual must always be used on one or more of the committees. Similarly, $\Delta^2 y_k = \Delta(\Delta y_k) = \Delta y_{k+1} - \Delta y_k$ gives the number of ways of selecting the committees from $(k + 2)$ individuals when two specific individuals must always be included on one or more committees.

In particular, $\Delta^2 y_0$ is the number of ways of selecting the committees from two individuals who must always be included on one or more of the committees. Generally, $\Delta^m y_0$ is the number of ways of selecting the committees from m individuals, each of whom must be included on one or more of the committees.

Since the m individuals can be chosen in $\binom{n}{m}$ ways, we see that the number of ways of choosing the committees so that exactly m members are called on is $\binom{n}{m}\Delta^m y_0$. The probability that exactly m individuals will be committee members is thus

$$\binom{n}{w}^{-r}\binom{n}{m}\Delta^m y_0 = \binom{n}{w}^{-r}\binom{n}{m}\sum_{k=w}^{m}(-1)^{m-k}\binom{m}{k}\binom{k}{w}^r. \tag{3.110}$$

(Note that $y_k = 0$ if $k < w$.) (See White, 1971.) From the above argument it also follows that:

1. When $w = 1$, as in the classical occupancy problem, the expression on the left-hand side becomes $n^{-r}\binom{n}{m}\Delta^m \mathbf{0}^r$. That is, with r balls tossed into n urns,

$$P\{\text{exactly } m \text{ urns are occupied}\} = n^{-r}\binom{n}{m}\Delta^m \mathbf{0}^r,$$

agreeing with (3.6′). This is clear because $\binom{n}{w} = \binom{n}{1} = n$, and $y_k = \binom{k}{w}^r = \binom{k}{1}^r = k^r$, giving $y_0 = \mathbf{0}^r$.

2. The number of ways of choosing the committees so that g specified persons are included and an additional s specified individuals are available [i.e., $(n - g - s)$ are excluded] is $\Delta^g y_s$.

3. In the case of varying committee size w_i ($i = 1, \ldots, r$), the only modification is to use

$$y_k = \prod_{i=1}^{r} \binom{k}{w_i}$$

in place of $\binom{k}{w}^r$. Thus the right-hand side of (3.110) becomes

$$\binom{n}{m} \sum_{k=0}^{m} (-1)^{m-k} \binom{m}{k} \prod_{i=1}^{r} \frac{\binom{k}{w_i}}{\binom{n}{w_i}}. \tag{3.111}$$

An alternative derivation of (3.111) was offered by Gittelsohn (1969), utilizing indicator random variables. It is outlined below.

Let A_i be the event that the ith person ($i = 1, \ldots, n$) is selected for *none* of the r committees. Let

$$I_i = \begin{cases} 1 & \text{if } A_i \text{ occurs,} \\ 0 & \text{otherwise.} \end{cases}$$

The sum, $X = \sum_{i=1}^{n} I_i$ is a random variable representing the number of persons never placed on any committee. The rest of the proof consists of finding the moments of X and expressing $P[X = m]$ in terms of its moments using, for example, (1.162).

Yet a further method of approach (see, for example, Sprott, 1969) is based on formula (1.85).

The probability that none of a specified set of j persons serve on any of the r committees is

$$\left[\frac{\binom{n-j}{w}}{\binom{n}{w}} \right]^r,$$

and so [in the notation of (1.85)]

$$S_j = \binom{n}{j} \left[\frac{\binom{n-j}{w}}{\binom{n}{w}} \right]^r. \tag{3.112}$$

Hence the probability that *exactly* $(n-m)$ persons do not serve on any committee is

$$\sum_{j=0}^{m}(-1)^j\binom{n-m+j}{j}S_{n-m+j}. \tag{3.113}$$

This is of course the same as the probability that exactly m persons serve on at least one committee. It is easy to establish the identity of (3.113) and (3.110).

Example 3.10. A court of discipline is drawn from six members, three of whom are chosen at random for any one sitting. What is the chance that all six will serve in five sittings?

Solution. This problem can be stated in terms of the committee problem, with $n = 6$, $w = 3$, $r = 5$, $m = 6$. In terms of differences [i.e., using the left-hand side of (3.110)], since $\binom{n}{m} = \binom{6}{6} = 1$, the desired probability is $\Delta^6 y_0 \Big/ \binom{6}{3}^5$, where $y_k = \binom{k}{3}^5$. We use the following table:

k	y_k	Δ^1	Δ^2	Δ^3	Δ^4	Δ^5	Δ^6
0	0	0	0	1	1,020	94,890	2,615,340
1	0	0	1	1,021	95,910	2,710,230	
2	0	1	1,022	96,931	2,806,140		
3	1	1,023	97,953	2,903,071			
4	1,024	98,976	3,001,024				
5	100,000	3,100,000					
6	3,200,000						

Hence the required probability is

$$\frac{2{,}615{,}340}{3{,}200{,}000} = .817.$$

The probability that *exactly* five will have served in five sittings is

$$\frac{\binom{6}{5}\Delta^5 y_0}{\binom{6}{5}^5} = \frac{6 \times 94{,}890}{3{,}200{,}000} = .178.$$

[This is a modification of a problem presented by Pearson (1931) and White (1971).]

An original scheme for investigating the more general question of the distribution of the random vector $\mathbf{f} = (f_0, f_1, \ldots, f_r)$, when the committee sizes may be unequal was developed by Mielke and Siddiqui (1965). Let the outcome of a committee selection be represented by an $r \times n$ matrix, say $\mathbf{\Omega}$, with the jth term in the ith row (i.e., the i, jth entry in the matrix) equal to 1, or 0, depending on whether the jth individual was chosen, or not chosen, respectively for the ith committee. Consider an example in the case $n = 7$, $r = 3, w_1 = 2, w_2 = 3, w_3 = 4$:

$$\begin{matrix} 0 & 0 & 1 & 0 & 1 & 0 & 0 \\ 0 & 1 & 1 & 0 & 0 & 0 & 1. \\ 0 & 1 & 1 & 1 & 1 & 0 & 0 \end{matrix}$$

Let

$n_j =$ the sum of terms in the jth column (the number of committees which include the jth individual).

Then

$$f_i = \text{the number of columns } (j\text{'s}) \text{ for which } n_j = i.$$

In the example above, $f_0 = 2, f_1 = 2, f_2 = 2, f_3 = 1$. Let

$$\mathscr{P} = \{(i_1, i_2, \ldots, i_k) | i_1 < i_2 < \cdots < i_k; \{i_1, i_2, \ldots, i_k\} \subset \{1, 2, \ldots, r\}\}.$$

Then each $\mathbf{p} \in \mathscr{P}$ represents a possible membership scheme for an individual. For example, when $r = 3$, $\mathbf{p} = (1, 2)$ represents membership in committees 1 and 2 but not in 3 ($\mathbf{p} = (1, 2)$ corresponds in the matrix $\mathbf{\Omega}$ to a column vector of the form $\begin{pmatrix} 1 \\ 0 \\ 0 \end{pmatrix}$). Now let $N_\mathbf{p}$ denote the number of individuals with membership scheme $\mathbf{p}$, and N_0 the number belonging to no committees. Note that each individual can have only one membership scheme, and thus $\sum_{\mathbf{p} \in \mathscr{P}} N_\mathbf{p} + N_0 = n$. In the above case, with $r = 3$, $f_0 = N_0$, $f_1 = N_1 + N_2 + N_3$, $f_2 = N_{12} + N_{13} + N_{23}$, $f_3 = N_{123}$. Since the total number of ways of choosing r committees of sizes $w_1, w_2, \ldots, w_r$ from n individuals is $\prod_{i=1}^{r} \binom{n}{w_i}$, it is easy to see that, in general

$$\Pr\{N_\mathbf{p} = j_\mathbf{p} \text{ for all } \mathbf{p} \in \mathscr{P}; N_0 = j_0\} = \left| \prod_{i=1}^{r} \binom{n}{w_i} \right|^{-1} \cdot \frac{n!}{j_0! \prod_{\mathbf{p} \in \mathscr{P}} j_\mathbf{p}!} \tag{3.113a}$$

(cf. Eicker et al., 1972).

Example 3.11. (Prepared by J. Seeger). The following numerical example demonstrates the need for caution in using this formula.

Let $n = 4, r = 3, w_1 = w_2 = w_3 = 2$. Consider

$$\Pr\{f_0 = 1, f_1 = 1, f_2 = 1, f_3 = 1\}.$$

A simple counting of possible cases yields the result that this probability is $\frac{72}{216} = \frac{1}{3}$. However, a direct attempt to use the distribution of the $N_{\mathbf{p}}$'s could lead to the following disconcerting result.

$$\begin{aligned}
&\Pr\{f_0 = 1, f_1 = 1, f_2 = 1, f_3 = 1\}\\
&\quad = \Pr\left\{N_0 = 1;\ \sum_{\mathbf{a}(1)} N_{\mathbf{a}(1)} = 1;\ \sum_{\mathbf{a}(2)} N_{\mathbf{a}(2)} = 1;\ N_{123} = 1\right\}\\
&\quad = \Pr\{\mathbf{N} = (1;1,0,0;1,0,0;1)\}\\
&\qquad + \Pr\{\mathbf{N} = (1;1,0,0;0,1,0;1)\} + \Pr\{\mathbf{N} = (1;1,0,0;0,0,1;1)\}\\
&\qquad + \Pr\{\mathbf{N} = (1;0,1,0;1,0,0;1)\} + \Pr\{\mathbf{N} = (1;0,1,0;0,1,0;1)\}\\
&\qquad + \Pr\{\mathbf{N} = (1;0,1,0;0,0,1;1)\} + \Pr\{\mathbf{N} = (1;0,0,1;1,0,0;1)\}\\
&\qquad + \Pr\{\mathbf{N} = (1;0,0,1;0,1,0;1)\} + \Pr\{\mathbf{N} = (1;0,0,1;0,0,1;1)\}\\
&\quad = \frac{4!}{\binom{4}{2}^3 (1)} \cdot 9 = 1
\end{aligned}$$

by formula (3.113a), where $\mathbf{N} = (N_0; N_1, N_2, N_3; N_{12}, N_{13}, N_{23}; N_{123})$.

The paradox is easily resolved by examining the $\mathbf{N}$ vector corresponding to each of the nine terms. Consider the first one (1; 1, 0, 0; 1, 0, 0; 1). It tells us one person belongs to no committees. His column vector in the matrix is

$$\begin{pmatrix}0\\0\\0\end{pmatrix}.$$

One person belongs to only the first committee. His column vector is

$$\begin{pmatrix}1\\0\\0\end{pmatrix}.$$

Similarly, the column vectors of the other two individuals are

$$\begin{pmatrix}1\\1\\0\end{pmatrix} \quad \text{and} \quad \begin{pmatrix}1\\1\\1\end{pmatrix}.$$

Putting these together yields the matrix

$$\boldsymbol{\Omega} = \begin{pmatrix} 0 & 1 & 1 & 1 \\ 0 & 0 & 1 & 1 \\ 0 & 0 & 0 & 1 \end{pmatrix}.$$

Recalling that each row of $\boldsymbol{\Omega}$ corresponds to a committee, we see that this means that the only committee having the correct size is the second. The first has three members, and the third only one. The only vectors that do not violate the rules ($w_1 = w_2 = w_3 = 2$) are the third, fifth, and seventh, $(1;1,0,0;0,0,1;1)$, $(1;0,1,0;0,1,0;1)$, and $(1;0,0,1;1,0,0;1)$ respectively. If we count only these as possibilities, we obtain a probability of

$$\frac{4!}{\binom{4}{2}^3(1)} \cdot 3 = \frac{1}{3}.$$

EXERCISE 3.39. Construct a table of $\Delta^r y_k$ when $y_k = \binom{k}{3}^4$ for $k = 0(1)7$, $r = 0(1)5$.

EXERCISE 3.40. In the situation described in Example 3.10 suppose that an extra person is added to the pool of members from whom the court of discipline may be chosen after its first sitting. What is the probability that each of the seven members will have served on at least one of the five sittings?

EXERCISE 3.41. In a certain city there is concern about a possible pollution problem causing increased frequency of asthma attacks on certain days. Three asthmatic patients are observed for a period of 120 days. It is assumed that, under normal "safe" external conditions, their attacks will fall randomly among the 120 days of observation. Under this assumption, two or more patients having an attack on the same day constitutes a rare event. The problem is to determine exactly how rare an event it is so that observation of such days could lead to information supporting the hypothesis of a pollution problem.

Assume the three patients had 6, 11, and 18 attacks, respectively, during the observation period. State the problem of finding the probability that under "safe" conditions two or more patients have an attack on the same day in terms of a committee problem and suggest a method for finding this probability. (*Source*: Eicker et al., 1972.)

Note: Mantel (1974) criticized the method of Eicker et al. (1972) in dealing with this problem because it provides no *simple* way of assessing the evidence for differences between days. Mantel suggests two purely statistical approaches to the problem by which, he claims, an extensive set of data for the above problem (with a significantly larger number of patients) can be assessed for evidence of day differences regardless of differences between individuals, and for differences between individuals regardless of day differences.

3.6.2. Grouped Membership

The members from whom a committee is to be chosen may themselves be organized in g disjoint groups, and we can consider the probability that a committee or committees contain no members from a specified group or groups. The following analysis is based on Walter (1975).

We denote our groups by the symbols $\mathscr{G}_1, \mathscr{G}_2, \ldots, \mathscr{G}_g$. The probability that no member of $\mathscr{G}_h$ is invited to join the ith committee is

$$\frac{\dbinom{n - n(\mathscr{G}_h)}{w_i}}{\dbinom{n}{w_i}} \qquad (h = 1, 2, \ldots, g),$$

where $n(\mathscr{G}_h)$ is the number of members in the group $\mathscr{G}_h$. The probability that no member from $\mathscr{G}_h$ is invited to be on any committee is

$$Q_h = \prod_{i=1}^{r} \left\{ \frac{\dbinom{n - n(\mathscr{G}_h)}{w_i}}{\dbinom{n}{w_i}} \right\}. \tag{3.114}$$

If we denote a set of k groups $\mathscr{G}_{h_1}, \mathscr{G}_{h_2}, \ldots, \mathscr{G}_{h_k}$ by $\mathscr{G}_{\mathbf{h}}^{(k)}$, then the probability that no member of any of these k groups is asked to serve on any committee is

$$Q_{\mathbf{h}}^{(k)} = Q_h,$$

with $n(\mathscr{G}_h)$ replaced by $\sum_{i=1}^{k} n(\mathscr{G}_{h_i})$. Writing $S_k = \sum \cdots \sum_{\mathbf{h}} Q_{\mathbf{h}}^{(k)}$ (summing being over all different sets of k h's), we have [see (1.85)]

Pr[exactly d groups are represented on at least one committee]

$$= \Pr[\text{exactly } (g-d) \text{ groups are not represented on any committee}]$$

$$= S_{g-d} - \binom{g-d+1}{g-d} S_{g-d+1} + \binom{g-d+2}{g-d} S_{g-d+2} - \cdots + (-1)^d \binom{g}{g-d} S_g. \tag{3.115}$$

EXERCISE 3.41. Suppose that in group $\mathscr{G}_h$ only $n_i(\mathscr{G}_h)$ members are eligible for the ith committee. Obtain a formula for S_k in this case.

We now consider a situation in which (1) the number of individuals in $\mathscr{G}_h$ eligible for the ith committee is $n_i(\mathscr{G}_h)$, with n_i denoting the total number of such individuals, *and* (2) there are r_i nominations for the ith committee, but these may include multiple nominations for some individuals and the final committee size w_i is variable. Then, on the assumption that nominations are made at random, the probability that a particular nomination for the ith committee is given to some member of $\mathscr{G}_{\mathbf{h}}^{(k)}$ is $p_i(\mathscr{G}_{\mathbf{h}}^{(k)}) \equiv n_i(\mathscr{G}_{\mathbf{h}}^{(k)})/n_i$. The probability that no member of $\mathscr{G}_{\mathbf{h}}^{(k)}$ is nominated in any of the r_i nominations is

$$\{q_i(\mathscr{G}_{\mathbf{h}}^{(k)})\}^{r_i} \qquad \text{where } q_i(\mathscr{G}_{\mathbf{h}}^{(k)}) = 1 - p_i(\mathscr{G}_{\mathbf{h}}^{(k)}).$$

Hence for this case we have, in (3.115),

$$S_k = \sum_{\mathbf{h}}\cdots\sum\left[\prod_{i=1}^{r}\{q_i(\mathscr{G}_{\mathbf{h}}^{(k)})\}^{r_i}\right] \tag{3.116}$$

(where the summation is over all possible sets $\mathbf{h}$ of k h's).

In the special case where each group has the same number n_i' of individuals eligible for the ith committee so that $n_i = gn_i'$, then

$$q_i(\mathscr{G}_{\mathbf{h}}^{(k)}) = 1 - \frac{kn_i'}{gn_i'}$$

$$= \frac{g-k}{g}, \tag{3.117}$$

and (3.115) becomes the formula for the classical occupancy problem [cf. (3.6′)].

3.6.3. Randomized Committee Problems

If we introduce a probability $(1 - p)$ of "falling through" in the committee problems described in the previous sections, this could correspond to a probability $(1 - p)$ that a person selected to serve on a committee refuses to do so. As before, we suppose there are n persons available and that each committee is of size w. We also suppose (perhaps unrealistically) that, if a person refuses to sit on a committee, no attempt is made to fill the vacant place.

Given j *specified* persons, suppose that α are selected for a *specified* committee and each refuses to serve on this committee, while the remaining $(j - \alpha)$ are not selected at all. The probability of this compound event is

$$\frac{\binom{j}{\alpha}(1-p)^{\alpha}\binom{n-j}{w-\alpha}}{\binom{n}{w}}.$$

The probability that none of the j specified persons serves on this specific committee is

$$P'_j = \sum_{\alpha=0}^{j} \frac{\binom{j}{\alpha}(1-p)^{\alpha}\binom{n-j}{w-\alpha}}{\binom{n}{w}} \tag{3.118}$$

The probability that none of the j specified persons serves on any of r committees is $(P'_j)^r$.

Using formula (1.85), but now with $S_j = \binom{n}{j}(P'_j)^r$, we see that the probability of exactly m people serving on at least one of the r committees is

$$\sum_{j=0}^{m}(-1)^j\binom{n-m+j}{j}\binom{n}{n-m+j}\frac{\left\{\sum_{\alpha}\binom{n-m+j}{\alpha}(1-p)^{\alpha}\binom{m-j}{w-a}\right\}^r}{\binom{n}{w}^r}$$

$$= \binom{n}{m}\binom{n}{w}^{-r}\sum_{j=0}^{m}(-1)^j\binom{m}{j}\left\{\sum_{\alpha=0}^{\min(w,n-m+j)}\binom{n-m+j}{\alpha}(1-p)^{\alpha}\binom{m-j}{w-\alpha}\right\}^r. \tag{3.119}$$

EXERCISE 3.43. Show that, if $w = 1$, we obtain formula (3.62).

EXERCISE 3.44. Assume that the probability of refusal $(1 - p)$ in the generalized committee problem remains constant, but the committee sizes are w_i $(i = 1, \ldots, r)$. Derive the distribution of the number of individuals that *do not* serve on any of the committees. [*Hint*: In (1.85) we now have $S_j = \binom{n}{j}\prod_{i=1}^{r} p''_{j,i}$, where $p''_{j,i} = \sum_{\alpha=0}^{j}\binom{j}{\alpha}\binom{n-j}{w_i-\alpha}p^{\alpha}\Big/\binom{n}{w_i}$.]

EXERCISE 3.45. Verify that the result obtained in Exercise 3.44 reduces to (3.111) when $p = 1$.

Some limit distributions related to committee problems are discussed in Section 6.2.7.

References

Arnold, B. C. (1972) The waiting time until first duplication, *J. Appl. Prob.*, **9**, 841–846.

Banjević, D. (1974) Generalization of waiting times model, *Mat. Vesn.*, **11**(26), 3–9.

Baticle, E. (1933, 1935) Le problème de la répartition, *C. R. Acad. Sci.* (*Paris*), **197**, 632–634; **201**, 862–864.

Baticle, E. (1946) Le problème des stocks, *C. R. Acad. Sci.* (*Paris*), **222**, 355–357.

Bolmarcich, J. J. and Belkin, B. (1974) *Sequential generalizing of the Bose-Einstein distribution*, Paoli, Pa.: D. H. Wagner Associates (preprint). Submitted to *J. Appl. Prob.*

Bolotnikov, Yu. V. (1968) Limit processes in a model of distribution of particles into cells with unequal probabilities, *Theor. Prob. Appl.*, **13**, 504–516.

Catcheside, D. G., Lea, D. E., and Thoday, J. M. (1945–1946) Types of chromosome structural change introduced by the irradiation of tradescantia microspores, *J. Genet.*, **47**, 113–149.

Decomps, B. and Kastler, A. (1963) Répartition de N particules entre g cellules; Loi de fluctuations, *C. R. Acad. Sci.* (*Paris*), **256**(5), 1087–1089.

Denning, P. J. and Schwartz, S. C. (1970) Properties of the working-set model, Princeton University Department of Electrical Engineering, no. 93. Reprinted in *Commun. ACM*, **15**(3) (1972), 191–198.

Driml, M. and Ullrich, M. (1967) Maximum likelihood estimate of the number of types, *Acta Techn. ČSAN*, **12**(3), 300–303.

Dwass, M. (1969) More birthday surprises, *J. Comb. Theory*, **7**(3), 258–261.

Eicker, P. J., Siddiqui, M. M. and Mielke, P. W. Jr. (1972) A matrix occupancy problem, *Ann. Math. Stat.*, **43**(3), 988–996.

El-Neweihi, E., Proschan, F. and Sethuraman, J. (1976). A simple model with applications in structural reliability, extinction of species, inventory depletion and urn sampling, I. Cut set failure. II. System lifelength, *Statistics Reports M391-2*, Florida State University.

Feller, W. (1957) *An Introduction to Probability Theory and Its Applications*, Vol. 1, 2nd ed., New York: John Wiley and Sons, (pp. 96 and 102, Exercise 16).

Feller, W. (1968) *An Introduction to Probability Theory and Its Applications*, Vol. 1, 3rd ed., New York: John Wiley and Sons.

Freund, J. E. and Pozner, A. N. (1956) Some results on restricted occupancy theory, *Ann. Math. Statist.*, **27**, 537–540.

Gittelsohn, A. M. (1969) An occupancy problem, *Am. Stat.*, **23**(4), 11–12.

Glasser, G. J. (1963) Random numbers, sample selection and occupancy problems, *J. Roy. Stat. Soc. Ser. A*, **126**, 115–119.

Harkness, W. L. (1970) The classical occupancy problem revisited, in *Random Counts in Physical Sciences*, G. P. Patil (Ed.), Pennsylvania State University Press, 107–126.

Harris, B. (1968) Statistical inference in the classical occupancy problem unbiased estimation of the number of classes, *J. Am. Stat. Assoc.*, **63**, 837–847.

Hill, B. M. (1974) The rank-frequency form of Zipf's law, *J. Am. Stat. Assoc.* **69**, 1017–1026.

Ijiri, Y. and Simon, H. A. (1975) Some distributions associated with Bose-Einstein statistics, *Proc. Nat. Acad. Sci. U.S.*, **72**(5), 1654–1657.

Ivchenko, G. I. and Medvedev, Yu. I. (1965) Some multi-dimensional theorems on a classical problem of permutations, *Theor. Prob. Appl.*, **10**, 144–149.

Johnson, N. L. and Kotz, Samuel (1976) On a multivariate generalized occupancy model, *J. Appl. Prob.*, **13**, 392–399.

Johnson, N. L., Kotz, S., and Srinivasan, R. (1974) Extended occupancy probability distribution; critical points, Mimeo Series No. 934, Institute of Statistics, University of North Carolina.

Johnson, N. L. and Young, D. H. (1960) Some applications of two approximations to the multinomial distribution, *Biometrika*, **47**, 463–468.

Jones, H. L. (1959) How many of a group of random numbers will be usable in selecting a particular sample?, *J. Am. Stat. Assoc.*, **54**, 102–122.

Klamkin, M. S. and Newman, D. J. (1967) Extensions of the birthday surprise, *J. Comb. Theory*, **3**, 279–282.

Mantel, N. (1974) Approaches to a health research occupancy problem, *Biometrics*, **30**, 355–362.

Mantel, N. and Pasternack, B. S. (1968) A class of occupancy problems, *Am. Stat.*, **22**(2), 23–24.

McCabe, B. (1970) Elementary Problem E 2263, *Am. Math. Mon.*, **77**, 1008 (correction: April 1971, **78**, 404.)

Mertz, D. B. and Davies, R. B. (1968) Cannibalism of the pupal stage by adult flour beetles, *Biometrics*, **24**, 247–275.

Mielke, P. W. and Siddiqui, M. M. (1965) A combinatorial test for independence of dichotomous responses, *J. Am. Stat. Assoc.*, **60**, 437–441.

Molenaar, W. (1970) *Approximations to the Poisson, Binomial and Hypergeometric Distribution Functions*, Mathematical Center Tracts No. 31, Amsterdam: Mathematisch Centrum.

Nath, H. B. (1973) Waiting time in the coupon-collector's problem, *Austr. J. Stat.*, **15**(2), 132–135.

Nath, H. B. (1974) On the collector's sequential sample size, *Trab. Estad.*, **25**, 85–88.

Nicholson, W. L. (1961) Occupancy distribution critical points, *Biometrika*, **48**, 175–180.

Nymann, J. E. (1975) Another generalization of the birthday problem, *Math. Mag.* **48**(1), 46–47.

Park, C. J. (1972) A note on the classical occupancy problem, *Ann. Math. Stat.*, **43**, 1698–1701.

Pearson, K. (1931) Introduction to *Tables for Statisticians and Biometricians*, Part II, Cambridge: Cambridge University Press, p. 237.

Pease, R. W. (1975) General solution to the occupancy problem with variably sized runs of adjacent cells occupied by single balls, *Math. Mag.* **48**, 131–134.

Pólya, G. (1930) Eine Wahrscheinlichkeitsaufgabe in der Kundenwerbung, *Z. Angew. Math. Mech.*, **10**(1), 96–97.

Price, G. B. (1946) Distributions derived from the multinomial expansion, *Am. Math. Mon.*, **53**, 59–74.

Richards, P. I. (1968) A generating function (Problem 67–18), *SIAM Rev.*, **10**, 455–456.

Samuel-Cahn, E. (1974) Asymptotic distributions for occupancy and waiting time problems with positive probability of falling through the cells, *Ann. Prob.*, **2**, 515–521.

Schelling, H. von (1951) Distribution of the ordinal number of simultaneous events which last during a finite time, *Ann. Math. Stat.*, **22**, 452–455.

Sprott, D. A. (1969) A note on a class of occupancy problems, *Am. Stat.*, **23**(4), 12–13.

Thomasian, A. (1969) *The Structure of Probability Theory with Applications*, New York: McGraw Hill.

4

Urn Models with Stochastic Replacements

4.1. INTRODUCTION

Urn models representing some form of contagion can be constructed in an unlimited variety of ways. Fortunately, it has been found that many cases of particular importance can be included within the scope of a few relatively simple methods of construction.

The development of these methods stems largely from a paper by Eggenberger and Pólya (1923), and the distributions as a group are known as Pólya's distributions. (There are differences in usage among different authors.)

The common feature of all these methods is that sampling with some form of replacement is used, but this is not simple replacement of the ball last chosen. The way in which the replacement is modified determines the particular subclass of distributions arising. In particular, the kind of replacement depends on the results of the sampling.

Eggenberger and Pólya (1923) supposed that an urn containing balls of two colors—say white and black—is used, and after each drawing, the chosen ball is returned, *together with s balls of the same color.*

Markov (1906) considered this problem (with $s = 1$) and (with general $s > 0$) in Markov (1917), and for this reason his name is coupled with Pólya's by some authors (e.g., Janardan and Schaeffer, 1975; Stancu, 1970a). (It appears, however, that the triple Markov–Eggenberger–Pólya is not used.) Also, a similar model has been proposed by Brillouin (1927). As pointed out by Fréchet (1943), this model with $s = 1$ was also considered by Chuprov before the appearance of the Eggenberger–Pólya (1923) paper. He presumably refers to Tschuprov (1922).

There are clearly many ways in which the model may be generalized. We mention a few of the most natural ones:

1. At each stage, s' balls of color opposite that chosen are added (as well as s balls of the same color).
2. s and s' may depend on the color of the chosen ball.
3. s (or s') may be negative. (Note that, if $s = -1$ and $s' = 0$, we have sampling without replacement.)
4. The values of s (and s') may vary with the number of the trial; that is, we have a prespecified value s_t for s (s'_t for s') on the tth trial.
5. The s_t (and s'_t) may be random variables.
6. There may be more than two colors for the balls in the urn. This leads to multivariate distributions.

Of course, modifications 1 through 5 may be combined with 6.

4.2. PÓLYA–EGGENBERGER DISTRIBUTIONS

The first model we study in detail is the one due to Eggenberger and Pólya (1923). In this scheme a single urn contains w white (W) balls and b black (B) balls. A ball is drawn at random and then replaced, together with s balls of the same color. We repeat the procedure n times and note the distribution of the random variable X representing the number of times a black ball is drawn. The distribution of X is given by

$$\Pr[X = k] = \binom{n}{k} \frac{b(b+s)\cdots\{b+(k-1)s\}w(w+s)\cdots\{w+(n-k-1)s\}}{(b+w)(b+w+s)\cdots\{(b+w+(n-1)s\}} \qquad (k = 0, 1, \ldots, n). \tag{4.1}$$

This expression may be established by the following argument. The probability of drawing a particular sequence containing k blacks, say BWWBWBB $\cdots$ W, for example, is

$$\frac{b}{b+w} \cdot \frac{w}{b+w+s} \cdot \frac{w+s}{b+w+2s} \cdot \frac{b+s}{b+w+3s} \cdot \frac{w+2s}{b+w+4s} \cdot \frac{b+2s}{b+w+5s} \times \frac{b+3s}{b+w+6s} \cdots \frac{w+(n-k-1)s}{b+w+(n-1)s}.$$

The numerator of this product merely consists of a rearrangement of the factors appearing in the numerator of (4.1). The denominator is the same as in (4.1). The factor $\binom{n}{k}$ accounts for the number of possible "k-success" sequences. If $s = 1$, then (4.1) can be written as

$$\Pr[X = k] = \binom{n}{k} \frac{w^{[k]}b^{[n-k]}}{(w + b)^{[n]}}. \tag{4.2}$$

We pause to take note of the fact that, in the special case $s = 0$, (4.1) becomes a simple *binomial* distribution (Section 2.1.1) with parameters n and $b/(b + w)$. When $b = w = s$, it becomes a discrete *rectangular* distribution (Section 2.1), and when $s = -1$, a *hypergeometric* (Section 2.3).

Pólya–Eggenberger distributions with $s > 0$ are, in a sense, on the side of the binomial ($s = 0$) "opposite" to the hypergeometric ($s = -1$), as pointed out by Kaiser and Stefansky (1972).

It is worthwhile to note that (4.1) can be written in many other ways. We can write the probability of a particular k-success sequence as

$$\begin{aligned}
&\frac{\alpha(\alpha + 1)\cdots[\alpha + (k - 1)]\beta(\beta + 1)\cdots(\beta + n - k - 1)}{(\alpha + \beta)(\alpha + \beta + 1)\cdots[\alpha + \beta + (n - 1)]}\\
&\qquad = \frac{\Gamma(\alpha + \beta)\Gamma(\alpha + k)\Gamma(\beta + n - k)}{\Gamma(\alpha)\Gamma(\beta)\Gamma(\alpha + \beta + n)}\\
&\qquad = \frac{1}{B(\alpha, \beta)} \times B(\alpha + k, \beta + n - k),
\end{aligned}$$

where $\alpha = b/s$, $\beta = w/s$; and (4.1) can be written as

$$\Pr[X = k] = \frac{\binom{n}{k}\alpha^{[k]}\beta^{[n-k]}}{(\alpha + \beta)^{[n]}}. \tag{4.1$'$}$$

Note that this depends only on n and the ratios $\alpha = b/s$, $\beta = w/s$.

The rth factorial moment of X is

$$\begin{aligned}
\mu_{(r)}(X) = E[X^{(r)}] &= \frac{\sum_k \left[k^{(r)}\binom{n}{k}\alpha^{[k]}\beta^{[n-k]}\right]}{(\alpha + \beta)^{[n]}}\\
&= \frac{n^{(r)}\alpha^{[r]} \sum_{k \geq r} \binom{n - r}{k - r}(\alpha + r)^{[k-r]}\beta^{[(n-r)-(k-r)]}}{(\alpha + \beta)^{[n]}}\\
&= \frac{n^{(r)}\alpha^{[r]}(\alpha + \beta + r)^{[n-r]}}{(\alpha + \beta)^{[n]}}\\
&= \frac{n^{(r)}\alpha^{[r]}}{(\alpha + \beta)^{[r]}}.
\end{aligned} \tag{4.3}$$

Putting $r = 1, 2$, we obtain

$$E[X] = \frac{n\alpha(\alpha+\beta+1)^{[n-1]}}{(\alpha+\beta)^{[n]}} = \frac{n\alpha}{\alpha+\beta} \tag{4.4}$$

and

$$E[X(X-1)] = n(n-1)\alpha(\alpha+1)(\alpha+\beta)^{-1}(\alpha+\beta+1)^{-1}.$$

Hence

$$\begin{aligned}\operatorname{var}(X) &= n(n-1)\alpha(\alpha+1)(\alpha+\beta)^{-1}(\alpha+\beta+1)^{-1} + n\alpha(\alpha+\beta)^{-1} \\ &\quad - n^2\alpha^2(\alpha+\beta)^{-2} \\ &= \frac{n\alpha}{\alpha+\beta}\left[\frac{(n-1)(\alpha+1)}{\alpha+\beta+1} + 1 - \frac{n\alpha}{\alpha+\beta}\right].\end{aligned} \tag{4.5}$$

Note that, if $(n-1)(\alpha+1)(\alpha+\beta+1)^{-1} = n\alpha(\alpha+\beta)^{-1}$, the expected value and variance of X are equal (as for the Poisson distribution).

The mean deviation of X is

$$2\theta \Pr[X = \theta](\alpha + n - \theta)(\alpha+\beta)^{-1}, \tag{4.6}$$

where θ is the integer part of $n\alpha/(\alpha+\beta)$ (Koźniewska, 1954).

Recurrence formulas connecting central moments ($\mu_j(X)$) of different orders (different values of j) have been obtained by Mühlbach (1972), Środka (1964), Stancu (1968), and Prete (1970), among others.

EXERCISE 4.1 (Prete, 1970). Obtain the following relationship among the central moments of the distribution (4.1′):

$$\mu_r = \sum_{j=0}^{r-2}\binom{r-1}{j}\{-(\alpha+\beta)^{-1}\mu_{j+1} + \delta\mu_{j+1} + \gamma\mu_j\},$$

where

$$\gamma = np(1-p)\{1 + n(\alpha+\beta)^{-1}\}$$

$$\delta = n(1-2p)(\alpha+\beta)^{-1} - p; \qquad p = \frac{\alpha}{\alpha+\beta}.$$

[*Hint*: (i) Use the ratio $\Pr[X = k]/\Pr[X = k-1]$ and (ii) note that $k - np = 1 + \{(k-1) - np\}$.]

EXERCISE 4.2 (Mühlbach, 1972). Denoting

$$\Pr[X = j] = \binom{n}{j} \frac{\alpha^{[j]}\beta^{[n-j]}}{(\alpha + \beta)^{[n]}}$$

by $P_{j,n}$, show that

(i) $\mu_r'(X) = \sum_{g=0}^{n} \binom{n}{g} P_{g,g} \Delta^g \mathbf{0}^r$.

(ii) $\mu_r(X) = \sum_{g=0}^{n} \binom{n}{g} P_{g,g} \Delta^g (\mathbf{0} - \mu_1'(X))^r$.

$\left[\textit{Hint}\text{: Use } \sum_{j=0}^{M} \binom{M}{j} \theta^{[j]} \phi^{[M-j]} = (\theta + \phi)^{[M]} \right]$.

EXERCISE 4.3. Denote by X_j the number of black balls drawn in the first n_j draws ($j = 1, 2; n_1 < n_2$). Show that the covariance between X_1 and X_2 is

$$\operatorname{var}(X_1) + \frac{n_1(n_2 - n_1)bws}{(b + w)^2(b + w + s)} = \operatorname{var}(X_1) + \frac{n_1(n_2 - n_1)\alpha\beta}{(\alpha + \beta)^2(\alpha + \beta + 1)}.$$

[*Hint*: Given X_1, $(X_2 - X_1)$ has the distribution (4.1) with b replaced by $(b + X_1 s)$, w replaced by $\{w + (n_1 - X_1)s\}$, and n replaced by $(n_2 - n_1)$.]

McFadden (1955) uses the special case $b = w$ as a basis for approximating orthant multinormal probabilities in the equiprobable case.

EXERCISE 4.4. Solve the problem set in Exercise 4.3, generalized by supposing that after the $(h - 1)$th drawing (and replacement) there are added s_{1h} balls of the first color drawn, s_{2h} of the second color drawn, ..., $s_{h-1,h}$ of the last color drawn, [*Hint*: See McFadden (1955) Section 3.]

Formula (4.1′) suggests some link with beta distributions. In fact, we can generate (4.1′) in the following way. Define the indicator random variables $\{Y_i\}_{i=1}^{n}$ such that

$$Y_i = \begin{cases} 1 & \text{if the } i\text{th drawing is black,} \\ 0 & \text{if the } i\text{th drawing is white.} \end{cases}$$

Note that the Y's are not independent. However, suppose they are, that is, that the n drawings are n independent binomial (Bernoulli) trials with probability of drawing black p and probability of drawing white $1 - p$ at each trial (Section 2.2.1). Then, given p, the probability of any particular ordered sequence containing just k blacks is $p^k(1 - p)^{n-k}$. If we now assign a prior beta distribution (with parameters α, β) to p, the overall probability of any *particular* k-blacks sequence is

$$\int_0^1 p^k(1-p)^{n-k}\frac{\Gamma(\alpha+\beta)}{\Gamma(\alpha)\Gamma(\beta)}p^{\alpha-1}(1-p)^{\beta-1}\,dp$$
$$=\frac{\Gamma(\alpha+\beta)}{\Gamma(\alpha)\Gamma(\beta)}\int_0^1 p^{\alpha+k-1}(1-p)^{\beta+n-k-1}\,dp$$
$$=\frac{B(\alpha+k,\beta+n-k)}{B(\alpha,\beta)}=\frac{\alpha^{[k]}\beta^{[n-k]}}{(\alpha+\beta)^{[n]}}\qquad\text{[cf. (4.1')].}\qquad(4.7)$$

There are $\binom{n}{k}$ different possible k-blacks sequences. Hence our model can also be regarded as being generated by a sequence of independent Bernoulli trials in which the parameter p has a beta distribution. [It can in fact be shown that (1) as $n \to \infty$ the ratio $X/(b + w + ns)$ tends to a limit P in any particular sequence and (2) P has a beta distribution with parameters α, β.] This representation has been quoted by numerous authors including Bricas (1949) and Pavlik (1971), and is implicit in the original paper of Eggenberger and Pólya (1923). (See also Section 6.3.3.)

Another approach, due to Morgenstern (1972), links Pólya–Eggenberger schemes with the study of order statistics. Suppose we are sampling at random from a continuous population obtaining values of a random variable X. Denote the *order statistics* from the first $(b + w - 1)$ values by

$$X_{(1)} \le X_{(2)} \le \cdots \le X_{(b+w-1)}.$$

The probability that the next value observed is less than or equal to $X_{(b)}$ is $b/(b + w)$, and this we can regard as *drawing a black ball.* If a black ball is drawn, the conditional probability of drawing a black ball (i.e., obtaining a value of X less than or equal to $X_{(b)}$) next time is $(b + 1)/(b + w + 1)$; if a white ball is drawn the first time (i.e., a value of X greater than $X_{(b)}$), then the conditional probability of drawing a black ball is $b/(b + w + 1)$. This corresponds to the Pólya–Eggenberger model in which a ball of the same color as that just drawn is added to the contents of the urn (i.e., $s = 1$).

Another completely different genesis of distribution (4.1) arises in terms of exceedances. (See Johnson and Kotz, 1969; Sarkadi, 1957.) We now briefly discuss this model.

Suppose we have t balls numbered $a_1, a_2, \ldots, a_t$, with $a_1 < a_2 < \cdots < a_t$, and we choose (without replacement) n of these balls. The random variable X is defined as the number of balls, among the n chosen, for which the corresponding a value is exceeded by the a values of at most $(m - 1)$ of the remaining $(t - n)$ balls.

If $X = k$, then (1) the kth greatest of the chosen a values must be one of $a_{t-m-k+2}, a_{t-m-k+3}, \ldots, a_{t-k+1}$, and (2) the least $(n - k)$ of the chosen a values must be selected from $a_1, a_2, \ldots, a_{t-m-k}$. Hence

$$\begin{aligned}\Pr[X = k] &= \binom{t}{n}^{-1}\binom{t-m-k}{n-k}\left[\binom{m+k-2}{k-1}\right.\\ &\qquad \left. + \binom{m+k-3}{k-1} + \cdots + \binom{k-1}{k-1}\right]\\ &= \binom{t}{n}^{-1}\binom{t-m-k}{n-k}\binom{m+k-1}{k}. \end{aligned} \tag{4.8}$$

[The expression in brackets is the coefficient of t^{k-1} in $(1 + t)^{m+k-2} + (1 + t)^{m+k-3} + \cdots + (1 + t) + 1 = t^{-1}\{(1 + t)^{m+k-1} - 1\}$.]

We can write (4.8) as

$$\begin{aligned}\Pr[X = k] &= \binom{n}{k}\frac{(t-m-k)!}{(t-m-n)!}\frac{(m+k-1)!}{(m-1)!}\cdot\frac{(t-n)!}{t!}\\ &= \binom{n}{k}\frac{m^{[k]}(t-m-n+1)^{[n-k]}}{(t-n+1)^{[n]}}. \end{aligned} \tag{4.8$'$}$$

This is identical with the distribution (4.1$'$), putting $\alpha = m$ and $\beta = t - n + 1$ (Sarkadi, 1957).

Example 4.1. The probability of drawing a black ball at the $(n + 1)$th trial is

$$\begin{aligned}\sum_{k=0}^{n}\frac{b+ks}{b+w+ns}\Pr[X = k] &= \sum_{k=0}^{n}\frac{\alpha+k}{\alpha+\beta+n}\binom{n}{k}\frac{\alpha^{[k]}\beta^{[n-k]}}{(\alpha+\beta)^{[n]}}\\ &= \sum_{k=0}^{n}\binom{n}{k}\frac{\alpha^{[k+1]}\beta^{[n-k]}}{(\alpha+\beta)^{[n+1]}} = \frac{\alpha}{\alpha+\beta}\sum_{k=0}^{n}\binom{n}{k}\frac{(\alpha+1)^{[k]}\beta^{[n-k]}}{(\alpha+\beta+1)^{[n]}}\\ &= \frac{\alpha}{\alpha+\beta} = \frac{b}{b+w}. \end{aligned} \tag{4.9}$$

Note that this probability does not depend on either n or s. This result was obtained by Jordan (1956).

Example 4.2 (The mode of the Pólya–Eggenberger distribution). From (4.1),

$$\frac{\Pr[X = k+1]}{\Pr[X = k]} = \frac{n-k}{k+1}\cdot\frac{b+ks}{w+(n-k-1)s}.$$

So $\Pr[X = k + 1] \gtrless \Pr[X = k]$ according as

$$(n - k)(b + ks) \gtrless (k + 1)\{w + (n - k - 1)s\},$$

that is, $(b + w - 2s)k \lessgtr n(b - s) - w + s$. The distribution therefore has at most one mode, or two neighboring and equal modal values.

If $(b - s) + (w - s) > 0$ and $n(b - s) - (w - s) > 0$, there is such a mode at

$$k = 1 + \frac{n(b - s) - w + s}{b + w - 2s} = \frac{(n + 1)(b - s)}{b + w - 2s}.$$

If this is an integer, there is an equal mode at

$$\frac{n(b - s) - w + s}{b + w - 2s}.$$

If $(b - s) + (w - s) > 0$ but $n(b - s) - (w - s) < 0$, the modal value is at $k = 0$. If $(b - s) + (w - s) < 0$ but $n(b - s) - (w - s) > 0$, there is an *antimode* (minimum value of $\Pr[X = k]$) at

$$k = \frac{(n + 1)(b - s)}{b + w - 2s}.$$

If this is an integer, there is an equal antimode at

$$\frac{n(b - s) - w + s}{b + w - 2s}.$$

It should be noted that a certain amount of confusion exists in the literature regarding the nomenclature of distribution (4.1). It is simply referred to as the "Pólya distribution" by Bosch (1963), and the "Pólya–Eggenberger distribution" discussed by Taguti (1952) turns out to be a limiting case ($n \to \infty$) of (4.1). [See (4.17).]

We refer to Johnson and Kotz (1969) and Chatfield and Theobald (1973) for a further discussion of ambiguous terminology. (See also Section 4.1.)

Kotz and Srinivasan (1974) calculated tables of the probabilities (4.1) and the cumulative distribution function (cdf) of X for $b = 2(2)10$, $w = 2(2)10$, $s = 1(1)8$, and $n = 2(1)10(5)35$. A portion of these tables is presented in the APPENDIX to this chapter.

By using these tables, the adequacy of the various asymptotic approximations to the distribution of X (Bosch, 1963; Johnson and Kotz, 1969; Taguti, 1952) was investigated. It was found that even for n as large as 35 the difference between the exact and approximate percentiles of X was never less than 5 and was as large as 10 in the tail areas, which, in our opinion, is far from satisfactory for practical purposes. It would therefore be of interest to derive more adequate approximations to the distribution of X and to study the rate of convergence in the case of the limiting distributions already known. Some of these limit distributions are discussed in Section 6.3.

4.3. GENERALIZATIONS OF PÓLYA–EGGENBERGER DISTRIBUTIONS

We first give an outline of the theory applicable to cases (1) through (3) in Section 4.1. We shall in fact consider the following slightly more general scheme.

TABLE 4.1

Generalized Pólya–Eggenberger Scheme

		Color of chosen ball	
		White	Black
Number of balls added to the urn	White	ω_w	ω_b
	Black	β_w	β_b

If a black (white) ball is chosen, we add β_b (β_w) black and ω_b (ω_w) white balls to the urn. Also, the chosen ball is returned to the urn. If there are B_n black balls and W_n white balls in the urn after n drawings, then (conditionally on $B_n = b_n$, $W_n = w_n$):

$$B_{n+1} = b_n + T_n\beta_b + (1 - T_n)\beta_w,$$

where T_n is a random variable with

$$\Pr[T_n = 0] = \frac{w_n}{b_n + w_n}$$

$$\Pr[T_n = 1] = \frac{b_n}{b_n + w_n}. \tag{4.10a}$$

Hence

$$E[B_{n+1} | b_n, w_n] = b_n + \frac{b_n\beta_b + w_n\beta_w}{b_n + w_n} \tag{4.10b}$$

$$\operatorname{var}[B_{n+1} | b_n, w_n] = (\beta_b - \beta_w)^2 \operatorname{var}[T_n] = \frac{(\beta_b - \beta_w)^2 b_n w_n}{(b_n + w_n)^2}. \tag{4.10c}$$

In the particular case where we know the value of $b_n + w_n$, as in the case where $\beta_b + \omega_b = \beta_w + \omega_w = \gamma$ (because the total number of balls added is the same, whether a black or a white ball is chosen), we can use (4.10) to obtain recurrence relations among the expected values and variances after n and after $(n + 1)$ drawings.

From (4.10a),

$$\begin{aligned} E[B_{n+1}] &= E[B_n + \{B_n\beta_b + (\alpha + n\gamma - B_n)\beta_w\}(\alpha + n\gamma)^{-1}] \\ &= \beta_w + \{1 + (\beta_b - \beta_w)(\alpha + n\gamma)^{-1}\}E[B_n], \end{aligned} \tag{4.11}$$

where $\alpha = \beta + \omega$ and β, ω are the *initial* numbers of black and white balls, respectively, in the urn.

From (4.10) we also obtain

$$\begin{aligned} \text{var}[B_{n+1}] &= (\beta_b - \beta_w)^2(\alpha + n\gamma)^{-2}E[B_n(\alpha + n\gamma - B_n)] \\ &\quad + \text{var}[B_n + \{B_n\beta_b + (\alpha + n\gamma - B_n)\beta_w\}(\alpha + n\gamma)^{-1}] \\ &= (\beta_b - \beta_w)^2(\alpha + n\gamma)^{-2}[(\alpha + n\gamma)E[B_n] - \text{var}[B_n] - \{[E[B_n]\}^2] \\ &\quad + \{1 + (\beta_b - \beta_w)(\alpha + n\gamma)^{-1}\}^2 \text{ var}(B_n) \\ &= \{1 + 2(\beta_b - \beta_w)(\alpha + n\gamma)^{-1}\}\text{var}(B_n) \\ &\quad + (\beta_b - \beta_w)^2E[(\alpha + n\gamma)^{-1}B_n]\{1 - E[(\alpha + n\gamma)^{-1}B_n]\}. \end{aligned} \tag{4.12}$$

Among special cases, in each of which the condition $\beta_b + \omega_b = \beta_w + \omega_w$ is satisfied, we note

1. $\beta_b = \omega_w$, $\beta_w = \omega_b$ (Bernstein 1940; Savkevitch, 1940; Friedman, 1949; Freedman, 1965), and the further specialization
2. $\beta_b = \omega_w = -1$, $\beta_w = \omega_b = 1$, which is the simplest Ehrenfest model of heat conduction. (See Section 4.7.2).
3. $\beta_w = \omega_b = 0$, $\beta_b = \omega_w$, which is the Pólya–Eggenberger model. (See Section 4.2.)
4. Woodbury (1949) has constructed a model in which $\beta_w = \omega_w = 0$ but $\beta_b = -\omega_b > 0$. This means that the drawing of a white (*uninfected*) ball has no effect on the composition of the urn, but a black (*infected*) ball causes β_b uninfected balls to become infected (since replacement of ω_b white balls by an equal number of black balls has the same effect as changing the color of ω_b balls from white to black). In this case the number of balls in the urn remains constant, and eventually all the balls will be black.
5. Wei (1976) has suggested the use of a model with $\beta_b = \omega_w = -1$, $\beta_w = \omega_b = \gamma$ to determine a sequential assignment of treatments in a clinical trial. (This does not use the results of previous treatments, but Wei claims advantages in that, (*a*) by taking $B_1 = W_1$, balance is obtained in early stages of the trial, while (*b*) for later stages complete randomization is approximated.)

In many cases it is possible to obtain useful limiting distributions (see Section 6.3).

Here we discuss some properties (in a finite number of drawings) of processes of type (1). Writing $\beta_b = \omega_w = \phi_1, \beta_w = \omega_b = \phi_2$ and noting that the number of balls in the urn just before the nth drawing is $b_0 + w_0 + (n-1)(\phi_1 + \phi_2) = b_0 + w_0 + (n-1)\phi$ (with $\phi = \phi_1 + \phi_2$) and letting X denote the number of black balls chosen in the first n drawings we have

$$\Pr[X = k] = \left[\prod_{i=1}^{n} \{b_0 + w_0 + (i-1)\phi\}\right]^{-1}$$
$$\times \sum_{j_1<j_2<\cdots<j_k}\sum\cdots\sum \prod_{i=1}^{k}\left[\{b_0 + (j_i - i)\phi_2 + (i-1)\phi_1\}\right.$$
$$\left.\times \prod_{h=1}^{j_i - j_{i-1}} \{w_0 + (j_{i-1} - i + h)\phi_1 + (i-1)\phi_2\}\right]$$
$$(\text{with } j_0 = 0). \quad (4.13)$$

Even for such a relatively simple scheme, we have quite a complicated formula for $\Pr[X = k]$.

The expected value of X can be obtained in the following way. We have

$$X = Y_1 + Y_2 + \cdots + Y_n,$$

where

$$Y_i = \begin{cases} 1 & \text{if the } i\text{th draw is black,} \\ 0 & \text{if the } i\text{th draw is white.} \end{cases}$$

Hence

$$E[X] = \sum_{i=1}^{n} E[Y_i] \quad (4.14)$$

and

$$E[Y_i] = \{b_0 + w_0 + (i-1)\phi\}^{-1}E[B_{i-1}], \quad (4.15)$$

where B_{i-1} = number of black balls in the urn at the moment of the ith drawing (i.e. after the $(i-1)$th draw).

From (4.11),

$$E[B_{h+1}] = \phi_2 + \{1 + (\phi_1 - \phi_2)(b_0 + w_0 + h\phi)^{-1}\}E[B_h]. \quad (4.16)$$

The expected value of B_n is then built up from (4.14) through (4.16).

EXERCISE 4.5. Find the expected value of X when there are initially 10 black and 10 white balls in the urn and $\phi_1 = 3, \phi_2 = 1$. Take $n = 1, 2, 3, 4, 5$.

If we suppose $\beta_b = \beta_w$ and $\omega_b = \omega_w$, we have a situation in which fixed numbers of black and white balls are added after each drawing, no matter what the result of the drawing. In this case of course we know the value of B_n exactly, and it is a constant $(\beta + n\beta_b = \beta + n\beta_w)$. This is not a Pólya-type scheme (see Section 4.7.2).

We now turn to some special cases in which the condition $\beta_b + \omega_b = \beta_w + \omega_w$ is not satisfied. If we take

$$\beta_b \neq \omega_w \qquad \text{and} \qquad \beta_w = \omega_b = 0,$$

we have a generalization of the Pólya–Eggenberger model proposed by Rosenblatt (1940), Bates and Neyman (1952), and Yntema (1954).

EXERCISE 4.6. Obtain a formula for the probability that there are exactly k white balls among the first n balls drawn, for the situation just described.

Styve (1965) considers the case

$$\beta_b = -1, \qquad \beta_w = 0, \qquad \omega_b = 1, \qquad \omega_w = 0.$$

This may serve as a model for inspection of items in a lot, when items found to be defective are replaced by nondefective items. In this case the constitution of the urn remains the same as long as white balls are chosen. If k black balls have been chosen $(k < \beta)$, then the probability of drawing a black ball at the next drawing is $(\beta - k)/(\beta + \omega)$.

The probability of drawing just k black balls in n drawings is the sum of $\binom{n}{k}$ terms like

$$(\beta + \omega)^{-n}[\beta(\beta - 1)\cdots(\beta - k + 1)\omega^{a_0}(\omega + 1)^{a_1}\cdots(\omega + k)^{a_k}],$$

corresponding to cases where the black balls occur at the $(a_0 + 1)$th, $(a_0 + a_1 + 2)$th, $(a_0 + a_1 + a_2 + 3)$th, $\ldots$, $(a_0 + a_1 + \cdots + a_{k-1} + k)$th draws, and $\sum_{j=0}^{k} a_j = n - k$. Hence the probability of k black balls in n drawings is

$$(\beta + \omega)^{-n}\beta^{(k)} \sum_{a_0}\cdots\sum_{a_k} \omega^{a_0}(\omega + 1)^{a_1}\cdots(\omega + k)^{a_k}. \tag{4.17}$$

The summation is equal to the coefficient of t^{n-k} in the expansion of

$$\prod_{j=0}^{k}\{1 - (\omega + j)t\}^{-1} = \frac{1}{k!}\sum_{j=0}^{k}(-1)^{k-j}\binom{k}{j}(\omega + j)^k\{1 - (\omega + j)t\}^{-1}$$

(see Section 1.1.3). This coefficient is

$$\frac{1}{k!}\sum_{j=0}^{k}(-1)^{k-j}\binom{k}{j}(\omega+j)^n = \frac{1}{k!}\Delta^k(\omega+\mathbf{0})^n.$$

So the required probability is

$$\frac{\beta^{(k)}}{(\beta+\omega)^n k!}\Delta^k(\omega+\mathbf{0})^n, \tag{4.18}$$

as given by Styve (1965) who determined this probability using a different method.

For the model $\beta_w = \beta_b = \omega_b = 0$, $\omega_w = -1$, (i.e., where white balls are not replaced when drawn, but black balls are), Kaminsky et al. (1975) have considered a *waiting-time type of distribution* which we discuss in Section 4.4.

EXERCISE 4.7. Taking $\beta_b = -1$, $\beta_w = \omega_w = 0$, $\omega_b = 2$, find the probability that exactly k balls are observed among the first n drawings.

Describe a situation for which this might be a reasonable model.

Rosenblatt (1940) has studied the case $\beta_b = \Delta$, $\omega_w = \Delta'$ (with $\Delta \neq \Delta'$), $\beta_w = \omega_b = 0$. In this case also, the number of balls in the urn after n drawings is not fixed. It depends on the number of times a black ball has been drawn. If this has happened r times, then the number of balls in the urn after n drawings is

$$\beta + \omega + r\Delta + (n-r)\Delta'.$$

Let X denote the number of black balls obtained in the first n drawings. It is easy to see that

$$\Pr[X=0] = \frac{\omega(\omega+\Delta')\cdots\{\omega+(n-1)\Delta'\}}{\tau(\tau+\Delta')\cdots\{\tau+(n-1)\Delta'\}}, \tag{4.19}$$

where $\tau = \beta + \omega$.

The expression for $\Pr[X = 1]$ is more complicated. We have to add up the probabilities that the drawing on which the black ball was observed was the first, second, ..., nth.

While the numerator of each term is $\beta\omega(\omega+\Delta')\cdots[\omega+(n-2)\Delta']$, the denominators are

$$\tau(\tau+\Delta)(\tau+\Delta+\Delta')\cdots\{\tau+\Delta+(n-2)\Delta'\},$$
$$\tau(\tau+\Delta')(\tau+\Delta+\Delta')\cdots\{\tau+\Delta+(n-2)\Delta'\},$$
$$\tau(\tau+\Delta')(\tau+2\Delta')(\tau+\Delta+2\Delta')\cdots\{\tau+\Delta+(n-2)\Delta'\},$$

and so on. [The last one is $\tau(\tau+\Delta')\cdots\{\tau+(n-2)\Delta'\}\{\tau+\Delta+(n-2)\Delta'\}$.]

Expressing our result in terms of $\rho = \beta/\tau$, $\sigma = \omega/\tau = 1 - \rho$, $\delta = \Delta/\tau$, $\delta' = \Delta'/\tau$, we have

$$\begin{aligned}\Pr[X = 1] &= \rho\sigma(\sigma + \delta')\cdots\{\sigma + (n-1)\delta'\}\\ &\quad\times\left[\frac{1}{(1+\delta)(1+\delta+\delta')\cdots\{1+\delta+(n-2)\delta'\}}\right.\\ &\quad+\frac{1}{(1+\delta')(1+\delta+\delta')\cdots\{1+\delta+(n-2)\delta'\}}\\ &\quad+\cdots\\ &\quad\left.+\frac{1}{(1+\delta')(1+2\delta')\cdots\{1+(n-2)\delta'\}\{1+\delta+(n-2)\delta'\}}\right].\end{aligned} \tag{4.20}$$

For general k, $\Pr[X = k]$ has a quite complicated expression. We suggest that the reader write out the appropriate general expression as an exercise.

Rosenblatt (1940) has shown that, if $n \to \infty$ with $n\delta = d$, $n\delta' = d'$, $n\rho = h$ constant, then

$$\lim_{n\to\infty} \Pr[X = k] = \frac{(h/d)^{[k]}}{k!}(1 + d')^{-h/d'}\{1 - (1 + d')^{-d/d'}\}^k. \tag{4.21}$$

Note that, although the number of drawings is supposed to increase without limit, we obtain nonzero limiting values for the probabilities that only 0, 1, 2, ... black balls are drawn. This is because the longer a black ball is not drawn, the smaller becomes the conditional probability that it will be obtained at the next drawing.

The rth factorial moment of this (limiting) distribution is

$$\begin{aligned}\mu_{(r)}(X) &= \sum_{k=r}^{\infty} k^{(r)}\frac{(h/d)^{[k]}}{k!}(1 + d')^{-h/d'}\{1 - (1 + d')^{-d/d'}\}^k\\ &= \left(\frac{h}{d}\right)^{[r]}(1 + d')^{-h/d'}\sum_{k=r}^{\infty}\frac{(hd^{-1} + r)^{[k-r]}}{(k-r)!}\{1 - (1 + d')^{-d/d'}\}^k\\ &= \left(\frac{h}{d}\right)^{[r]}\{(1 + d')^{d/d'} - 1\}^r.\end{aligned} \tag{4.22}$$

(Note that the summation is the *negative binomial expansion* of

$$[1 - \{1 - (1 + d')^{-d/d'}\}]^{-(hd^{-1}+r)};$$

see Section 2.4.)

Thorp, E. O. (1964) Repeated independent trials and a class of dice problems, *Am. Math. Mon.*, **71**, 778–781.

Uppuluri, V. R. R. and Carpenter, J. A. (1971) A generalization of the classical occupancy problem, *J. Math. Anal. Appl.*, **34**, 316–324.

Vantilborgh, H. (1974) On the working set size and its normal approximation, *BIT, Sweden*, **14**, 240–251.

von Mises, R. (1939) Über Aufteilungs- und Besetzungs-Wahrscheinlichkeiten, *Rev. Fac. Sci. Univ. Istanbul*, No. 5, **4**, 145–163. Reprinted in *Selected Papers of Richard von Mises*, Vol. 2, Providence, R.I.: American Mathematical Society, 1964, pp. 313–334.

von Mises, R. (1964) *Mathematical Theory of Probability and Statistics*, H. Geiringer (Ed.), New York: Academic Press.

Walter, S. D. (1975) *Some Generalizations of the Committee Problem*, Department of Biometry, School of Medicine, Yale University (manuscript).

White, C. (1971) The committee problem, *Am. Stat.*, **25**(4), 25–26.

Young, D. H. (1961) Quota fulfillment using unrestricted random sampling, *Biometrika*, **48**, 333–342.

Putting $d = d'$, we obtain the limiting distribution (and its factorial moments) for the original Pólya–Eggenberger distribution:

$$\lim_{n\to\infty} \Pr[X = k] = \frac{(h/d)^{[k]}}{k!}(1 + d)^{-k-h/d}\, d^k \tag{4.23}$$

$$\mu_{(r)}(X) = \left(\frac{h}{d}\right)^{[r]} d^k. \tag{4.24}$$

Rosenblatt applied this limiting distribution to data on bubonic plague in Peru, for 102 months during the period 1932–1940. There were 926 cases in all. Rosenblatt fitted the distribution (of months having 0, 1, 2, ... cases of plague) by equating the sample mean and standard deviation to the population values, h and $h(1 + d)$, respectively. Numerical results are shown in Table 4.2 [based on data in Rosenblatt (1940)].

The distribution (4.23) which we call *limiting Pólya–Eggenberger* is called simply "Pólya–Eggenberger" by Taguti (1952) [and simply "Pólya" by Rosenblatt (1940) and Lundberg (1964)]. In the latter reference the distribution has been fitted to numbers of claims in certain classes of life insurance. The distribution is in fact a negative binomial distribution (Section 2.4). Taguti (1952) gives tables of h such that the cdf

$$\sum_{j=0}^{k} \frac{(h/d)^{[j]}}{j!}(1 + d)^{-j-h/d} d^j$$

TABLE 4.2

Data Fitted by Limiting Pólya–Eggenberger (P–E) Distribution

Cases per month	Number of months	Fitted limiting P–E	Cases per month	Number of months	Fitted limiting P–E
0	3	5.21	13	3	3.17
1	7	6.86	14	2	2.82
3	8	7.53	15	1	2.51
4	6	7.31	16	3	2.21
5	7	6.92	18	1	3.53 (17–18)
6	7	6.45	19	3	2.61 (19–20)
7	10	5.94	22	1	2.00 (21–22)
8	2	5.42	23	5	2.42 (23–25)
9	4	4.91	28	1	1.77 (26–28)
10	6	4.42	33	1	1.83 (>28)
11	4	3.98	44	1	

is equal to .95 or .99 for $h/d = .5(.5)15, 20, 30, 60, \infty$ and $k = 1(1)25(5)40(10)60$, 75, 100, 200, 500, ∞ to three significant figures.

The distribution (4.23) can be generated in the following way. Consider a discrete random process [representing a discrete random variable with a discrete (or continuous) parameter λ] such that

1. The probability that the event E does not occur during the interval $(t, t + \Delta t)$ is $1 - \lambda\Delta t$.
2. The probability that E occurs once during the interval $(t, t + \Delta t)$ is $\lambda\Delta t$.
3. The probability that E occurs twice or more in the interval $(t, t + \Delta t)$ is $o(\Delta t)$.

The basic equation governing this process is

$$P_k(t + \Delta t) = P_k(t)(1 - \lambda\Delta t) + P_{k-1}(t)\lambda\Delta t + o(\Delta t),$$

where $P_k(t + \Delta t)$ is the probability that the event appears exactly k times during the time interval $(0, t + \Delta t)$. Letting $\Delta t \to 0$ we obtain the difference-differential equation

$$\frac{dP_k(t)}{dt} = \lambda[P_{k-1}(t) - P_k(t)]. \tag{4.25}$$

The solution of (4.25) with initial conditions $P_0(0) = 1$ and $P_k(0) = 0$ for $k \geq 1$ is the Poisson distribution

$$P_k(t) = e^{-\lambda t}\frac{(\lambda t)^k}{k!} \qquad (k = 0, 1 \ldots). \tag{4.26}$$

We have so far assumed that λ does not depend on k or t. If we assume that

$$\lambda = \frac{\alpha + k}{\beta + t} \qquad (\alpha \geq 0, \beta \geq 0),$$

that is, that the probability of the occurrence of E in the interval $(t, t + \Delta t)$ increases linearly with k—the number of occurrences of E in the interval $(0, t)$—and also for given k that this probability decreases as t increases, then (with the initial conditions as above) it can be shown that

$$P(t) = \frac{\alpha^{[k]}}{k!}\left(1 + \frac{t}{\beta}\right)^{-\alpha-k} \cdot t^k. \tag{4.27}$$

On putting $\beta = 1$, $\alpha = h/d$, $t = d$, we obtain (4.23).

Example 4.3. Another possible genesis is a mixture of Poisson distributions. If we define

$$p^*(k) = \int_0^\infty p(k; \lambda)p(\lambda)\, d\lambda,$$

where $p(k;\lambda) = e^{-\lambda}(\lambda^k/k!)\,(k = 0, 1, \ldots)$ and $p(\lambda) = \{\gamma^\alpha/\Gamma(\alpha)\}e^{-\gamma\lambda}\lambda^{\alpha-1}$ $(0 < \lambda)$, we find

$$p^*(k) = \frac{\gamma^\alpha}{\Gamma(\alpha)k!}\int_0^\infty e^{-(\gamma+1)\lambda}\lambda^{k-\alpha-1}\,d\lambda$$

$$= \left(\frac{\gamma}{\gamma+1}\right)^\alpha \frac{\alpha^{[k]}}{k!}\left(\frac{1}{\gamma+1}\right)^k = \frac{\alpha^{[k]}}{k!}(1+\gamma)^{-\alpha-k}\gamma^k \qquad \text{[cf. (4.27)]}.$$

[This classical derivation, given by Greenwood and Yule (1920), exemplifies how quite different models can give rise to identical distributions corresponding to observable variates. Note that λ has a gamma distribution, as defined in Section 2.2.4.]

4.4. INVERSE PÓLYA DISTRIBUTIONS

Just as the negative binomial and negative hypergeometric distributions were obtained (Sections 2.4 and 2.5) from waiting-time problems in situations where a fixed number of observations would produce binomial and hypergeometric distributions, so we can derive waiting-time distributions for situations where Pólya-type distributions are obtained from fixed numbers of observations. These distributions are called *inverse Pólya* distributions.

They are the distributions of the numbers of drawings needed to observe a prespecified number, say k, of black balls when replacements follow the schedule shown in Table 4.1. The *inverse Pólya–Eggenberger* distribution, for example, is the distribution of the number N of drawings needed to obtain k black balls under the conditions $\beta_b = \omega_w = s$, $\beta_w = \beta_b = 0$.

Since $\Pr[N > n] = \Pr[X < k] = P_n$ say, where X has the Pólya–Eggenberger distribution (4.1′), we have

$$\Pr[N = n] = P_{n-1} - P_n,$$

with P_n given by (4.1′). From this we find

$$\Pr[N = k + x] = \binom{k+x-1}{x}\frac{\alpha^{[k]}\beta^{[x]}}{(\alpha+\beta)^{[k+x]}} \qquad (x = 0, 1, 2, \ldots). \tag{4.28}$$

EXERCISE 4.8. Using (4.28), write down the distribution of the number of white balls drawn before k black balls are drawn.

Kaminsky et al. (1975) considered the distribution of the number Y of black balls drawn before g white balls have been drawn when $\beta_w = \beta_b = \omega_b = 0$, $\omega_w = -1$ (i.e., white balls are not replaced when drawn, while black balls are). This could correspond to an industrial sampling situation when defective (white) items are discarded.

We use a different method and consider the somewhat more general case $\beta_w = \beta_b = \omega_b = 0$, $\omega_w = -s$. We have

$$\Pr[Y = y] = \Pr[(g-1) \text{ white balls in } (y-1) \text{ draws}]$$
$$\times \Pr[\text{white ball on } y\text{th drawing} \mid (g-1) \text{ white balls drawn in } (y-1) \text{ draws}]$$
$$= \frac{w(w-s)\cdots\{w-(g-1)s\}}{(b+w)(b+w-s)\cdots\{(b+w-(g-1)s\}}$$
$$\times \sum\cdots\sum \prod_{i=0}^{g-1}\left(\frac{b}{b+w-is}\right)^{a_i}, \tag{4.29}$$

where the summation is over $\sum_{i=0}^{g-1} a_i = (y-1)-(g-1) = y-g$ (and of course $a_i \geq 0$). A set $a_0, a_1, \ldots, a_{g-1}$ corresponds to occurrences of a white ball at the (a_0+1)th, (a_0+a_1+2)th, $\ldots$, $(a_0+a_1+\cdots+a_{g-2}+g-1)$th drawings.

We can write (4.29) in the form

$$\Pr[Y = y] = \frac{(w/s)^{(g)}}{[(b+w)/s]^{(g)}}\sum\cdots\sum \prod_{i=0}^{g-1}\left(\frac{b/s}{[(b+w)/s]-i}\right)^{a_i}. \tag{4.30}$$

The summation in (4.30) is equal to the coefficient of t^{y-g} in the expansion of

$$\prod_{i=0}^{g-1}\left\{1-\frac{(b/s)t}{[(b+w)/s]-i}\right\}^{-1} = \sum_{i=0}^{g-1} A_i\left\{1-\frac{(b/s)t}{[(b+w)/s]-i}\right\}^{-1},$$

with

$$A_i = \frac{[(b+w)/s]^{(g)}}{(g-1)!}(-1)^{g-1-i}\binom{g-1}{i}\left(\frac{b+w}{s}-i\right)^{-1},$$

as may be established using the method of undetermined coefficients (Section 1.1.3).

Hence

$$\Pr[Y = y] = \frac{(w/s)^{(g)}}{[(b+w)/s]^{(g)}}\sum_{i=0}^{g-1}\frac{[(b+w)/s]^{(g)}}{(g-1)!}(-1)^{g-1-i}$$
$$\times \binom{g-1}{i}\left(\frac{b}{s}\right)^{y-g}\left(\frac{b+w}{s}-i\right)^{-y+g-1}$$
$$= \left(\frac{b}{s}\right)^{y-g}\frac{(w/s)^{(g)}}{(g-1)!}\sum_{i=0}^{g-1}(-1)^{g-1-i}\binom{g-1}{i}\left(\frac{b+w}{s}-i\right)^{-y+g-1}$$
$$= \left(\frac{b}{s}\right)^{y-g}\frac{(w/s)^{(g)}}{(g-1)!}\Delta^{g-1}\left(\frac{b+w}{s}-\mathbf{0}\right)^{-y+g-1}. \tag{4.31}$$

EXERCISE 4.9. Find the distribution of the number of drawings needed to obtain g white balls for

(i) Styve's (1965) model (see Section 4.3).
(ii) The model in Exercise 4.7.

EXERCISE 4.10. Let an urn contain n red balls and b black balls. Balls are drawn at random, red balls are retained, and black balls are replaced. Let $T(r)$ denote the number of black balls selected before r red ones are removed.

(i) Show that

$$P(T(r) < m) = \sum \prod_{i=0}^{r-1} \left(\frac{b}{b+n-i}\right)^{\alpha_i} \left(\frac{n-i}{b+n-i}\right),$$

where the sum is over all nonnegative integers $k, \alpha_0, \alpha_1, \ldots, \alpha_{r-1}$, satisfying $k \le m-1$ and $\sum_{i=0}^{r-1} \alpha_i = k$. (*Hint*: $T(r)$ is the sum of independent *geometric* random variables [see (2.30)] with varying probability of success.)

(ii) Show that

$$E(T(r)) = b \sum_{j=0}^{r-1} (n-j)^{-1}$$

$$\operatorname{var}(T(r)) = b \sum_{j=0}^{r-1} (b+n-j)(n-j)^{-2}.$$

(*Source*: Kaminsky et al., 1975.)

4.5. MULTIVARIATE PÓLYA DISTRIBUTIONS

4.5.1. Multivariate Pólya–Eggenberger Distributions

The simplest multivariate generalization of the Pólya–Eggenberger distribution is constructed by supposing that we initially have c_j balls of color $\mathscr{C}_j$ ($j = 0, 1, \ldots, k$) in the urn. At each drawing (of a single ball), the ball is returned together with s balls of the same color. After n draws, the joint distribution of the numbers $X_0, X_1, \ldots, X_k$ of times the balls drawn were of color $\mathscr{C}_0, \mathscr{C}_1, \ldots, \mathscr{C}_k$, respectively, is

$$\Pr\left[\bigcap_{j=0}^{k} (X_j = x_j)\right] = \binom{n}{x_0 \; x_1 \cdots x_k} \frac{\prod_{i=0}^{k} c_i(c_i+s)\cdots(c_i+\overline{x_i-1}s)}{c(c+s)(c+2s)\cdots(c+\overline{n-1}s)}, \tag{4.32}$$

where

$$c = \sum_{j=0}^{k} c_j \qquad \text{and} \qquad \sum_{j=0}^{k} x_j = n.$$

As for the multinomial distribution (Section 2.7.1), this is not a $(k+1)$-dimensional, but a k-dimensional distribution, even though there are $(k+1)$ random variables involved.

(The notation c_j for the initial number of balls of color $\mathscr{C}_j$ is really more suitable than the w, b used in the univariate case. Although we could have used c_0, c_1 in that case, we felt that the clearly different symbols w and b may be helpful on a first reading.)

Putting $c_j/s = \alpha_j$, $c/s = \alpha = \sum_{j=0}^k \alpha_j$, we can write (4.32) as

$$\Pr\left[\bigcap_{j=0}^{k} (X_j = x_j)\right] = \frac{n!}{\prod_{j=0}^{k} x_j!} \cdot \frac{\prod_{j=0}^{k} \alpha_j^{[x_j]}}{\alpha^{[n]}}. \tag{4.32$'$}$$

The joint factorial moment formula is

$$\mu_{(r_0, r_1, \ldots, r_k)} = E\left[\prod_{j=0}^{k} X_j^{(r_j)}\right] = \frac{n^{(r)} \prod_{j=0}^{k} \alpha_j^{[r_j]}}{\alpha^{[r]}}, \tag{4.33}$$

where $r = \sum_{j=0}^k r_j$.

In particular, $E[X_i X_j] = n(n-1)\alpha_i \alpha_j \alpha^{-1}(\alpha+1)^{-1}$,

$$E[X_i] = \frac{n\alpha_i}{\alpha} \qquad \operatorname{var}(X_i) = \frac{n\alpha_i(\alpha - \alpha_i)(\alpha + n)}{\alpha^2(\alpha + 1)},$$

$$\operatorname{cov}(X_i, X_j) = -\frac{n\alpha_i\alpha_j(\alpha + n)}{\alpha^2(\alpha + 1)}, \qquad \operatorname{corr}(X_i, X_j) = -\sqrt{\frac{\alpha_i\alpha_j}{(\alpha - \alpha_i)(\alpha - \alpha_j)}}. \tag{4.34}$$

The correlation between two variables is thus given by

$$\operatorname{corr}(X_i, X_j) = -\sqrt{\frac{\alpha_i\alpha_j}{(\alpha - \alpha_i)(\alpha - \alpha_j)}} \tag{4.35}$$

$$= -\sqrt{\frac{p_i p_j}{q_i q_j}}, \tag{4.35$'$}$$

where $p_i = \alpha_i/\alpha$ and $q_i = 1 - p_i$. This is the form usually given in the literature (e.g., Steyn, 1955; Janardan and Patil, 1970). (Cf. similar formula for the multinomial distribution in Section 2.7.1.)

The marginal distribution of any one of the X's is a Pólya–Eggenberger distribution. This can easily be seen (e.g., for X_i) by treating $\mathscr{C}_i$ as one color and all the others as another color. Similarly, any subset of the X's has a joint distribution of the same form as (4.32).

The marginal joint distribution of X_0 and X_1, for example, is

$$\Pr[(X_0 = x_0) \cap (X_1 = x_1)] = \frac{n!\alpha_0^{[x_0]}\alpha_1^{[x_1]}(\alpha - \alpha_0 - \alpha_1)^{[n-x_0-x_1]}}{x_0!x_1!(n - x_0 - x_i)!\alpha^{[n]}}. \quad (4.36)$$

Given $X_0 = x_0$, the joint distribution of $X_1, \ldots, X_k$ is

$$\Pr\left[\bigcap_{j=1}^{k} (X_j = x_j) | X_0 = x_0\right] = \frac{(n - x_0)!}{\prod_{j=1}^{k} x_j!} \cdot \frac{\prod_{j=1}^{k} \alpha_j^{[x_j]}}{(\alpha - \alpha_0)^{[n-x_0]}} \quad \left(\sum_{j=0}^{k} x_j = n - x_0\right). \quad (4.37)$$

(Note that

$$\Pr[X_0 = x_0] = \frac{n!}{x_0!(n - x_0)!} \cdot \frac{\alpha_0^{[x_0]}(\alpha - \alpha_0)^{[n-x_0]}}{\alpha^{[n]}}.)$$

This conditional distribution does not depend on α_0. Although α_0 appears in (4.37), we observe that $\alpha - \alpha_0 = \alpha_1 + \alpha_2 + \cdots + \alpha_k$. In particular, the conditional distribution of X_1, given $X_0 = x_0$, is

$$\Pr[X_1 = x_1 | X_0 = x_0] = \frac{(n - x_0)!}{x_1!(n - x_0 - x_1)!} \cdot \frac{\alpha_1^{[x_1]}(\alpha - \alpha_0 - \alpha_1)^{[n-x_0-x_1]}}{(\alpha - \alpha_0)^{[n-x_0]}}. \quad (4.38)$$

Mosimann (1962), Santacroce (1967), Stancu (1970b), and Dyczka (1973) have demonstrated a connection between multivariate Pólya–Eggenberger distributions and Dirichlet distributions similar to that between univariate Pólya–Eggenberger distributions and beta distributions. [See (4.7).]

We can write (4.32) in the form

$$\Pr\left[\bigcap_{j=0}^{k} (X_j = x_j)\right] = \frac{n!}{\prod_{j=0}^{k} x_j!} \cdot \frac{\Gamma(\alpha)}{\Gamma(\alpha + n)} \prod_{j=0}^{k} \frac{\Gamma(\alpha_j + x_j)}{\Gamma(\alpha_j)}. \quad (4.32'')$$

The distribution can be obtained as a mixture of multinomial distributions:

$$\Pr\left[\bigcap_{j=0}^{k} (X_j = x_j)\right] = \frac{n!}{\prod_{j=0}^{k} x_j!} \prod_{j=0}^{k} p_j^{x_j},$$

with the p_j's having a joint Dirichlet distribution (Section 2.7.6) with density function

$$\frac{\Gamma(\alpha)}{\prod_{j=0}^{k} \Gamma(\alpha_j)} \prod_{j=0}^{k} p_j^{\alpha_j - 1} \quad \left(0 < p_j; \sum_{j=0}^{k} p_j = 1\right).$$

As in most situations, it is possible to construct a model for the distribution (4.32) without using urns (or "boxes," or other terms which are synonymous for our purposes). Lumel'skiĭ (1973) has expressed the model in terms of a random walk in $(k + 1)$ dimensions with steps each of unit length parallel to an axis. The probability of choosing the ith axis when the coordinates are $(x_0, x_1, \ldots, x_k)$ is $(c_i + sx_i)/(c + s\sum_{j=0}^{k} x_j)$.

4.5.2. Other Multivariate Pólya Distributions

Kriz (1972) has considered parallel, simultaneous series of drawings from m urns, so generating m independent multivariate Pólya–Eggenberger distributions like (4.32). Each distribution has the same values for c (the initial number of balls in the urn) and s, but the sets of colors $(\mathscr{C}_{i1}, \ldots, \mathscr{C}_{ik_i})$ in the ith urn $(i = 1, \ldots, m)$ are disjoint, one from another, so that there are $\sum_{i=1}^{m} k_i$ different colors in all. We denote by c_{ij} the number of balls with color $\mathscr{C}_{ij}$ in the ith urn. Necessarily, $\sum_{j=1}^{k_i} c_{ij} = c$ for all i. Kriz considers the joint distribution of the $\prod_{i=1}^{m} k_i$ random variables $Y_{j_1 j_2 \cdots j_m}$ which represent the numbers of sets of m balls (out of n sets in all), one from each urn, in which color $\mathscr{C}_{ij_i}$ is drawn from the ith urn (for $i = 1, 2, \ldots, m$), that is, the number of times the event $\bigcap_{i=1}^{m} \mathscr{C}_{ij_i}$ occurs.

The joint distribution of the Y's is

$$\Pr\left[\bigcap_{j_1=1}^{k_1} \bigcap_{j_2=1}^{k_2} \cdots \bigcap_{j_m=1}^{k_m} (Y_{j_1 j_2 \cdots j_m} = y_{j_1 j_2 \cdots j_m})\right]$$

$$= \frac{n!}{\prod_{j_1=1}^{k_1} \prod_{j_2=1}^{k_2} \cdots \prod_{j_m=1}^{k_m} y_{j_1 j_2 \cdots j_m}!} \prod_{i=1}^{m} \left[\frac{\prod_{j=1}^{k_i} \prod_{r=1}^{t_{ij}} \{c_{ij} + (r-1)s\}}{\prod_{r=1}^{n} \{c + (r-1)s\}}\right] \tag{4.39}$$

where c_{ij} = sum of $y_{j_1 j_2 \cdots j_m}$ over all $j_1, j_2, \ldots, j_m$, subject to $j_i = j$ (i.e., the total number of times color $\mathscr{C}_{ij}$ is drawn).

Kriz calls the distribution (4.39) the *polytomous multivariate Pólya* (*PMP*) *distribution.* This distribution can be specialized in several ways.

If there is only one urn, we obtain a Pólya–Eggenberger distribution (which of course further specializes to multinomial, multivariate hypergeometric, binomial, and hypergeometric distributions—see Section 4.2). For $m \geq 2$ we obtain distributions which arise in the analysis of contingency tables of second and higher orders (Fisher, 1922; Freeman and Halton, 1949).

Although Kriz's model is rather general, it is itself specialized in several ways. Not only are the initial numbers of balls in each urn the same, but they remain the same, increasing by s at each stage. Furthermore, there is no overlap in the sets of colors for each urn.

More general multivariate distributions can be obtained (using a single urn) by supposing it possible for each ball to have more than one color. A special, and relatively simple case arises when there are two sets of colors, $\mathscr{C}'_1, \ldots, \mathscr{C}'_{k'}$ and $\mathscr{C}''_1, \ldots, \mathscr{C}''_{k''}$, and each ball has one color from each set. After drawing a ball with the color combination $\mathscr{C}'_{j'}, \mathscr{C}''_{j''}$ it is returned to the urn with s additional balls of the same color combination.

One can regard this as leading to "regular" multivariate Pólya–Eggenberger distributions with k', k'' variables (number of drawings of each of the k', k'' possible color combinations) provided each is present in the urn initially. (If some are not present, we still obtain a distribution of the same type with fewer variables.)

However, we can consider the joint distribution of the number of times sets of *single* colors appear in n drawings. Although the univariate distributions (for a single color such as $\mathscr{C}'_i$) are Pólya–Eggenberger distributions, the joint distributions are different from (4.32). If both colors come from the same set, we again obtain multivariate Pólya distributions of type (4.32).

The joint distribution of the numbers of times $\mathscr{C}'_{i'}$ and $\mathscr{C}''_{i''}$ appear (not necessarily together) is the joint distribution of

$$X'_{i'} = \sum_{h=1}^{k''} Y_{i'h} \qquad \text{and} \qquad X''_{i''} = \sum_{h=1}^{k'} Y_{hi''},$$

where Y_{ab} represents the number of times $\mathscr{C}'_a$ and $\mathscr{C}''_b$ appear *on the same ball.* The $k'k''$ Y's have a joint multivariate Pólya distribution of type (4.32). Kaiser and Stefansky (1972) recommend the use of distributions of this kind for teaching purposes.

4.5.3. Multivariate Inverse Pólya Distributions

As in the univariate case, we can regard this class of distributions as waiting-time distributions (e.g., in inverse sampling).

In the sampling situation described in Section 4.4, suppose that we continue to draw balls from the urn until we observe a prespecified number, say g_0, of balls of color $\mathscr{C}_0$. Then the joint distribution of the numbers of balls $X_1, X_2, \ldots, X_k$ of colors $\mathscr{C}_1, \mathscr{C}_2, \ldots, \mathscr{C}_k$, respectively, drawn when sampling ceases is

$$\Pr\left[\bigcap_{j=1}^{k}(X_j = x_j)\right] = \frac{(x + g_0 - 1)!}{(g_0 - 1)!\prod_{j=1}^{k} x_j!} \, \frac{\alpha_0^{[g_0]}\prod_{j=1}^{k}\alpha_j^{[x_j]}}{\alpha^{[x+g_0]}}, \qquad (4.40)$$

where $x = \sum_{j=1}^{k} x_j$ and α's are defined in 4.5.1.

This formula can be derived as

[Probability of obtaining $(g_0 - 1)$ balls of color $\mathscr{C}_0$ and x_j of color $\mathscr{C}_j$ for $j = 1, \ldots, k$ among the first $(x + g_0 - 1)$ balls chosen] $\times$ [conditional probability of choosing a $\mathscr{C}_0$ on the next draw].

We note that the sum of (4.40) over all nonnegative integer values of $x_1, x_2, \ldots, x_k$ must be equal to 1.

The joint factorial moment is

$$\mu_{(r_1, \ldots, r_k)} = E\left[\prod_{j=1}^{k} X_j^{(r_j)}\right] = \sum_{x_1} \cdots \sum_{x_k} \left\{ \Pr\left[\bigcap_{j=1}^{k} (X_j = x_j)\right] \prod_{j=1}^{k} x_j^{(r_j)} \right\}.$$

Using the relationships (Section 1.1.2, eq. (1.10′)),

$$\frac{x_j^{(r_j)}}{x_j!} = \{(x_j - r_j)!\}^{-1} \qquad \{(g_0 - 1)!\}^{-1} = \frac{g_0^{[r]}}{(g_0 + r - 1)!},$$

$$\alpha_0^{[g_0]} = \frac{(\alpha_0 - r)^{[g_0 + r]}}{(\alpha_0 - 1)^{(r)}},$$

we have [putting $y_j = x_j - r_j$, $y = \sum_{j=1}^{k} (x_j - r_j) = x - r$]

$$\mu_{(r_1, \ldots, r_k)} = \frac{g_0^{[r]} \prod_{j=1}^{k} \alpha_j^{[r_j]}}{(\alpha_0 - 1)^{(r)}} \sum_{y_1} \cdots \sum_{y_k} \frac{(y + g_0 + r - 1)!}{(g_0 + r - 1)! \prod_{j=1}^{k} y_j!}$$

$$\times \frac{(\alpha_0 - r)^{[g_0 + r]} \prod_{j=1}^{k} (\alpha_j + r_j)^{[y_j]}}{\alpha^{[y + g_0 + r]}}$$

$$= \frac{g_0^{[r]} \prod_{j=1}^{k} \alpha_j^{[r_j]}}{(\alpha_0 - 1)^{(r)}} \qquad (\text{for } \alpha_0 > r). \tag{4.41}$$

The summation in (4.41) is of the same form as $\sum_{x_1} \cdots \sum_{x_k} \Pr[\bigcap_{j=1}^{k} (X_j = x_j)]$ [with α_0 changed to $\alpha_0 - r$, α_j to $(\alpha_j + r_j)(j = 1, \ldots, k)$, and g_0 to $(g_0 + r)$—note that $\alpha = (\alpha_0 - r) + \sum_{j=1}^{k} (\alpha_j + r_j)$] and, as we have already pointed out, this sum is equal to 1. For $r \geq \alpha_0$ the joint factorial moment is infinite.

By giving the r's specific values, we obtain, from (4.41),

$$E[X_i] = \frac{g_0 \alpha_i}{\alpha_0 - 1}, \tag{4.42a}$$

$$\operatorname{var}(X_i) = (\alpha_0 + g_0 - 1) g_0 \alpha_i (\alpha_0 + \alpha_i - 1)(\alpha_0 - 1)^{-2}(\alpha_0 - 2)^{-1}, \tag{4.42b}$$

$$\operatorname{cov}(X_i, X_j) = (\alpha_0 + g_0 - 1) g_0 \alpha_i \alpha_j (\alpha_0 - 1)^{-2}(\alpha_0 - 2)^{-1}, \tag{4.42c}$$

hence

$$\operatorname{corr}(X_i, X_j) = \left\{\frac{\alpha_i \alpha_j}{(\alpha_0 + \alpha_i - 1)(\alpha_0 + \alpha_j - 1)}\right\}^{1/2}.$$

Note that the correlation is always positive.

In Section 4.5.1 we have noted that a multivariate Pólya–Eggenberger distribution (4.32) can be regarded as a mixture of multinomial distributions with the parameters $p_0, p_1, p_2, \ldots, p_k$ having a joint Dirichlet distribution. It follows that the waiting-time distribution corresponding to (4.32) is also a mixture of the waiting time distributions corresponding to multinomial distributions (which, as we have seen in Section 2.7.3 are negative multinomial distributions) again with a Dirichlet mixing distribution.

Much additional detailed information on the properties of multivariate inverse Pólya distributions is given in Janardan and Patil (1971, 1974).

4.6. STAGEWISE LINKAGE *(How Long Before Everyone Is the Same?)*

In this model, one urn is filled with n initially distinguishable balls. At each stage two balls are drawn. The probability of drawing a specified pair is $(1/n)[1/(n-1)]$, that is, the balls have equal likelihood of being drawn. After a single drawing, the two balls drawn become *equivalently related* in the following way. Let the balls be initially labeled $1, 2, \ldots, n$. Suppose at the first drawing balls i and j are drawn. Then ball i is labeled with an i and a j, and so is ball j, and the balls are replaced. This will subsequently be called the *linkage*. So after the first stage we have two balls labeled $\{i, j\}$ and $(n-2)$ balls each labeled with one of the remaining $(n-2)$ elements of $1, 2, \ldots, i-1, i+1, \ldots, j-1, j+1, \ldots, n$. The process is repeated. Each time a pair of balls is drawn with respective labels $\{i_1, i_2, \ldots, i_k\}$ and $\{j_1, j_2, \ldots, j_l\}$, they are then relabeled $\{i_1, \ldots, i_k\} \cup \{j_1, \ldots, j_l\}$. The question is, How many drawings must we perform before all the balls carry the same label? (i.e., all carry every one of the numbers $1, 2, \ldots, n$).

A related problem was originally posed by A. Boyd [as quoted by Tijdeman (1971)] in terms of n persons each possessing a piece of news not known to any of the others. How many telephone calls must be made before they all know the news? The exact minimum number of calls needed to spread the news has been computed independently by Hajnal et al. (1971), Tijdeman (1971), and Bumby (quoted by Tijdeman). For $n \geq 4$, the minimum number of needed calls is $2n - 4$.

Moon (1972) posed and partially answered another question: What is the expected waiting time for everybody to have the same information when the *linkages* are made randomly? Our model is relevant to this problem.

Let C denote the number of drawings (linkages) needed to make the balls identically labeled. Moon (1972) showed that:

If ε denotes any positive constant, then

$$(1-\varepsilon)n \ln n < E[C] < (2+\varepsilon)n(\ln n)^2 \qquad [\text{for all } n \geq N(\varepsilon)].$$

We derive *explicit* expressions for lower and upper bounds for $E[C]$. (We do not have an explicit expression for $E[C]$.)

We first obtain a lower bound. Let X denote the number of drawings (linkages) needed for all n balls to have labels that include 1. Clearly, $E[X] \leq E[C]$.

For all i $(1 \leq i \leq n-1)$, let X_i denote the number of drawings (waiting time) needed after the drawing on which the ith ball has 1 included on its label until (and including) the next drawing on which a further $(i+1)$th ball has 1 added to its label (i.e., until one more ball, not yet having a 1, is drawn with one that does have a 1). Then,

$$X = X_1 + X_2 + \cdots + X_{n-1}.$$

Suppose that, at a certain stage, exactly i balls have a 1 on their label. The probability that $(i+1)$ balls will have 1's after the next drawing is

$$p_i = \frac{\binom{i}{1}\binom{n-i}{1}}{\binom{n}{2}} = \frac{2i(n-i)}{n(n-1)}. \tag{4.43}$$

Hence

$$\Pr[X_i = k] = q_i^{k-1} p_i \qquad (k = 1, 2, \ldots),$$

where $q_i = 1 - p_i$. The random variable X_i has a geometric distribution (see Section 2.4), and so $E[X_i] = p_i^{-1}$. Hence

$$\begin{aligned} E[X] &= \sum_{i=1}^{n-1} E[X_i] = \sum_{i=1}^{n-1} \frac{n(n-1)}{2i(n-i)} = \frac{1}{2} n(n-1) \sum_{i=1}^{n-1} \{i(n-i)\}^{-1} \\ &= (n-1) \sum_{i=1}^{n-1} i^{-1}. \end{aligned} \tag{4.44}$$

From a standard result in integral calculus,

$$(n-1)\{1 + \log_e(n-1)\} > E[X] > (n-1)\log_e n. \tag{4.45}$$

Since $E[X] \leq E[C]$, we have $E[C] > (n-1)\log_e n$.

EXERCISE 4.11. In (4.44) we used the relation

$$\frac{1}{2}n\sum_{i=1}^{n-1}\{i(n-i)\}^{-1} = \sum_{i=1}^{n-1} i^{-1}.$$

Prove this. [*Hint*: Use $i^{-1}(n-i)^{-1} = n^{-1}\{i^{-1} + (n-i)^{-1}\}$.]

Now for the upper bound. Let

$$k_i = \alpha p_i^{-1}\log_e n, \qquad i = 1, \ldots, n-1,$$

and denote by $[k_i]$ the integer part of k_i. (The analysis would be simpler if k_i were an integer, but it may not be possible to choose α to make every one of $k_1, k_2, \ldots, k_{n-1}$ integers.)

Then,

$$\begin{aligned}\Pr[X_i \le k_i] = 1 - \Pr[X_i > [k_i]] &= 1 - \sum_{j=[k_i]+1}^{\infty} p_i q_i^j \\ &= 1 - q_i^{[k_i]} \\ &> 1 - \exp(-p_i^{[k_i]}) \\ &> 1 - e^{p_i}e^{-p_i k_i} \\ &> 1 - en^{-\alpha},\end{aligned} \tag{4.46}$$

since $e^{p_i} < e$.

If $X_i \le k_i$ for each i, then certainly

$$X \le \sum_{i=1}^{n-1} k_i = \alpha\left(\sum_{i=1}^{n-1} p_i^{-1}\right)\log_e n = \alpha E[X]\log_e n.$$

The random variables $X_1, \ldots, X_{n-1}$ are mutually independent, so

$$\begin{aligned}\Pr[X \le \alpha E[X]\log_e n] \ge \prod_{i=1}^{n-1}\Pr[X_i \le k_i] &> (1 - en^{-\alpha})^{n-1} \\ &> (1 - en^{-\alpha})^n, \\ &> 1 - en^{1-\alpha},\end{aligned}$$

and so

$$\Pr[X > \alpha E[X]\log_e n] < en^{1-\alpha}. \tag{4.47}$$

The same inequality is valid if the variate X represents the number of drawings needed to obtain "i" on every ball for any i $(=1, 2, \ldots, n)$. Hence, from (4.47),

$$\Pr[C > \alpha E[X]\log_e n] < nen^{1-\alpha} = en^{2-\alpha} \tag{4.48}$$

since, for any γ, $(C > \gamma) \equiv \bigcap_{i=1}^{n} (X_i > \gamma)$. From (1.110′),

$$E[C] = \sum_{j=0}^{\infty} \Pr[C > j] = \sum_{i=0}^{\infty} \sum_{j=iA}^{(i+1)A-1} \Pr[C > j]$$

$$\leq A \sum_{i=0}^{\infty} \Pr[C > iA] \qquad \text{(for any positive integer } A\text{)},$$

since $\Pr[C > j]$ is nonincreasing with respect to j. Using (4.48), we find

$$E[C] \leq A\left\{1 + \sum_{i=1}^{\infty} en^{2-iA/\phi}\right\} \qquad (\text{with } \phi = E[X]\log_e n)$$

$$= A\left\{1 + \frac{en^{2-A/\phi}}{1 - n^{-A/\phi}}\right\} \qquad \text{(for any positive integer } A\text{)}. \qquad (4.49)$$

Choosing A equal to the integer part of $\{(2 + \beta)E[X]\log_e n + 1\} = [(2 + \beta)\phi + 1]$ for some $\beta > 0$, and noting that the expression in braces on the right-hand side of (4.49) is a decreasing function of A, we find

$$E[C] \leq [\{(2 + \beta)E[X]\log_e n + 1\}]\left(1 + \frac{en^{-\beta}}{1 - n^{-2-\beta}}\right).$$

Using (4.45), we have

$$E[C] \leq [\{(2 + \beta)(n - 1)\{1 + \log_e(n - 1)\}\log_e n + 1\}]\left(1 + \frac{en^{-\beta}}{1 - n^{-2-\beta}}\right)$$

$$\leq (2 + \beta)n(\log_e n)^2\{1 + en^{-\beta}(1 - n^{-2})\} \qquad (\text{for any } \beta > 0). \qquad (4.50)$$

4.7. RANDOMIZED SCHEMES

4.7.1. Randomized Pólya Schemes

In other schemes so far discussed the numbers and colors of balls to be added to the urn are determined by the outcome(s) of previous drawing(s). If these quantities are random variables, we have *randomized* Pólya schemes.

A simple scheme of this type is obtained from the Pólya–Eggenberger model (4.1) by changing from a constant to a random variable the number of additional balls s of the same color as the drawn ball. Thus, if a black (white) ball is chosen, it is returned to the urn together with S additional black (white) balls, where S is a random variable with known distribution. More generally, the distribution of S may depend on the color of the drawn ball.

Further generalizations can be obtained by (1) replacing β_b, β_w, ω_b, ω_w in Table 4.1 (Section 4.3) by random variables B_b, B_w, W_b, W_w with known joint distributions for B_b and W_b, and for B_w and W_w, and (2) considering situations with balls of more than two colors in the urn.

Athreya (1969) has considered a situation with k different colors $\mathscr{C}_1, \mathscr{C}_2, \ldots, \mathscr{C}_k$. If a ball of color $\mathscr{C}_i$ is drawn, it is returned to the urn together with A_i additional balls of color $\mathscr{C}_i$, where A_i is a random variable with a distribution depending only on i (and not on the number of the drawing or on the results of other drawings).

Suppose there are initially $a_{01}, a_{02}, \ldots, a_{0k}$ balls of colors $\mathscr{C}_1, \mathscr{C}_2, \ldots, \mathscr{C}_k$, respectively, in the urn. Let $A_{n1}, A_{n2}, \ldots, A_{nk}$ denote the corresponding numbers of balls in the urn after adding the balls consequent on the nth drawing. Further, we write

$$q_{ij} = \Pr[A_i = j], \qquad E[A_i] = \theta_i$$

and suppose

1. $\sum_j q_{ij} j \log j$ converges (for each $i = 1, 2, \ldots, k$).
2. $\theta_1, \theta_2, \ldots, \theta_k$ are finite.

Athreya shows that, if $\theta_1 = \theta_2 = \cdots = \theta_r > \theta_{r+1} \geq \theta_{r+2} \geq \cdots \geq \theta_k$, then the vector of proportions $\mathbf{P}_{ni} = A_{ni}/\sum_{j=1}^{k} A_{nj}$ $(i = 1, \ldots, k)$ tends to a limiting (random) vector $(P_1, P_2, \ldots, P_k)$ as $n \to \infty$, with probability 1, and $P_{r+1} = P_{r+2} = \cdots = P_k = 0$.

Athreya also points out that in the special case where $A_1, A_2, \ldots, A_r$ are not random variables but each equal to a fixed value θ $(>E[A_i]$ for $i = r + 1, \ldots, k)$, then the joint limiting distribution of $P_1, P_2, \ldots, P_r$ is Dirichlet, with parameters $a_{0i}\theta^{-1}$ $(i = 1, 2, \ldots, r)$.

This extends the result for entirely nonrandomized procedures to cases where there are other colors present with randomized additions. The expected value of the number of additions in each case is less than the common value for colors with a fixed number of additions (when they are drawn). In a nonprecise sense, one could say that, as the number of draws increases, the proportion of balls having colors with randomized numbers of additions decreases and has less and less effect on the distribution.

4.7.2. Other Randomized Schemes

We now suppose that the numbers of balls of various colors added to the urn after each drawing are still random variables, but with distributions that do not depend on the results of previous drawings. We did not explicitly consider the corresponding situations when these numbers are fixed. The situations correspond to sequences of independent drawings with probabilities of the possible outcomes changing from drawing to drawing, but in a prespecified way. The relevant theory is quite straightforward, though the algebra may become complicated.

When the numbers added are random variables, the properties of the resultant process will be those of a mixture, with appropriate weights, of processes with prespecified numbers of balls added after each drawing.

As an example we consider the following model, constructed by Arnold (1973). After each drawing from an urn (and whatever the color of the drawn ball) the value of a random variable Y is observed. This value in turn determines the numbers of balls of various colors to be added to the urn before the next drawing. If the value obtained is g, then q_{ig} balls of color $\mathscr{C}_i$ are added to the urn. We denote the random variable representing the value of Y obtained immediately after the nth drawing by Y_n. (Note that no Y is observed before the first drawing.)

Initially the urn contains c_i balls of color $\mathscr{C}_i$ $(i = 1, \ldots, k)$. The conditional probability that the ball drawn on the nth drawing, given $\bigcap_{j=1}^{n-1} (Y_j = y_j)$ is of color $\mathscr{C}_i$, is

$$\frac{c_i + \sum_{j=1}^{n-1} q_{iy_j}}{\sum_{h=1}^{k} \left(c_h + \sum_{j=1}^{n-1} q_{iy_h} \right)}. \tag{4.51}$$

Arnold generalizes this model by causing the probability that a ball is drawn to vary with the number of drawings since the ball was placed in the urn. He divides the contents of the urn into groups $\mathscr{G}_1, \mathscr{G}_2, \ldots, \mathscr{G}_{n-1}$ and $\mathscr{G}_0$. The group $\mathscr{G}_i$ $(i = 1, \ldots, n-1)$ consists of balls added after the $(n-i)$th drawing (i.e., corresponding to Y_{n-i}); $\mathscr{G}_0$ consists of all balls in the urn not included in $\mathscr{G}_1, \mathscr{G}_2, \ldots, \mathscr{G}_{n-1}$. The nth drawing is then done in two stages—first a group is chosen, and then a ball is chosen at random from the selected group. It is supposed that the probability of selecting group $\mathscr{G}_i$ is δ_i (with of course $\sum_{i=0}^{n-1} \delta_i = 1$ and each $\delta_i > 0$). (Drawings are still with replacement.)

This model is further discussed in Chapters 5 and 6.

4.8. URN TRANSFER MODELS

4.8.1. Some Historical Remarks

D. Bernoulli (1768–1769) proposed one of the earliest models of this type. In his model, there are two urns. Urn I contains k white balls and urn II contains k black balls. The process begins by drawing (at random) a ball from urn I and putting it in urn II. Then a ball is drawn (at random) from urn II and placed in urn I. The process continues in this way, balls being drawn from each urn in turn and placed in the other urn. Bernoulli sought to find the expected number (say, E_r) of white balls in urn I after r balls have been drawn from each urn.

(We encountered a problem of this kind in Exercise 1.17 and we use difference equations to solve the present problem in the way described in Chapter 1.)

Example 4.4. If there are G_r white [and $(k - G_r)$ black] balls in urn I [and so $(k - G_r)$ white and G_r black balls in urn II] after r balls have been drawn from each urn, then after the next two drawings (first from urn I and then from urn II) the number of white balls will be

(i) Increased by 1, with probability $\dfrac{k - G_r}{k} \cdot \dfrac{k - G_r}{k + 1}$.

(ii) Decreased by 1, with probability $\dfrac{G_r}{k} \cdot \dfrac{G_r}{k + 1}$.

(iii) Unchanged, with probability $1 - \dfrac{(k - G_r)^2 + G_r^2}{k(k + 1)}$.

Hence the expected increase in the number of white balls in urn I is

$$\{k(k + 1)\}^{-1}\{(k - G_r)^2 - G_r^2\} = (k + 1)^{-1}(k - 2G_r),$$

and so (noting that E_r is the expected value of G_r)

$$E_{r+1} - E_r = \frac{k - 2E_r}{k + 1};$$

that is,

$$E_{r+1} - \frac{k - 1}{k + 1} E_r = \frac{k}{k + 1}. \tag{4.52}$$

Putting $E_r = E_{r+1} = c$, we obtain the particular solution

$$E_r = \tfrac{1}{2}k.$$

The general solution is

$$E_r = \alpha\left(\frac{k - 1}{k + 1}\right)^r + \tfrac{1}{2}k.$$

When $r = 0$, $E_0 = k$; so $\alpha = \frac{1}{2}k$, and we obtain

$$E_r = \tfrac{1}{2}k\left\{\left(\frac{k - 1}{k + 1}\right)^r + 1\right\}. \tag{4.53}$$

Laplace (1812), in his classical *Théorie Analytique des Probabilités*, discusses the same model as Bernoulli but with the more general initial conditions of g_0 white balls and $(k - g_0)$ black balls in urn I and $(k - g_0)$ white and g_0 black balls in urn II.

EXERCISE 4.12. Obtain the value of E_r for this case.

$$\left[\textit{Answer}:\ (g_0 - \tfrac{1}{2}k)\left(\frac{k-1}{k+1}\right)^r + \tfrac{1}{2}k.\right]$$

Laplace obtained an approximate formula for the distribution of G_r. He also allowed for a possible prior distribution of g_0.

There is a closely related process in which balls are taken *simultaneously* from one urn and transferred to the other urn (see Exercise 1.17). Although the process is not the same as that studied by D. Bernoulli and Laplace, its properties are very similar, especially when the number of balls $(2k)$ is large.

Either of these processes can be generalized by supposing

1. There are unequal numbers of balls in the two urns.
2. The total numbers of black and white balls are not equal.

Markov (1915) and Steklov (1915) have studied the properties of the Bernoulli-Laplace process in this more general situation.

In the processes so far described, the numbers of balls in each urn must remain effectively constant. P. and T. Ehrenfest (1907) introduced a model, representing heat exchange between two isolated bodies, in which this is not so. (They represented the bodies by urns and molecules by balls.)

In the Ehrenfest model it is supposed that at each stage a ball is selected at random from among *all* the balls in *both* urns, and transferred to the *other* urn than that wherein it was found. In Section 4.8.2 we discuss models of this kind in some detail.

4.8.2. Ehrenfest Urn Models

We denote by k' (k'') the initial number of balls in urn I (II). The model is in fact equivalent to a Pólya-Eggenberger model of the kind discussed in Section 4.2. To see this consider a single urn (say, A) containing k' white and k'' black balls. A ball is drawn at random from urn A and is replaced by a ball of opposite color. This corresponds to $\beta_b = \omega_w = -1$, $\beta_w = \omega_b = 1$ in our general scheme (Table 4.1). By identifying white balls with balls in urn I and black balls with balls in urn II, the effective identity of the two models can be seen.

We are interested in the distribution of K'_n, the number of white balls (or balls in urn I) after n drawings.

Putting $\Pr[K'_n = k] = P_n(k)$, we have

$$k_0 P_n(k) = (k+1)P_{n-1}(k+1) + (k_0 - k + 1)P_{n-1}(k-1), \quad (4.54)$$

where $k_0 = k' + k''$ (the total number of balls) with initial conditions $P_0(k') = 1$, $P_0(k) = 0$ $(k \neq k')$. As $n \to \infty$, if $P_n(k) \to P(k)$, then

$$k_0 P(k) = (k+1)P(k+1) + (k_0 - k + 1)P(k-1). \quad (4.55)$$

Equation (4.55) can be rearranged as

$$k_0\{P(k) - P(k-1)\} = (k+1)P(k+1) - (k-1)P(k-1), \quad (4.56)$$

whence

$$\frac{P(k+1)}{P(k)} = \frac{k_0 - k}{k+1}$$

and

$$P(k) = \binom{k_0}{k} 2^{-k_0} \qquad (0 \le k \le k_0). \quad (4.57)$$

We have shown that the limiting distribution of K'_n as $n \to \infty$ is binomial with parameters $k_0, \frac{1}{2}$.

EXERCISE 4.13. Using methods similar to those in Example 4.4, show that the expected number of balls in urn I after n transfers is

$$\tfrac{1}{2}k_0 + (k' - \tfrac{1}{2}k_0)(1 - 2k_0^{-1})^n.$$

Kac (1964) describes experimental simulations of an Ehrenfest process in which the initial numbers of balls were $k' = 16{,}384$, $k'' = 0$ (i.e., the second urn was empty). In each realization there were 200,000 drawings. Figure 4.1 shows the number of balls in the first urn after every 1000 drawings. Kac notes

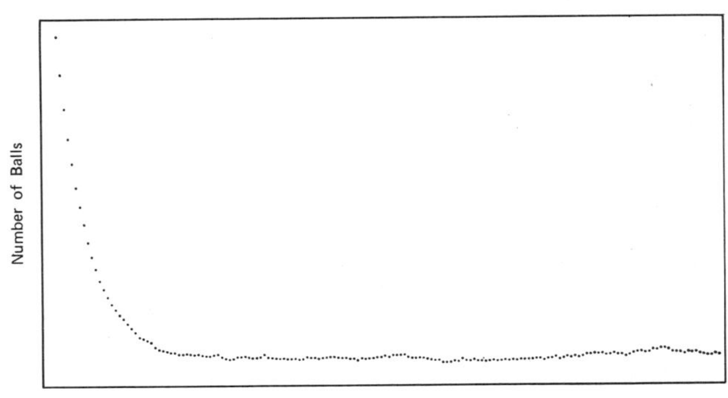

FIGURE 4.1

Results of simulation of an Ehrenfest process (Kac, 1964).

that, as expected, the number of balls in the first urn declined almost exactly exponentially until the equilibrium number ($8{,}192 = \frac{1}{2} \times 16{,}384$) was nearly reached. After that point "the curve becomes wiggly, moving randomly up and down around that number" (i.e., 8,192). Results of a smaller-scale simulation also have been exhibited by Kohlrausch and Schrödinger (1926).

4.8.3. Modifications of the Ehrenfest Model

Generalizations of the Pólya model along the lines described in Section 4.3 also apply to the Ehrenfest model. In particular, we may suppose that there is a probability p that the chosen ball is replaced by one of opposite color (and $q = 1 - p$ that it is returned, unchanged, to the urn). Then formula (4.54) becomes

$$k_0 P_n(k) = (k + 1)pP_{n-1}(k + 1) + k_0 qP_{n-1}(k) + (k_0 - k + 1)pP_{n-1}(k - 1). \tag{4.58}$$

If $P_n(k) \to P(k)$ as $n \to \infty$, then

$$k_0 P(k) = [(k + 1)P(k + 1) + (k_0 - k + 1)P(k - 1)]p + k_0 qP(k). \tag{4.59}$$

This is in fact identical with (4.55) (as can be seen by dividing by p), and so the limiting distribution is again *binomial* with parameters $k_0, \frac{1}{2}$. Note that this distribution does not depend on p.

If we denote the expected number of white balls *after* the nth stage by E_n, then, since each white ball present after the $(n - 1)$th stage contributes, on the average, $k_0^{-1}q + (1 - k_0^{-1})p = 1 - k_0^{-1}p$ to E_n, and each black ball contributes $k_0^{-1}p$ to E_n, we have

$$E_n = (1 - k_0^{-1}p)E_{n-1} + k_0^{-1}p(k_0 - E_{n-1});$$

that is,

$$E_n - (1 - 2k_0^{-1}p)E_{n-1} = p.$$

This difference equation has the general solution

$$E_n = A(1 - 2k_0^{-1}p)^n,$$

and the special solution $E_n = \frac{1}{2}k_0$. The *general solution* is therefore

$$E_n = A(1 - 2k_0^{-1}p)^n + \tfrac{1}{2}k_0,$$

with A such that $E_0 = k'$. This leads to

$$A = k' - \tfrac{1}{2}k_0$$

and

$$E_n = \tfrac{1}{2}k_0 + (k' - \tfrac{1}{2}k)(1 - 2k_0^{-1}p)^n. \tag{4.60}$$

The convergence of E_n to $\frac{1}{2}k_0$ as $n \to \infty$ is clear from this formula.

Another approach, which allows us also to determine the variance of K'_n, uses indicator variables

$$Y_{ij} = \begin{cases} 1 & \text{if the } j\text{th ball is white at the conclusion of the } i\text{th stage,} \\ 0 & \text{otherwise.} \end{cases}$$

Then, $K'_n = \sum_{j=1}^{k_0} Y_{nj}$. If a ball experiences an even number of color changes in the first n stages, then $Y_{nj} = Y_{0j}$. If the number of color changes is odd, then $Y_{nj} = Y_{0j} + 1$ if $Y_{0j} = 0$; $Y_{nj} = Y_{0j} - 1$ if $Y_{0j} = 1$.

We define

$$Y_{0j} = \begin{cases} 1 & \text{for } 1 \le j \le k', \\ 0 & \text{for } k' + 1 \le j \le k_0 \end{cases} \tag{4.61}$$

(which is consistent with the definition of Y_{ij}). We define further indicator variables Z_{nj} such that

$$\begin{aligned} Y_{nj} &= Y_{0j} + Z_{nj} \qquad \text{if} \quad Y_{0j} = 0, \\ Y_{nj} &= Y_{0j} - Z_{nj} \qquad \text{if} \quad Y_{0j} = 1. \end{aligned} \tag{4.62}$$

Then the Z_{nj}'s are identically (but not independently) distributed. In fact, $Z_{nj} = 0$ or 1, according as the number of color changes is even or odd.

For any one ball, the number of times M it is chosen in n stages is distributed binomially with parameters n, k_0^{-1}. Given it is chosen M times, the conditional probability that it experiences an even number of color changes is

$$P_{(M)} = \sum_{j=0}^{[M/2]} \binom{M}{2j} p^{2j}(1-p)^{M-2j} = \tfrac{1}{2}[1 + (1-2p)^M], \tag{4.63}$$

where $[M/2]$ is the integral part of $M/2$. Hence

$$\begin{aligned} E[Z_{nj}] = \Pr[Z_{nj} = 1] = E[1 - P_{(M)}] &= \tfrac{1}{2} - \tfrac{1}{2}E[(1-2p)^M] \\ &= \tfrac{1}{2} - \tfrac{1}{2}\{k_0^{-1}(1-2p) + 1 - k_0^{-1}\}^n \\ &= \tfrac{1}{2}\{1 - (1 - 2k_0^{-1}p)^n\}, \end{aligned} \tag{4.64a}$$

and

$$\begin{aligned} \operatorname{var}(Z_{nj}) &= \tfrac{1}{2}\{1 - (1 - 2k_0^{-1}p)^n\} - [\tfrac{1}{2}\{1 - (1 - 2k_0^{-1}p)^n\}]^2 \\ &= \tfrac{1}{4}\{1 - (1 - 2k_0^{-1}p)^{2n}\}. \end{aligned} \tag{4.64b}$$

For any *two balls*, the probability that *both* experience an odd number of color changes is

$$E[(1 - P_{(M_1)})(1 - P_{(M_2)})],$$

where $M_1, M_2, (n - M_1 - M_2)$ have a joint multinomial distribution with parameters $n; k_0^{-1}, k_0^{-1}, (1 - 2k_0^{-1})$.

Hence (with $j \neq j'$)

$$\begin{aligned} \Pr[Z_{nj}Z_{nj'} = 1] &= \tfrac{1}{4}E[1 - (1-2p)^{M_1} - (1-2p)^{M_2} + (1-2p)^{M_1+M_2}] \\ &= \tfrac{1}{4}\{1 - 2(1 - 2k_0^{-1}p)^n + (1 - 4k_0^{-1}p)^n\}, \end{aligned} \tag{4.65}$$

and so

$$\text{cov}(Z_{nj}, Z_{nj'}) = \tfrac{1}{4}\{(1 - 4k_0^{-1}p)^n - (1 - 2k_0^{-1}p)^{2n}\}. \tag{4.66}$$

[Note that $(M_1 + M_2)$ has a binomial distribution with parameters n, $2k_0^{-1}$.]

Now

$$K'_n = k' - \sum_{j=1}^{k'} Z_{nj} + \sum_{j=k'+1}^{k_0} Z_{nj}.$$

Hence

$$\begin{aligned} E_n = E[K'_n] &= k' - k'E[Z_{nj}] + k''E[Z_{nj}] = k' + (k'' - k')E[Z_{nj}] \\ &= k' + \tfrac{1}{2}(k'' - k')\{1 - (1 - 2k_0^{-1}p)^n\} \\ &= \tfrac{1}{2}[k_0 + (k' - k'')(1 - 2k_0^{-1}p)^n]. \end{aligned} \tag{4.67a}$$

Also

$$\begin{aligned} V_n = \text{var}[K'_n] &= k_0 \,\text{var}[Z_{nj}] + \{k'(k' - 1) + k''(k'' - 1) - 2k'k''\} \\ &\quad \times \text{cov}[Z_{nj}, Z_{nj'}] \\ &= \tfrac{1}{4}k_0\{1 - (1 - 2k_0^{-1}p)^{2n}\} + \tfrac{1}{4}\{(k' - k'')^2 - k_0\} \\ &\quad \times \{(1 - 4k_0^{-1}p)^n - (1 - 2k_0^{-1}p)^{2n}\}. \end{aligned} \tag{4.67b}$$

The method can be extended, in a straightforward manner, to the calculation of higher moments. For example (for $j \neq j' \neq j''$),

$$\begin{aligned} \Pr[Z_{nj}Z_{nj'}Z_{nj''} = 1] = \tfrac{1}{8}[1 &- 3(1 - 2k_0^{-1}p)^n + 3(1 - 4k_0^{-1}p)^n \\ &- (1 - 6k_0^{-1}p)^n], \end{aligned}$$

and so

$$E[Z_{nj}^{\beta} Z_{nj'}^{\beta'} Z_{nj''}^{\beta''}] = \tfrac{1}{8}[1 - 3(1 - 2k_0^{-1}p)^n + 3(1 - 4k_0^{-1}p)^n - (1 - 6k_0^{-1}p)^n], \tag{4.68}$$

for any positive integers β, β', β''.

More generally,

$$G_s = E\left[\prod_{r=1}^{s} Z_{nj_r}^{\beta_r}\right] = 2^{-s}\nabla^s(1 + 2k_0^{-1}p\mathbf{0})^n, \tag{4.69}$$

where ∇ is the backward difference operator (cf. (1.27)).

Hence

$$E\left[\left(\sum_{j=1}^{k} Z_{nj}\right)\right] = kG_1, \tag{4.70a}$$

$$E\left[\left(\sum_{j=1}^{k} Z_{nj}\right)^2\right] = kG_1 + \binom{k}{2} 2\, G_2, \tag{4.70b}$$

$$E\left[\left(\sum_{j=1}^{k} Z_{nj}\right)^3\right] = kG_1 + \binom{k}{2} 6G_2 + \binom{k}{3} 6G_3, \tag{4.70c}$$

and generally (for α a positive integer),

$$E\left[\left(\sum_{j=1}^{k} Z_{nj}\right)^{\alpha}\right] = \sum_{j=1}^{k} \binom{k}{j}(\Delta^j \mathbf{0}^{\alpha})(\tfrac{1}{2}\nabla)^j(1 + 2k_0^{-1}p\mathbf{0})^n, \tag{4.71}$$

because $\Delta^j\mathbf{0}^{\alpha}$ is equal to the sum of the coefficients of the terms in the expansion of $(x_1 + x_2 + \cdots + x_j)^{\alpha}$ that contain each x at least once (see Exercise 3.2).

Also, by similar argument,

$$\begin{aligned} E\left[\left(\sum_{j=1}^{k'} Z_{nj}\right)^{\alpha'}\left(\sum_{j=k'+1}^{k_0} Z_{nj}\right)^{\alpha''}\right] &= \sum_{j'=1}^{k'} \binom{k'}{j'}(\Delta^{j'}\mathbf{0}^{\alpha'}) \sum_{j''=1}^{k''} \binom{k''}{j''}(\Delta^{j''}\mathbf{0}^{\alpha''})G_{j'+j''} \\ &= \sum_{j=1}^{k_0}\left\{\sum_{i=0}^{\min(k',j)} \binom{k'}{i}\binom{k''}{j-i}(\Delta^i\mathbf{0}^{\alpha'})(\Delta^{j-i}\mathbf{0}^{\alpha''})\right\}G_j. \end{aligned} \tag{4.72}$$

In principle we can evaluate all moments of K_n' using this formula.

If we suppose that the "stages" occur at randomly varying points in time, such that T_i, the time between the $(i-1)$th and ith stages, is independent of all other T_j's $(j \neq i)$, then the number of stages completed in a fixed time t will be a random variable, say $N_{(t)}$, and the distribution of the number of white balls after time t, say $K_{(t)}'$, will be a mixture of distributions of $\{K_{N_{(t)}}'\}$ with weights $\Pr[N_{(t)} = n]$ $(n = 0, 1, \ldots)$.

In particular, if each T_i has the same exponential distribution (cf. (2.18))

$$\Pr[T_i > \tau] = e^{-\lambda\tau} \qquad (\lambda, \tau > 0),$$

then the distribution of $N_{(t)}$ will be Poisson with the expected value λt. Hence

$$\begin{aligned} E[K_{(t)}'] &= E_{N_{(t)}}[E[K_{N_{(t)}} \mid N_{(t)}]] \qquad \text{(using (4.67a))} \\ &= E_{N_{(t)}}[\tfrac{1}{2}k_0 + (k' - \tfrac{1}{2}k_0)(1 - 2k_0^{-1}p)^{N(t)}] \\ &= \tfrac{1}{2}k_0 + (k' - \tfrac{1}{2}k_0)\exp(-2\lambda t k_0^{-1}p). \end{aligned} \tag{4.73}$$

Similarly, we can derive

$$\operatorname{var}(K_{(t)}') = \tfrac{1}{4}k_0(1 - e^{-4\lambda t k_0^{-1}}p).$$

This is evaluated from the formula

$$\operatorname{var}(K_{(t)}') = E_{N_{(t)}}[\operatorname{var}(K_{N_{(t)}}) \mid N_{(t)}] + E_{N_{(t)}}[\{E[K_{N_{(t)}} \mid N_t] - E[K_{(t)}']\}^2\}. \tag{4.74}$$

This model has been considered by Karlin and McGregor (1965).

Vincze (1964) has considered a modified Ehrenfest model in which there is at each stage a probability, say ω, that no drawing of a ball (and *a fortiori* no transfer) will be made. Hence, after n stages, the effective number of drawings is distributed binomially with parameters n, $1 - \omega$. The distribution of the number of balls X in urn I (or the number of white balls, if the alternative model is used) is a mixture of the appropriate Ehrenfest distributions. In particular, after n stages

$$P_{(n)}(k) = \Pr[X = k] = \sum_{j=0}^{n} \binom{n}{j}(1 - \omega)^j \omega^{n-j} P_j(k). \tag{4.75}$$

If $\lim_{n\to\infty} P_n(k) = P(k)$, then it will also be true that $\lim_{n\to\infty} P_{(n)}(k) = P(k)$, though the approach to the limit is usually slower in this case.

To see the validity of the last sentence observe that

$$P_{(n)}(k) = E[P_{J_n}(k)],$$

where J_n has a binomial n, $1 - \omega$ distribution. For $j > j(\varepsilon)$, $|P_j(k) - P(k)| < \varepsilon$, and for $n > n(j(\varepsilon))$, $\Pr[J_n < j(\varepsilon)] < \varepsilon$. Hence, for $n > n(j(\varepsilon))$,

$$|P_{(n)}(k) - P(k)| < \Pr[J_n < j(\varepsilon)] + \varepsilon \Pr[J_n > j(\varepsilon)] < \varepsilon + \varepsilon = 2\varepsilon$$

[since $|P_{(n)}(k) - P(k)| \leq 1$ for any n and k].

George (1961) has generalized the Ehrenfest model by supposing that, as in Section 4.8.2, at each stage a ball is changed to the opposite color, and that the probability that the chosen ball is white depends on the number of white balls k' in the urn, but this is not necessarily equal to k'/k_0 (as in the original case). George supposes that the probability of the chosen ball being white is a linear function of k', say $a_1'k' + a_2'$. He also supposes that "stages" occur randomly in time [as do Karlin and McGregor (1965)], though his modification is not essentially linked to this concept. (He also considers situations of the kind described in Section 4.8.1, where balls are exchanged simultaneously between two urns.)

George obtains difference-differential equations for the distribution of the number of white balls (or number of balls in urn I) after time t, but the solutions are rather complicated. A typical difference-differential equation is discussed in Exercise 4.14.

EXERCISE 4.14. Let X_t denote the number of white balls after time t. Assuming that, if there are k_t' white and k_t'' black balls at time t, then, in the time interval $(t, t + \delta t)$, the probability that

(i) Some one white ball changes to black is $(a_1'k_t' + a_2')(\delta t)$, and

(ii) Some one black ball changes to white is $(a_1''k_t'' + a_2'')(\delta t)$ and no other changes are possible in time t,

obtain a difference-differential equation expressing $d\Pr[X_t = k]/dt$ in terms of $\Pr[X_t = k - 1]$, $\Pr[X_t = k]$, and $\Pr[X_t = k + 1]$.

References

Arnold, B. C. (1973) Response distributions of a generalized urn scheme under non-contingent reinforcement, *J. Math. Psychol.*, **10**, 232–239.

Athreya, K. B. (1969) On a characteristic property of Pólya's urn, *Stud. Sci. Math. Hung.*, **4**, 31–35.

Athreya, K. B. and Karlin, S. (1968) Embedding of urn schemes into continuous time Markov branching processes and related limit theorems, *Ann. Math. Stat.*, **39**, 1801–1818.

Bates, G. E. and Neyman, J. (1952) Contributions to the theory of accident proneness II. *Univ. of Calif. Publ. Stat.* **1**, 255–275.

Bellman, R. and Harris, T. E. (1951) Recurrence times for Ehrenfest model, *Pacific J. Math.*, **1**, 184–188.

Bernoulli, D. (1768) De usu algorithmi infinitesimalis in arte conjectandi specimen, *Novi Comment Acad. Sci. Imp. Petropolitanae, 1766–1767*, **12**, 87–98.

Bernoulli, D. (1769) Disquisitiones analyticae de novo problemate conjecturall, *Novi Comment. Acad. Sci. Imp. Petropolitanae*, **14**, par 1, 1769 (1770), 3–25.

Bernstein, S. (1940) Sur un problème du schéma des urnes à composition variable, *C. R. Dokl. Acad. Sci. URSS*, **28**, 5–7.

Binet, F. E. (1974) Letter to the Editor, *Austr. J. Stat.*, **16**(3), 174.

Blackwell, D. and Kendall, D. (1964) The Martin boundary for Pólya's urn scheme and an application to stochastic population growth, *J. Appl. Prob.*, **1**, 284–296.

Blackwell, D. and MacQueen, J. B. (1973) Ferguson distributions via Pólya urn schemes, *Ann. Stat.*, **1**, 353–355.

Bosch, A. J. (1963) The Pólya distribution, *Stat. Neerl.*, **17**, 201–213.

Bricas, M. A. (1949) *Le Système de Courbes de Pearson et le Schéma d'Urne de Pólya*, Athens: Cristou.

Brillouin, L. (1927) Comparaison des différentes statistiques appliquées aux problèmes de quanta, *Ann. Phys.*, **7**, 315–331.

Chatfield, C. and Theobald, C. M. (1973) Mixtures and random sums, *Statistician*, **22**, 281–287.

David, F. N. and Barton, D. E. (1962) *Combinatorial Chance*, London: Griffin.

Dyczka, W. (1969) The moments of Pólya distribution. Special case, *Comment. Math.*, **13**, 129–139.

Dyczka, W. (1972) Pólya distribution connected with the problem of Bayes, *Demonst. Math.*, **4**, 145–165.

Dyczka, W. (1973) On the multidimensional Pólya distribution, *Ann. Soc. Math. Polonae, Ser. I: Comment. Math.*, **17**, 43–63.

Eggenberger, F. and Pólya, G. (1923). Über die Statistik verketteter Vorgänge, *Z. Angew. Math. Mech.*, **1**, 279–289.

Ehrenfest, P. and Ehrenfest, T. (1907) Über zwei bekannte Einwände gegen das Boltzmannsche H-Theorem, *Physik. Z.*, **8**, 311–314.

Fisher, R. A. (1922) On the interpretation of χ^2 from contingency tables, and the calculation of P, *J. R. Stat. Soc. Ser. A*, **85**, 87–94.

Fréchet, M. (1943) *Les Probabilités Associées a un Système d'Événements Compatibles et Dépendants*, Paris: Hermann.

Freedman, D. (1963) L'urne de Bernard Friedman, *C. R. Acad. Sci.*, (*Paris*) **257**, 3809.

Freeman, G. H. and Halton, J. H. (1949) Note on an exact treatment of contingency, goodness of fit and other problems of significance, *Biometrika*, **36**, 141–149.

Friedman, B. (1949) A simple urn model, *Commun. Pure Appl. Math.*, **2**, 59–70.

George, Cl. (1961) Les urnes d'Ehrenfest et les processus stochastiques du premier ordre, *Acad. Roy. Belg. Bull. Cl. Sci.* (5), **47**, 92–110.

Gerstenkorn, T. (1975) The multivariate truncated Polya distribution, Fourth International Symposium on Multivariate Analysis, Dayton, Ohio.

Greenwood, M. and Yule, G. U. (1920) An inquiry into the nature of frequency distributions representative of multiple happenings, *J. R. Stat. Soc.*, **83**, 255–279.

Hajnal, A., Milner, E. C., and Szemeredi, E. (1972) A cure of the telephone disease, *Can. Math. Bull.*, **15**, 447–450.

Janardan, K. G. and Patil, G. P. (1970) On the multivariate Pólya distribution: a model of contagion for data with multiple counts, in *Random Counts in Scientific Work*, **3**, G. P. Patil (Ed.), Pennsylvania State Univeristy Press, pp. 143–161.

Janardan, K. G. and Patil, G. P. (1971) The multivariate inverse Pólya distribution: a model of contagion for data with multiple counts in inverse sampling, in *Studi di Probabilit, Statistica e Ricerca Operativa in Onore di Giuseppe Pompilj*, Gubbio: Oderisi, pp. 1–15.

Janardan, K. G. and Patil, G. P. (1974) On multivariate modified Pólya and inverse Pólya distributions and their properties, *Ann. Inst. Stat. Math.* (*Tokyo*), **26**, 271–276.

Janardan, K. G. and Schaeffer, D. J. (1975) A generalization of Markov-Pólya distribution; its extensions and applications, Sangamon State University (preprint).

Johnson, N. L. and Kotz, S. (1969) *Distributions in Statistics: Discrete Distributions*, New York: John Wiley and Sons.

Johnson, N. L. and Kotz, S. (1976) Two variants of Pólya's urn models, *Am. Stat.*, **30**(4), 186–188.

Jordan, K. (1973) *Treatise on Probability Theory*, Budapest: Hungarian Academy of Sciences (Hungarian edition, 1956).

Kac, M. (1947) Random walk and the theory of Brownian motion, *Am. Math. Mon.*, **54**, 369–391.

Kac, M. (1964) Probability, *Sci. Am.*, Vol. 211, 102–118.

Kaiser, H. F. and Stefansky, W. (1972) A Pólya distribution for teaching, *Am. Stat.*, **26**, 40–43.

Kaminsky, K. S., Luks, E. M., and Nelson, P. I. (1975) An urn model and the prediction of order statistics, *Comm. Stat.*, **4**, 245–250.

Karlin, S. and McGregor, J. (1965) Ehrenfest urn models, *J. Appl. Prob.*, **2**, 352–376.

Kohlrausch, K. W. F. and Schrödinger, E., (1926) Das Ehrenfestsche Modell der *H*-kurve, *Phys. Z.*, **27**, 306–313.

Kotz, S. and Srinivasan, R. (1974) *Tables of the Pólya-Eggenberger Distribution*, Technical Report 74-6, Department of Mathematics, Temple University.

Koźniewska, J. (1954) The first absolute central moment for Pólya's distribution (in Polish) *Zastosow. Mat.*, **1**, 206–211.

Kriz, J. (1972) Die PMP-Verteilung, *Stat. Hefte*, **13**(3), 211–224.

Laplace, P. S. (1812) *Théorie Analytique des Probabilités*, *Oeuvres complètes*, Vol. 7, Nos. 1–2 (1886), Livre 2, pp. 413–414.

Lumel'skiĭ, Ja. P. (1973) Random walks related to generalized urn schemes, *Dokl. Akad. Nauk SSR*, **209**(6), *Sov. Math. Dokl.*, **14**(2), 628–631.

Lundberg, O. (1964) *On Random Processes and Their Application to Sickness and Accident Statistics*, Uppsala: Almquist and Wiksells.

Markov, A. A. (1906) Extension of the law of large numbers to dependent variables, *Izv. Fiz-Mat. Obschch. Kazan Univ. Ser. 2*, **15**(4), 135–256. (in Russian). See also *Collected Works*,

Theory of Numbers and Probability Theory, Moscow: Izdatel'stvo Akademii Nauk (1951) p. 351 (from a 1907 reprint of the 1906 paper).

Markov, A. A. (1915) Application of the mathematical expectation method to variables connected into a series (in Russian), *Izv. Akad. Nauk (Petersburg) Ser. 6*, **9**(14), 1453–1484.

Markov, A. A. (1917) A generalization of a problem on a sequential exchange of balls (in Russian), (*Collected Works*, p. 589, presented at a meeting of Physico-Mathematical Section of the Academy of Sciences, October 8, 1917).

McFadden, J. A. (1955) Urn models of correlation and a comparison with the multivariate normal integral, *Ann. Math. Stat.*, **26**, 478–489.

Molina, E. C. (1930) The theory of probability: some comments on Laplace's Théorie Analytique, *Bull. Am. Math. Soc.*, **36**, 369–392.

Moon, J. W. (1972) Random exchanges of information, *Nieuw Arch. Wisk.*, **20**(3), 246–249.

Morgenstern, D. (1972) Überschreitungswahrscheinlichkeiten, das Polyasche Urnenmodel und ein Wartezeitproblem bei Urnenziehungen, *Math.-Phys. Semest., Göttingen*, **19**(2), 213–215.

Mosimann, J. E. (1962) On the compound multinomial distribution, the multivariate β-distribution, and correlations among proportions, *Biometrika*, **49**, 65–82.

Mosimann, J. E. (1963) On the compound negative multinomial distribution and correlations among inversely sampled pollen counts, *Biometrika*, **50**, 47–54.

Mühlbach, G. von. (1972) Rekursionsformulen für die zentralen Momente der Pólya- und der Beta-Verteilung, *Metrika*, **19**, 171–177.

Ondar, H. O. (1970) On a paper by V. A. Steklov in probability theory, *Istor. Metodol. Estestv. Nauk*, **9**, 262–266.

Pavlik, M. (1970) Approximate construction of a two-dimensional confidence region, *Apl. Mat.*, **15**(5), 305–309.

Pavlik, M. (1971) Die Betaverteilung als Wahrscheinlichkeitscharacteristik einiger Zufallsfunktionen, *Wiss. Z. Techn. Hochsch. Karl-Marx-Stadt.*, **13**, Hefte 1, 43–50.

Pólya, G. (1931) Sur quelques points de la théorie des probabilites, *Ann. Inst. H. Poincaré*, **1**, 117–161.

Prete, P. del (1970) Una formula ricorrente per i momenti di alcune distribuzioni probabilistiche, *Stud. Econ. (Naples)*, **25**(2), 175–182.

Rosenblatt, A. (1940) Sur le concept de contagion de M. G. Pólya dans le Calcul des probabilités. Divers schèmes. Application à la peste bubonique au Pérou, *Proc. Acad. Nac. Cien. Exactas, Fis. Nat., Peru (Lima)*, **3**, 186–204.

Santacroce, G. (1967) Sú i momenti fattoriali di una distribuzione del Pólya a più variabili, *Giornale dell'Istituto Italiano degli Attuari*, **30**, 90–94.

Sarkadi, K. (1957) On the distribution of the number of exceedances, *Ann. Math. Stat.*, **28**, 1021–1023.

Sarkadi, K. (1957) Generalized hypergeometric distributions, *Publ. Math. Inst. Hung.*, **2**, 59–69.

Savkevitch, V. (1940) Sur le schèma des urnes à composition variable, *C. R. Dokl. Acad. Sci. URSS*, **28**, 8–12.

Severo, N. C. (1970) A note on the Ehrenfest multiurn model. *J. Appl. Prob.*, **7**, 444–445.

Sheynin, O. B. (1973) D. Bernoulli's work in probability, *RETE*, **1**, 273–300.

Siegert, A. J. F. (1950) On the approach to statistical equilibrium, *Phys. Rev.*, **76**, 1708–1714.

Śródka, T. (1964) Functions limites et degrés de convergence dans la distribution de Pólya, *Bull. Soc. Sci. Lettres, Lódź*, **15**(2), 1–9.

Stancu, D. D. (1970a) Relatii de recurenta pentru momentele centrate ale unor legi discrete de probabilitati, *Stud. Univ. Babes-Bolyai, Ser. Math. Mech.*, **1**, 55–62.

Stancu, D. D. (1970b) Probabilistic methods in the theory of approximation of functions of several variables by linear positive operators, in *Approximation Theory*, A. Talbot (Ed.), New York: Academic Press.

Steklov, V. A. (1915) On a problem of Laplace, *Izv. Akad. Nauk*, **9** (14), 1515–1537 (in Russian).

Steyn, H. S. (1951) On discrete multivariate probability functions of hypergeometric type, *Indag. Math.* **17**, 588–595.

Styve, B. (1965) A variant of Pólya urn model, *Nord. Mat. Tidskr.*, **13**, 147–152 (in Norwegian).

Taguti, G. (1952) Tables of the 5% and 1% points for the Pólya-Eggenberger's distribution function, *Rep. Stat. Appl. Res., JUSE*, **2**, 27–32.

Tijdeman, R. (1971) On a telephone problem, *Nieuw Arch. Wisk.*, **19**(3), 188–192.

Tschuprov, A. (1922) Ist die normale stabilität empirisch nachweisbar?, *Nord. Stat. Tidskr.*, B/I, 373–378.

Vincze, I. (1964) Über das Ehrenfestsche Modell der Wärmeübertragung *Arch. Math.*, **15**, Fasc. 4/5, 394–400.

Wei, Lee-Jen (1976) *A Class of Designs for Sequential Clinical Trials*, (manuscript). Department of Mathematics, University of South Carolina.

Woodbury, M. A. (1949) On a probability distribution, *Ann. Math. Stat.*, **20**, 311–313.

Yntema, L. (1954) Einiges zur Wahrscheinlichkeitsansteckung, *Verzeker. Arch. Actuar. Bijvoegsel*, **31**, 86–91.

APPENDIX. Tables of Pólya-Eggenberger Distribution

In this appendix we present tables of the Pólya–Eggenberger cumulative distribution function, that is,

values of $\Pr[X \leq k] = \sum_{j=0}^{k} \Pr[X = j]$, with

$$\Pr[X = j] = \binom{n}{j} \frac{\left(\dfrac{A}{S}\right)^{[j]} \left(\dfrac{B}{S}\right)^{[n-j]}}{\left(\dfrac{A+B}{S}\right)^{[n]}}$$

for $n = 2(1)10(5)25$, $k = 0(1)n$, $A = 2(2)10$, $B = A(2)10$, $S = 1(1)5$.

Using these tables as well as the tables for n up to 35 not presented here, we investigated the adequacy of the various asymptotic approximations to the distribution of X (cf. Bosch, 1963; Johnson and Kotz, 1969; Taguti, 1952). It was found that even for n as large as 35 the difference between the exact and approximate percentile of X was never less than 5 and was as large as 10 in the tail areas, which, in our opinion, is far from satisfactory for practical purposes. It would therefore be of interest to derive more adequate approximations to the distribution of X, and to study the rate of convergence in the case of the limiting distributions that are already known. (Cf. Section 6.3.)

n	k	A= 2 B= 2				
		S=1	S=2	S=3	S=4	S=5
2	0	.3000	.3333	.3571	.3750	.3889
	1	.7000	.6667	.6429	.6250	.6111
	2	1.0000	1.0000	1.0000	1.0000	1.0000
3	0	.2000	.2500	.2857	.3125	.3333
	1	.5000	.5000	.5000	.5000	.5000
	2	.8000	.7500	.7143	.6875	.6667
	3	1.0000	1.0000	1.0000	1.0000	1.0000
4	0	.1429	.2000	.2418	.2734	.2982
	1	.3714	.4000	.4176	.4297	.4386
	2	.6286	.6000	.5824	.5703	.5614
	3	.8571	.8000	.7582	.7266	.7018
	4	1.0000	1.0000	1.0000	1.0000	1.0000
5	0	.1071	.1667	.2115	.2461	.2734
	1	.2857	.3333	.3626	.3828	.3977
	2	.5000	.5000	.5000	.5000	.5000
	3	.7143	.6667	.6374	.6172	.6023
	4	.8929	.8333	.7885	.7539	.7266
	5	1.0000	1.0000	1.0000	1.0000	1.0000
6	0	.0833	.1429	.1893	.2256	.2545
	1	.2262	.2857	.3229	.3486	.3677
	2	.4048	.4286	.4422	.4512	.4577
	3	.5952	.5714	.5578	.5488	.5423
	4	.7733	.7143	.6771	.6514	.6323
	5	.9167	.8571	.8107	.7744	.7455
	6	1.0000	1.0000	1.0000	1.0000	1.0000

n	k	A= 2 B= 4				
		S=1	S=2	S=3	S=4	S=5
2	0	.4762	.5000	.5185	.5333	.5455
	1	.8571	.8333	.8148	.8000	.7879
	2	1.0000	1.0000	1.0000	1.0000	1.0000
3	0	.3571	.4000	.4321	.4571	.4773
	1	.7143	.7000	.6914	.6857	.6818
	2	.9286	.9000	.8765	.8571	.8409
	3	1.0000	1.0000	1.0000	1.0000	1.0000
4	0	.2778	.3333	.3745	.4063	.4318
	1	.5952	.6000	.6049	.6095	.6136
	2	.8333	.8000	.7778	.7619	.7500
	3	.9603	.9333	.9095	.8889	.8712
	4	1.0000	1.0000	1.0000	1.0000	1.0000
5	0	.2222	.2857	.3329	.3694	.3986
	1	.5000	.5238	.5409	.5541	.5647
	2	.7381	.7143	.7010	.6926	.6871
	3	.8968	.8571	.8290	.8081	.7920
	4	.9762	.9524	.9296	.9091	.8910
	5	1.0000	1.0000	1.0000	1.0000	1.0000
6	0	.1818	.2500	.3012	.3410	.3729
	1	.4242	.4643	.4914	.5115	.5272
	2	.6515	.6429	.6400	.6394	.6397
	3	.8247	.7857	.7619	.7459	.7344
	4	.9329	.8929	.8625	.8392	.8207
	5	.9848	.9643	.9430	.9231	.9051
	6	1.0000	1.0000	1.0000	1.0000	1.0000

n	k	A= 2 B= 6				
		S=1	S=2	S=3	S=4	S=5
2	0	.5833	.6000	.6136	.6250	.6346
	1	.9167	.9000	.8864	.8750	.8654
	2	1.0000	1.0000	1.0000	1.0000	1.0000
3	0	.4667	.5000	.5260	.5469	.5641
	1	.8167	.8000	.7890	.7813	.7756
	2	.9667	.9500	.9351	.9219	.9103
	3	1.0000	1.0000	1.0000	1.0000	1.0000
4	0	.3818	.4286	.4641	.4922	.5151
	1	.7212	.7143	.7116	.7109	.7113
	2	.9121	.8857	.8653	.8516	.8400
	3	.9848	.9714	.9580	.9453	.9337
	4	1.0000	1.0000	1.0000	1.0000	1.0000
5	0	.3182	.3750	.4177	.4512	.4783
	1	.6364	.6429	.6497	.6562	.6622
	2	.8485	.8214	.8044	.7930	.7848
	3	.9545	.9286	.9076	.8906	.8768
	4	.9924	.9821	.9706	.9590	.9479
	5	1.0000	1.0000	1.0000	1.0000	1.0000
6	0	.2692	.3333	.3814	.4189	.4493
	1	.5629	.5833	.5993	.6123	.6232
	2	.7832	.7619	.7506	.7441	.7402
	3	.9138	.8810	.8592	.8418	.8294
	4	.9749	.9524	.9322	.9150	.9005
	5	.9959	.9881	.9783	.9678	.9574
	6	1.0000	1.0000	1.0000	1.0000	1.0000

7	0	.0667	.1250	.1721	.2095	.2396	.1515	.2222	.2761	.3183	.3522	.2308	.3000	.3520	.3928	.4256
	1	.1833	.2500	.2925	.3223	.3444	.3636	.4167	.4518	.4774	.4972	.5000	.5333	.5574	.5760	.5912
	2	.3333	.3750	.3988	.4146	.4259	.5758	.5833	.5905	.5967	.6022	.7203	.7083	.7041	.7029	.7033
	3	.5000	.5000	.5000	.5000	.5000	.7525	.7222	.7060	.6962	.6897	.8671	.8333	.8127	.7991	.7895
	4	.6667	.6250	.6012	.5854	.5741	.8788	.8333	.8038	.7832	.7680	.9487	.9167	.8924	.8738	.8594
	5	.8167	.7500	.7075	.6777	.6556	.9545	.9167	.8860	.8615	.8418	.9854	.9667	.9482	.9315	.9170
	6	.9333	.8750	.8279	.7905	.7604	.9899	.9722	.9525	.9333	.9156	.9977	.9917	.9833	.9738	.9641
	7	1.0000	1.0000	1.0000	1.0000	1.0000	1.0000	1.0000	1.0000	1.0000	1.0000	1.0000	1.0000	1.0000	1.0000	1.0000
8	0	.0545	.1111	.1583	.1964	.2273	.1282	.2000	.2556	.2995	.3350	.2000	.2727	.3278	.3709	.4053
	1	.1515	.2222	.2684	.3011	.3256	.3147	.3778	.4192	.4493	.4724	.4462	.4909	.5220	.5455	.5642
	2	.2788	.3333	.3648	.3857	.4008	.5105	.5333	.5494	.5616	.5715	.6615	.6606	.6636	.6677	.6720
	3	.4242	.4444	.4555	.4626	.4677	.6845	.6667	.6590	.6552	.6534	.8182	.7879	.7715	.7617	.7554
	4	.5758	.5556	.5445	.5374	.5323	.8205	.7778	.7531	.7371	.7260	.9161	.8788	.8539	.8365	.8236
	5	.7212	.6667	.6352	.6143	.5992	.9138	.8667	.8343	.8109	.7932	.9683	.9394	.9155	.8963	.8808
	6	.8485	.7778	.7316	.6989	.6744	.9681	.9333	.9032	.8784	.8580	.9911	.9758	.9591	.9433	.9290
	7	.9455	.8889	.8417	.8036	.7727	.9930	.9778	.9595	.9412	.9239	.9986	.9939	.9867	.9782	.9691
	8	1.0000	1.0000	1.0000	1.0000	1.0000	1.0000	1.0000	1.0000	1.0000	1.0000	1.0000	1.0000	1.0000	1.0000	1.0000
9	0	.0455	.1000	.1470	.1855	.2169	.1099	.1818	.2386	.2838	.3204	.1750	.2500	.3073	.3524	.3889
	1	.1273	.2000	.2488	.2837	.3099	.2747	.3455	.3920	.4257	.4515	.4000	.4545	.4916	.5193	.5411
	2	.2364	.3000	.3372	.3622	.3803	.4545	.4909	.5147	.5321	.5456	.6077	.6182	.6282	.6371	.6450
	3	.3636	.4000	.4198	.4327	.4419	.6224	.6182	.6188	.6208	.6231	.7692	.7455	.7344	.7288	.7259
	4	.5000	.5000	.5000	.5000	.5000	.7622	.7273	.7092	.6983	.6913	.8794	.8409	.8179	.8028	.7924
	5	.6364	.6000	.5802	.5673	.5581	.8671	.8182	.7883	.7682	.7537	.9455	.9091	.8828	.8634	.8486
	6	.7636	.7000	.6628	.6378	.6197	.9371	.8909	.8573	.8322	.8129	.9797	.9545	.9318	.9127	.8969
	7	.8727	.8000	.7512	.7163	.6901	.9770	.9455	.9164	.8916	.8709	.9944	.9818	.9668	.9520	.9382
	8	.9545	.9000	.8530	.8145	.7831	.9950	.9818	.9649	.9474	.9305	.9991	.9955	.9892	.9815	.9730
	9	1.0000	1.0000	1.0000	1.0000	1.0000	1.0000	1.0000	1.0000	1.0000	1.0000	1.0000	1.0000	1.0000	1.0000	1.0000
10	0	.0385	.0909	.1375	.1762	.2081	.0952	.1667	.2241	.2703	.3079	.1544	.2308	.2897	.3364	.3742
	1	.1084	.1818	.2323	.2689	.2966	.2418	.3182	.3687	.4054	.4335	.3603	.4231	.4653	.4966	.5210
	2	.2028	.2727	.3144	.3426	.3631	.4066	.4545	.4849	.5067	.5235	.5588	.5804	.5970	.6104	.6215
	3	.3147	.3636	.3905	.4080	.4205	.5664	.5758	.5841	.5912	.5973	.7217	.7063	.7010	.6996	.7000
	4	.4371	.4545	.4638	.4697	.4739	.7063	.6818	.6708	.6651	.6619	.8405	.8042	.7845	.7725	.7648
	5	.5629	.5455	.5362	.5303	.5261	.8182	.7727	.7475	.7316	.7207	.9182	.8776	.8512	.8331	.8200
	6	.6853	.6364	.6095	.5920	.5795	.8998	.8485	.8155	.7926	.7758	.9636	.9301	.9038	.8836	.8677
	7	.7972	.7273	.6856	.6574	.6369	.9530	.9091	.8752	.8492	.8288	.9866	.9650	.9438	.9252	.9093
	8	.8916	.8182	.7677	.7311	.7034	.9830	.9545	.9267	.9023	.8914	.9963	.9860	.9726	.9587	.9454
	9	.9615	.9091	.8625	.8238	.7919	.9963	.9848	.9692	.9524	.9359	.9994	.9965	.9911	.9840	.9760
	10	1.0000	1.0000	1.0000	1.0000	1.0000	1.0000	1.0000	1.0000	1.0000	1.0000	1.0000	1.0000	1.0000	1.0000	1.0000

n	k	A= 2 B= 2					A= 2 B= 4					A= 2 B= 6				
		S=1	S=2	S=3	S=4	S=5	S=1	S=2	S=3	S=4	S=5	S=1	S=2	S=3	S=4	S=5
15	0	.0196	.0625	.1061	.1445	.1772	.0526	.1176	.1751	.2233	.2635	.0909	.1667	.2291	.2799	.3218
	1	.0564	.1250	.1784	.2192	.2510	.1404	.2279	.2894	.3349	.3703	.2273	.3137	.3722	.4153	.4488
	2	.1078	.1875	.2401	.2773	.3050	.2487	.3309	.3823	.4187	.4461	.3780	.4424	.4836	.5134	.5365
	3	.1716	.2500	.2955	.3277	.3503	.3661	.4265	.4629	.4885	.5078	.5231	.5539	.5755	.5921	.6056
	4	.2451	.3125	.3496	.3737	.3908	.4835	.5147	.5348	.5495	.5610	.6512	.6495	.6532	.6582	.6633
	5	.3260	.3750	.4007	.4170	.4285	.5942	.5956	.5999	.6045	.6088	.7568	.7304	.7198	.7151	.7132
	6	.4118	.4375	.4506	.4589	.4646	.6935	.6691	.6594	.6548	.6526	.8390	.7978	.7769	.7648	.7573
	7	.5000	.5000	.5000	.5000	.5000	.7786	.7353	.7140	.7016	.6936	.8994	.8529	.8259	.8085	.7967
	8	.5882	.5625	.5494	.5411	.5354	.8483	.7941	.7643	.7455	.7325	.9412	.8971	.8676	.8471	.8322
	9	.6740	.6250	.5993	.5830	.5715	.9025	.8456	.8105	.7869	.7699	.9683	.9314	.9027	.8811	.8645
	10	.7549	.6875	.6504	.6263	.6092	.9422	.8897	.8528	.8262	.8062	.9845	.9571	.9318	.9109	.8938
	11	.8284	.7500	.7035	.6723	.6497	.9693	.9265	.8913	.8638	.8420	.9934	.9755	.9554	.9368	.9205
	12	.8922	.8125	.7599	.7227	.6950	.9861	.9559	.9258	.8998	.8778	.9977	.9877	.9737	.9588	.9446
	13	.9436	.8750	.8216	.7808	.7490	.9951	.9779	.9561	.9344	.9144	.9994	.9951	.9870	.9770	.9662
	14	.9804	.9375	.8939	.8555	.8228	.9990	.9926	.9814	.9677	.9533	.9999	.9988	.9957	.9910	.9850
	15	1.0000	1.0000	1.0000	1.0000	1.0000	1.0000	1.0000	1.0000	1.0000	1.0000	1.0000	1.0000	1.0000	1.0000	1.0000
20	0	.0119	.0476	.0880	.1254	.1580	.0333	.0909	.1463	.1945	.2356	.0598	.1304	.1926	.2448	.2885
	1	.0344	.0952	.1477	.1897	.2232	.0913	.1775	.2423	.2918	.3308	.1556	.2490	.3149	.3642	.4027
	2	.0665	.1429	.1984	.2392	.2703	.1664	.2597	.3209	.3648	.3981	.2692	.3563	.4117	.4514	.4819
	3	.1073	.1905	.2442	.2816	.3093	.2522	.3377	.3895	.4256	.4526	.3878	.4529	.4932	.5222	.5445
	4	.1553	.2381	.2871	.3199	.3436	.3434	.4113	.4511	.4788	.4995	.5024	.5392	.5638	.5823	.5971
	5	.2095	.2857	.3279	.3555	.3750	.4356	.4805	.5075	.5266	.5412	.6072	.6160	.6258	.6348	.6428
	6	.2688	.3333	.3674	.3892	.4045	.5252	.5455	.5596	.5705	.5793	.6988	.6838	.6807	.6813	.6834
	7	.3320	.3810	.4059	.4216	.4326	.6095	.6061	.6081	.6113	.6146	.7760	.7431	.7295	.7231	.7200
	8	.3981	.4286	.4438	.4533	.4599	.6866	.6623	.6533	.6495	.6478	.8387	.7945	.7729	.7608	.7534
	9	.4658	.4762	.4813	.4845	.4867	.7551	.7143	.6958	.6856	.6793	.8879	.8385	.8115	.7949	.7840
	10	.5342	.5238	.5187	.5155	.5133	.8143	.7619	.7356	.7198	.7095	.9251	.8758	.8457	.8260	.8122
	11	.6019	.5714	.5562	.5467	.5401	.8640	.8052	.7729	.7526	.7386	.9522	.9068	.8758	.8542	.8384
	12	.6680	.6190	.5941	.5784	.5674	.9043	.8442	.8079	.7839	.7669	.9710	.9322	.9022	.8798	.8628
	13	.7312	.6667	.6326	.6108	.5955	.9360	.8788	.8407	.8141	.7945	.9835	.9526	.9251	.9030	.8854
	14	.7905	.7143	.6721	.6445	.6250	.9597	.9091	.8712	.8431	.8218	.9913	.9684	.9446	.9239	.9065
	15	.8447	.7619	.7129	.6801	.6564	.9765	.9351	.8995	.8713	.8488	.9958	.9802	.9610	.9425	.9261
	16	.8927	.8095	.7558	.7184	.6907	.9877	.9567	.9254	.8985	.8760	.9983	.9887	.9743	.9589	.9442
	17	.9334	.8571	.8016	.7608	.7297	.9945	.9740	.9489	.9249	.9035	.9994	.9944	.9848	.9731	.9609
	18	.9656	.9048	.8523	.8103	.7768	.9981	.9870	.9696	.9506	.9321	.9998	.9977	.9925	.9848	.9760
	19	.9881	.9524	.9120	.8746	.8420	.9996	.9957	.9871	.9756	.9628	1.0000	.9994	.9975	.9940	.9893
	20	1.0000	1.0000	1.0000	1.0000	1.0000	1.0000	1.0000	1.0000	1.0000	1.0000	1.0000	1.0000	1.0000	1.0000	1.0000

25	0	.0079	.0385	.0761	.1123	.1446	.0230	.0741	.1271	.1746	.2159	.0423	.1071	.1679	.2202	.2648
	1	.0232	.0769	.1275	.1696	.2039	.0640	.1453	.2106	.2620	.3030	.1129	.2063	.2755	.3282	.3698
	2	.0452	.1154	.1710	.2134	.2464	.1188	.2137	.2793	.3275	.3644	.2005	.2979	.3616	.4075	.4428
	3	.0733	.1538	.2102	.2508	.2814	.1833	.2792	.3395	.3820	.4140	.2964	.3822	.4349	.4722	.5006
	4	.1068	.1923	.2467	.2843	.3119	.2543	.3419	.3939	.4298	.4565	.3942	.4594	.4992	.5275	.5493
	5	.1453	.2308	.2813	.3151	.3396	.3289	.4017	.4438	.4728	.4943	.4889	.5299	.5565	.5762	.5918
	6	.1880	.2692	.3145	.3441	.3652	.4046	.4587	.4902	.5122	.5287	.5773	.5940	.6080	.6197	.6297
	7	.2344	.3077	.3468	.3718	.3895	.4792	.5128	.5336	.5487	.5605	.6572	.6520	.6546	.6591	.6639
	8	.2839	.3462	.3782	.3985	.4127	.5512	.5641	.5745	.5830	.5901	.7277	.7042	.6969	.6950	.6952
	9	.3358	.3846	.4091	.4244	.4351	.6192	.6125	.6131	.6154	.6182	.7881	.7509	.7354	.7279	.7241
	10	.3895	.4231	.4396	.4499	.4570	.6822	.6581	.6496	.6462	.6449	.8388	.7924	.7704	.7583	.7510
	11	.4444	.4615	.4699	.4750	.4786	.7395	.7009	.6843	.6756	.6705	.8802	.8291	.8023	.7863	.7760
	12	.5000	.5000	.5000	.5000	.5000	.7905	.7407	.7172	.7037	.6951	.9133	.8611	.8312	.8122	.7994
	13	.5556	.5385	.5301	.5250	.5214	.8352	.7778	.7485	.7308	.7190	.9391	.8889	.8573	.8363	.8215
	14	.6105	.5769	.5604	.5501	.5430	.8736	.8120	.7782	.7569	.7423	.9585	.9127	.8809	.8585	.8422
	15	.6642	.6154	.5909	.5756	.5649	.9057	.8433	.8064	.7821	.7650	.9728	.9328	.9020	.8791	.8617
	16	.7161	.6538	.6218	.6015	.5873	.9319	.8718	.8331	.8066	.7873	.9829	.9496	.9208	.8980	.8802
	17	.7656	.6923	.6532	.6282	.6105	.9528	.8974	.8583	.8303	.8094	.9898	.9634	.9373	.9155	.8976
	18	.8120	.7308	.6855	.6559	.6348	.9688	.9202	.8821	.8533	.8312	.9943	.9744	.9518	.9314	.9140
	19	.8547	.7692	.7187	.6849	.6604	.9805	.9402	.9044	.8758	.8530	.9970	.9829	.9642	.9459	.9295
	20	.8932	.8077	.7533	.7157	.6881	.9888	.9573	.9252	.8977	.8748	.9986	.9893	.9747	.9590	.9440
	21	.9267	.8462	.7898	.7492	.7186	.9942	.9715	.9444	.9191	.8969	.9994	.9939	.9833	.9706	.9575
	22	.9548	.8846	.8290	.7866	.7536	.9974	.9829	.9618	.9400	.9196	.9998	.9969	.9901	.9806	.9701
	23	.9768	.9231	.8725	.8304	.7961	.9991	.9915	.9773	.9604	.9432	.9999	.9988	.9951	.9891	.9816
	24	.9921	.9615	.9239	.8877	.8554	.9998	.9972	.9903	.9804	.9688	1.0000	.9997	.9984	.9957	.9917
	25	1.0000	1.0000	1.0000	1.0000	1.0000	1.0000	1.0000	1.0000	1.0000	1.0000	1.0000	1.0000	1.0000	1.0000	1.0000

n	k	A= 2 B= 8				
		S=1	S=2	S=3	S=4	S=5
2	0	.6545	.6667	.6769	.6857	.6933
	1	.9455	.9333	.9231	.9143	.9067
	2	1.0000	1.0000	1.0000	1.0000	1.0000
3	0	.5455	.5714	.5923	.6095	.6240
	1	.8727	.8571	.8462	.8381	.8320
	2	.9818	.9714	.9615	.9524	.9440
	3	1.0000	1.0000	1.0000	1.0000	1.0000
4	0	.4615	.5000	.5300	.5541	.5741
	1	.7972	.7857	.7794	.7758	.7738
	2	.9483	.9286	.9130	.9004	.8902
	3	.9930	.9857	.9777	.9697	.9619
	4	1.0000	1.0000	1.0000	1.0000	1.0000
5	0	.3956	.4444	.4818	.5115	.5358
	1	.7253	.7222	.7227	.7246	.7272
	2	.9051	.8810	.8644	.8525	.8436
	3	.9770	.9603	.9453	.9324	.9213
	4	.9970	.9921	.9858	.9790	.9721
	5	1.0000	1.0000	1.0000	1.0000	1.0000
6	0	.3429	.4000	.4432	.4774	.5052
	1	.6593	.6667	.6745	.6820	.6889
	2	.8571	.8333	.8190	.8099	.8037
	3	.9530	.9286	.9097	.8951	.8836
	4	.9890	.9762	.9632	.9510	.9402
	5	.9986	.9952	.9904	.9846	.9785
	6	1.0000	1.0000	1.0000	1.0000	1.0000

n	k	A= 2 B=10				
		S=1	S=2	S=3	S=4	S=5
2	0	.7051	.7143	.7222	.7292	.7353
	1	.9615	.9524	.9444	.9375	.9314
	2	1.0000	1.0000	1.0000	1.0000	1.0000
3	0	.6044	.6250	.6420	.6562	.6684
	1	.9066	.8929	.8827	.8750	.8690
	2	.9890	.9821	.9753	.9687	.9626
	3	1.0000	1.0000	1.0000	1.0000	1.0000
4	0	.5238	.5556	.5808	.6016	.6189
	1	.8462	.8333	.8254	.8203	.8170
	2	.9670	.9524	.9400	.9297	.9210
	3	.9963	.9921	.9871	.9818	.9764
	4	1.0000	1.0000	1.0000	1.0000	1.0000
5	0	.4583	.5000	.5324	.5586	.5803
	1	.7857	.7778	.7744	.7734	.7737
	2	.9368	.9167	.9013	.8906	.8820
	3	.9872	.9762	.9655	.9557	.9470
	4	.9985	.9960	.9925	.9883	.9838
	5	1.0000	1.0000	1.0000	1.0000	1.0000
6	0	.4044	.4545	.4930	.5237	.5489
	1	.7279	.7273	.7296	.7332	.7371
	2	.9013	.8788	.8641	.8540	.8469
	3	.9724	.9545	.9396	.9272	.9171
	4	.9946	.9870	.9785	.9700	.9619
	5	.9994	.9978	.9952	.9919	.9882
	6	1.0000	1.0000	1.0000	1.0000	1.0000

n	k	A= 4 B= 4				
		S=1	S=2	S=3	S=4	S=5
2	0	.2778	.3000	.3182	.3333	.3462
	1	.7222	.7000	.6818	.6667	.6538
	2	1.0000	1.0000	1.0000	1.0000	1.0000
3	0	.1667	.2000	.2273	.2500	.2692
	1	.5000	.5000	.5000	.5000	.5000
	2	.8333	.8000	.7727	.7500	.7308
	3	1.0000	1.0000	1.0000	1.0000	1.0000
4	0	.1061	.1429	.1738	.2000	.2224
	1	.3485	.3714	.3877	.4000	.4097
	2	.6515	.6286	.6123	.6000	.5903
	3	.8939	.8571	.8262	.8000	.7776
	4	1.0000	1.0000	1.0000	1.0000	1.0000
5	0	.0707	.1071	.1390	.1667	.1906
	1	.2475	.2857	.3128	.3333	.3495
	2	.5000	.5000	.5000	.5000	.5000
	3	.7525	.7143	.6872	.6667	.6505
	4	.9293	.8929	.8610	.8333	.8094
	5	1.0000	1.0000	1.0000	1.0000	1.0000
6	0	.0490	.0833	.1149	.1429	.1675
	1	.1795	.2262	.2599	.2857	.3062
	2	.3834	.4048	.4186	.4286	.4362
	3	.6166	.5952	.5814	.5714	.5638
	4	.8205	.7738	.7401	.7143	.6938
	5	.9510	.9167	.8851	.8571	.8325
	6	1.0000	1.0000	1.0000	1.0000	1.0000

7	0	.3000	.3636	.4116	.4493	.4799	.3595	.4167	.4601	.4946	.5227	.0350	.0667	.0972	.1250	.1499
	1	.6000	.6182	.6332	.6459	.6567	.6740	.6818	.6902	.6982	.7057	.1329	.1833	.2209	.2500	.2733
	2	.8077	.7879	.7777	.7722	.7693	.8627	.8409	.8282	.8204	.8155	.2960	.3333	.3576	.3750	.3883
	3	.9231	.8939	.8741	.8600	.8496	.9526	.9293	.9119	.8988	.8887	.5000	.5000	.5000	.5000	.5000
	4	.9755	.9545	.9364	.9214	.9090	.9872	.9735	.9603	.9486	.9394	.7040	.6667	.6424	.6250	.6117
	5	.9944	.9848	.9738	.9629	.9526	.9975	.9924	.9857	.9785	.9713	.8671	.8167	.7791	.7500	.7267
	6	.9993	.9970	.9931	.9882	.9828	.9997	.9987	.9968	.9942	.9910	.9650	.9333	.9028	.8750	.8501
	7	1.0000	1.0000	1.0000	1.0000	1.0000	1.0000	1.0000	1.0000	1.0000	1.0000	1.0000	1.0000	1.0000	1.0000	1.0000
8	0	.2647	.3333	.3850	.4257	.4586	.3216	.3846	.4322	.4699	.5005	.0256	.0545	.0838	.1111	.1359
	1	.5471	.5758	.5975	.6148	.6292	.6244	.6410	.6553	.6677	.6785	.1002	.1515	.1910	.2222	.2475
	2	.7588	.7455	.7404	.7390	.7393	.8230	.8042	.7948	.7899	.7875	.2308	.2788	.3104	.3333	.3508
	3	.8891	.8586	.8399	.8277	.8193	.9290	.9021	.8840	.8713	.8622	.4048	.4242	.4362	.4444	.4506
	4	.9570	.9293	.9083	.8923	.8800	.9763	.9565	.9398	.9262	.9151	.5952	.5758	.5638	.5556	.5494
	5	.9865	.9697	.9533	.9389	.9265	.9937	.9837	.9726	.9621	.9524	.7692	.7212	.6896	.6667	.6492
	6	.9970	.9899	.9807	.9709	.9613	.9988	.9953	.9901	.9840	.9776	.8998	.8485	.8090	.7778	.7525
	7	.9996	.9980	.9949	.9907	.9858	.9999	.9992	.9978	.9956	.9929	.9744	.9455	.9162	.8889	.8641
	8	1.0000	1.0000	1.0000	1.0000	1.0000	1.0000	1.0000	1.0000	1.0000	1.0000	1.0000	1.0000	1.0000	1.0000	1.0000
9	0	.2353	.3077	.3624	.4054	.4403	.2895	.3571	.4082	.4485	.4813	.0192	.0455	.0733	.1000	.1246
	1	.5000	.5385	.5662	.5878	.6054	.5789	.6044	.6243	.6407	.6545	.0769	.1273	.1676	.2000	.2266
	2	.7118	.7063	.7068	.7094	.7129	.7833	.7692	.7638	.7621	.7623	.1818	.2364	.2731	.3000	.3207
	3	.8529	.8238	.8077	.7981	.7921	.9025	.8741	.8567	.8454	.8378	.3287	.3636	.3851	.4000	.4111
	4	.9344	.9021	.8801	.8646	.8533	.9621	.9371	.9181	.9037	.8927	.5000	.5000	.5000	.5000	.5000
	5	.9751	.9510	.9308	.9145	.9014	.9876	.9720	.9571	.9441	.9331	.6713	.6364	.6149	.6000	.5889
	6	.9924	.9790	.9646	.9511	.9390	.9968	.9895	.9804	.9710	.9621	.8182	.7636	.7259	.7000	.6793
	7	.9983	.9930	.9853	.9766	.9677	.9994	.9970	.9929	.9877	.9820	.9231	.8727	.8324	.8000	.7734
	8	.9998	.9986	.9961	.9925	.9881	.9999	.9995	.9984	.9966	.9943	.9808	.9545	.9267	.9000	.8754
	9	1.0000	1.0000	1.0000	1.0000	1.0000	1.0000	1.0000	1.0000	1.0000	1.0000	1.0000	1.0000	1.0000	1.0000	1.0000
10	0	.2105	.2857	.3428	.3878	.4242	.2619	.3333	.3873	.4298	.4644	.0147	.0385	.0649	.0909	.1152
	1	.4582	.5055	.5337	.5640	.5843	.5376	.5714	.5966	.6167	.6332	.0600	.1084	.1487	.1818	.2093
	2	.6672	.6703	.6764	.6830	.6894	.7444	.7363	.7352	.7368	.7396	.1448	.2028	.2430	.2727	.2958
	3	.8158	.7902	.7777	.7711	.7676	.8741	.8462	.8305	.8211	.8153	.2682	.3147	.3435	.3636	.3787
	4	.9087	.8741	.8527	.8336	.8288	.9450	.9161	.8961	.8819	.8715	.4194	.4371	.4475	.4545	.4598
	5	.9601	.9301	.9075	.8906	.8778	.9791	.9580	.9401	.9256	.9140	.5806	.5629	.5525	.5455	.5402
	6	.9851	.9650	.9463	.9304	.9171	.9933	.9814	.9685	.9565	.9458	.7318	.6853	.6565	.6364	.6213
	7	.9955	.9850	.9724	.9599	.9484	.9983	.9930	.9855	.9773	.9691	.8552	.7972	.7570	.7273	.7042
	8	.9990	.9950	.9885	.9807	.9725	.9997	.9980	.9947	.9903	.9852	.9400	.8916	.8513	.8182	.7907
	9	.9999	.9990	.9969	.9938	.9898	1.0000	.9997	.9988	.9973	.9953	.9853	.9615	.9351	.9091	.8848
	10	1.0000	1.0000	1.0000	1.0000	1.0000	1.0000	1.0000	1.0000	1.0000	1.0000	1.0000	1.0000	1.0000	1.0000	1.0000

n	k	A = 2, B = 8: S=1	S=2	S=3	S=4	S=5
15	0	.1304	.2105	.2739	.3248	.3664
	1	.3083	.3860	.4383	.4770	.5074
	2	.4862	.5304	.5607	.5836	.6020
	3	.6403	.6478	.6571	.6661	.6743
	4	.7620	.7417	.7347	.7327	.7329
	5	.8513	.8155	.7976	.7876	.7818
	6	.9125	.8723	.8485	.8334	.8232
	7	.9519	.9149	.8895	.8717	.8588
	8	.9755	.9458	.9219	.9036	.8894
	9	.9886	.9675	.9472	.9299	.9157
	10	.9953	.9819	.9663	.9514	.9382
	11	.9983	.9910	.9802	.9685	.9572
	12	.9995	.9961	.9897	.9816	.9729
	13	.9999	.9987	.9957	.9910	.9853
	14	1.0000	.9997	.9988	.9971	.9945
	15	1.0000	1.0000	1.0000	1.0000	1.0000
20	0	.0887	.1667	.2317	.2850	.3293
	1	.2200	.3116	.3742	.4208	.4571
	2	.3640	.4368	.4835	.5175	.5439
	3	.5023	.5440	.5723	.5938	.6111
	4	.6247	.6352	.6465	.6569	.6662
	5	.7268	.7120	.7092	.7104	.7130
	6	.8081	.7760	.7625	.7563	.7535
	7	.8700	.8287	.8079	.7961	.7890
	8	.9153	.8715	.8464	.8307	.8203
	9	.9471	.9058	.8790	.8609	.8482
	10	.9684	.9327	.9063	.8872	.8731
	11	.9822	.9534	.9291	.9101	.8953
	12	.9905	.9689	.9477	.9298	.9150
	13	.9953	.9802	.9628	.9466	.9326
	14	.9979	.9881	.9746	.9608	.9480
	15	.9991	.9934	.9837	.9726	.9615
	16	.9997	.9967	.9903	.9821	.9731
	17	.9999	.9986	.9950	.9894	.9828
	18	1.0000	.9995	.9979	.9949	.9906
	19	1.0000	.9999	.9994	.9983	.9965
	20	1.0000	1.0000	1.0000	1.0000	1.0000

n	k	A = 2, B = 10: S=1	S=2	S=3	S=4	S=5
15	0	.1692	.2500	.3125	.3622	.4028
	1	.3808	.4474	.4928	.5269	.5538
	2	.5739	.6009	.6216	.6384	.6525
	3	.7261	.7183	.7186	.7217	.7258
	4	.8348	.8063	.7931	.7865	.7833
	5	.9065	.8709	.8505	.8379	.8297
	6	.9506	.9170	.8944	.8788	.8677
	7	.9757	.9489	.9276	.9114	.8989
	8	.9891	.9702	.9522	.9371	.9246
	9	.9955	.9837	.9700	.9571	.9456
	10	.9984	.9919	.9824	.9723	.9625
	11	.9995	.9964	.9906	.9834	.9758
	12	.9999	.9986	.9956	.9912	.9859
	13	1.0000	.9996	.9984	.9962	.9932
	14	1.0000	.9999	.9996	.9989	.9978
	15	1.0000	1.0000	1.0000	1.0000	1.0000
20	0	.1183	.2000	.2657	.3189	.3627
	1	.2814	.3667	.4244	.4673	.5009
	2	.4475	.5043	.5422	.5704	.5928
	3	.5951	.6170	.6348	.6497	.6624
	4	.7157	.7082	.7095	.7135	.7184
	5	.8084	.7811	.7703	.7660	.7647
	6	.8759	.8387	.8201	.8097	.8037
	7	.9229	.8835	.8606	.8464	.8371
	8	.9542	.9178	.8936	.8772	.8657
	9	.9740	.9435	.9202	.9031	.8904
	10	.9860	.9623	.9414	.9248	.9116
	11	.9929	.9758	.9581	.9427	.9299
	12	.9966	.9851	.9710	.9575	.9455
	13	.9985	.9913	.9807	.9694	.9588
	14	.9994	.9953	.9878	.9789	.9699
	15	.9998	.9976	.9928	.9862	.9790
	16	.9999	.9989	.9961	.9917	.9863
	17	1.0000	.9996	.9982	.9956	.9920
	18	1.0000	.9999	.9993	.9981	.9961
	19	1.0000	1.0000	.9998	.9995	.9987
	20	1.0000	1.0000	1.0000	1.0000	1.0000

n	k	A = 4, B = 4: S=1	S=2	S=3	S=4	S=5
15	0	.0048	.0196	.0401	.0625	.0848
	1	.0207	.0564	.0924	.1250	.1535
	2	.0536	.1078	.1521	.1875	.2163
	3	.1069	.1716	.2167	.2500	.2758
	4	.1816	.2451	.2848	.3125	.3333
	5	.2755	.3260	.3552	.3750	.3895
	6	.3839	.4118	.4273	.4375	.4449
	7	.5000	.5000	.5000	.5000	.5000
	8	.6161	.5882	.5727	.5625	.5551
	9	.7245	.6740	.6448	.6250	.6105
	10	.8184	.7549	.7152	.6875	.6667
	11	.8931	.8284	.7833	.7500	.7242
	12	.9464	.8922	.8479	.8125	.7837
	13	.9793	.9436	.9076	.8750	.8465
	14	.9952	.9804	.9599	.9375	.9152
	15	1.0000	1.0000	1.0000	1.0000	1.0000
20	0	.0020	.0119	.0282	.0476	.0680
	1	.0089	.0344	.0652	.0952	.1229
	2	.0239	.0666	.1075	.1429	.1728
	3	.0496	.1073	.1538	.1905	.2200
	4	.0878	.1553	.2029	.2381	.2653
	5	.1392	.2095	.2542	.2857	.3094
	6	.2036	.2688	.3072	.3333	.3525
	7	.2792	.3320	.3615	.3810	.3951
	8	.3638	.3981	.4165	.4286	.4372
	9	.4539	.4658	.4721	.4762	.4791
	10	.5461	.5342	.5279	.5238	.5209
	11	.6362	.6019	.5835	.5714	.5628
	12	.7208	.6680	.6385	.6190	.6049
	13	.7964	.7312	.6928	.6667	.6475
	14	.8608	.7905	.7458	.7143	.6906
	15	.9122	.8447	.7971	.7619	.7347
	16	.9504	.8927	.8462	.8095	.7800
	17	.9761	.9334	.8925	.8571	.8272
	18	.9911	.9656	.9348	.9048	.8771
	19	.9980	.9881	.9718	.9524	.9320
	20	1.0000	1.0000	1.0000	1.0000	1.0000

25	0	.0642	.1379	.2027	.2570	.3026	.0873	.1667	.2333	.2882	.3339	.0010	.0079	.0213	.0385	.0572
	1	.1644	.2611	.3293	.3806	.4208	.2157	.3103	.3755	.4242	.4623	.0044	.0232	.0494	.0769	.1033
	2	.2809	.3706	.4280	.4696	.5016	.3557	.4335	.4836	.5201	.5486	.0122	.0452	.0817	.1154	.1451
	3	.3999	.4674	.5098	.5406	.5645	.4900	.5384	.5708	.5952	.6147	.0258	.0733	.1170	.1538	.1845
	4	.5128	.5526	.5795	.6001	.6166	.6090	.6272	.6430	.6567	.6685	.0469	.1068	.1548	.1923	.2223
	5	.6144	.6272	.6398	.6512	.6612	.7090	.7017	.7037	.7083	.7137	.0764	.1453	.1944	.2308	.2589
	6	.7021	.6920	.6924	.6958	.7001	.7895	.7639	.7551	.7524	.7524	.1148	.1880	.2356	.2692	.2946
	7	.7755	.7480	.7384	.7351	.7346	.8519	.8152	.7986	.7903	.7861	.1622	.2344	.2779	.3077	.3297
	8	.8349	.7960	.7787	.7700	.7655	.8987	.8572	.8355	.8231	.8155	.2181	.2839	.3212	.3462	.3643
	9	.8816	.8368	.8141	.8012	.7933	.9327	.8912	.8668	.8516	.8415	.2814	.3358	.3653	.3846	.3985
	10	.9174	.8712	.8451	.8291	.8186	.9567	.9184	.8932	.8763	.8645	.3508	.3895	.4099	.4231	.4325
	11	.9440	.8998	.8722	.8540	.8415	.9730	.9399	.9154	.8978	.8849	.4243	.4444	.4549	.4615	.4663
	12	.9632	.9234	.8957	.8763	.8624	.9838	.9566	.9338	.9163	.9029	.5000	.5000	.5000	.5000	.5000
	13	.9766	.9425	.9159	.8962	.8814	.9906	.9693	.9491	.9323	.9189	.5757	.5556	.5451	.5385	.5337
	14	.9857	.9579	.9333	.9140	.8988	.9948	.9789	.9615	.9461	.9331	.6492	.6105	.5901	.5769	.5675
	15	.9917	.9699	.9481	.9297	.9146	.9973	.9860	.9716	.9577	.9455	.7186	.6642	.6347	.6154	.6015
	16	.9954	.9792	.9605	.9435	.9290	.9986	.9910	.9795	.9676	.9564	.7819	.7161	.6788	.6538	.6357
	17	.9976	.9861	.9707	.9556	.9420	.9994	.9944	.9858	.9758	.9659	.8378	.7656	.7221	.6923	.6703
	18	.9988	.9912	.9791	.9660	.9536	.9997	.9968	.9905	.9825	.9740	.8852	.8120	.7644	.7308	.7054
	19	.9995	.9947	.9857	.9749	.9640	.9999	.9982	.9940	.9878	.9809	.9236	.8547	.8056	.7692	.7411
	20	.9998	.9971	.9908	.9824	.9732	1.0000	.9991	.9964	.9920	.9866	.9531	.8932	.8452	.8077	.7777
	21	.9999	.9985	.9945	.9884	.9812	1.0000	.9996	.9981	.9952	.9912	.9742	.9267	.8830	.8462	.8155
	22	1.0000	.9994	.9971	.9932	.9879	1.0000	.9999	.9991	.9974	.9948	.9878	.9548	.9183	.8846	.8549
	23	1.0000	.9998	.9988	.9966	.9934	1.0000	1.0000	.9997	.9989	.9975	.9956	.9768	.9506	.9231	.8967
	24	1.0000	1.0000	.9997	.9989	.9975	1.0000	1.0000	.9999	.9997	.9992	.9990	.9921	.9787	.9515	.9428
	25	1.0000	1.0000	1.0000	1.0000	1.0000	1.0000	1.0000	1.0000	1.0000	1.0000	1.0000	1.0000	1.0000	1.0000	1.0000

n	k	A= 4 B= 6 S=1	S=2	S=3	S=4	S=5
2	0	.3818	.4000	.4154	.4286	.4400
	1	.8182	.8000	.7846	.7714	.7600
	2	1.0000	1.0000	1.0000	1.0000	1.0000
3	0	.2545	.2857	.3115	.3333	.3520
	1	.6364	.6286	.6231	.6190	.6160
	2	.9091	.8857	.8654	.8476	.8320
	3	1.0000	1.0000	1.0000	1.0000	1.0000
4	0	.1762	.2143	.2460	.2727	.2957
	1	.4895	.5000	.5083	.5152	.5210
	2	.7832	.7571	.7379	.7229	.7110
	3	.9510	.9286	.9079	.8892	.8723
	4	1.0000	1.0000	1.0000	1.0000	1.0000
5	0	.1259	.1667	.2012	.2308	.2563
	1	.3776	.4048	.4248	.4406	.4534
	2	.6573	.6429	.6335	.6270	.6223
	3	.8671	.8333	.8074	.7869	.7702
	4	.9720	.9524	.9330	.9148	.8979
	5	1.0000	1.0000	1.0000	1.0000	1.0000
6	0	.0923	.1333	.1690	.2000	.2270
	1	.2937	.3333	.3622	.3846	.4027
	2	.5455	.5476	.5500	.5524	.5548
	3	.7692	.7381	.7170	.7016	.6899
	4	.9161	.8810	.8526	.8295	.8103
	5	.9832	.9667	.9491	.9318	.9154
	6	1.0000	1.0000	1.0000	1.0000	1.0000

n	k	A= 4 B= 8 S=1	S=2	S=3	S=4	S=5
2	0	.4615	.4762	.4889	.5000	.5098
	1	.8718	.8571	.8444	.8333	.8235
	2	1.0000	1.0000	1.0000	1.0000	1.0000
3	0	.3297	.3571	.3802	.4000	.4171
	1	.7253	.7143	.7062	.7000	.6952
	2	.9451	.9286	.9136	.9000	.8877
	3	1.0000	1.0000	1.0000	1.0000	1.0000
4	0	.2418	.2778	.3078	.3333	.3553
	1	.5934	.5952	.5975	.6000	.6025
	2	.8571	.8333	.8148	.8000	.7879
	3	.9744	.9603	.9465	.9333	.9210
	4	1.0000	1.0000	1.0000	1.0000	1.0000
5	0	.1813	.2222	.2565	.2857	.3109
	1	.4835	.5000	.5130	.5238	.5330
	2	.7582	.7381	.7243	.7143	.7068
	3	.9231	.8968	.8752	.8571	.8419
	4	.9872	.9762	.9643	.9524	.9407
	5	1.0000	1.0000	1.0000	1.0000	1.0000
6	0	.1387	.1818	.2185	.2500	.2773
	1	.3946	.4242	.4465	.4643	.4790
	2	.6613	.6515	.6460	.6429	.6410
	3	.8552	.8247	.8025	.7857	.7725
	4	.9570	.9329	.9115	.8929	.8767
	5	.9932	.9848	.9749	.9643	.9535
	6	1.0000	1.0000	1.0000	1.0000	1.0000

n	k	A= 4 B=10 S=1	S=2	S=3	S=4	S=5
2	0	.5238	.5357	.5462	.5556	.5639
	1	.9048	.8929	.8824	.8730	.8647
	2	1.0000	1.0000	1.0000	1.0000	1.0000
3	0	.3929	.4167	.4370	.4545	.4699
	1	.7857	.7738	.7647	.7576	.7519
	2	.9643	.9524	.9412	.9307	.9211
	3	1.0000	1.0000	1.0000	1.0000	1.0000
4	0	.3004	.3333	.3610	.3846	.4051
	1	.6702	.6667	.6650	.6643	.6644
	2	.9013	.8810	.8645	.8508	.8394
	3	.9853	.9762	.9668	.9574	.9483
	4	1.0000	1.0000	1.0000	1.0000	1.0000
5	0	.2337	.2727	.3054	.3333	.3574
	1	.5675	.5758	.5831	.5897	.5957
	2	.8242	.8030	.7877	.7762	.7673
	3	.9526	.9329	.9156	.9005	.8874
	4	.9935	.9870	.9795	.9716	.9635
	5	1.0000	1.0000	1.0000	1.0000	1.0000
6	0	.1845	.2273	.2633	.2941	.3208
	1	.4796	.5000	.5161	.5294	.5408
	2	.7431	.7273	.7172	.7104	.7057
	3	.9053	.8788	.8583	.8420	.8289
	4	.9763	.9600	.9443	.9298	.9167
	5	.9969	.9924	.9866	.9799	.9729
	6	1.0000	1.0000	1.0000	1.0000	1.0000

7	0	.0692	.1091	.1449	.1765	.2043	.1078	.1515	.1894	.2222	.2509	.1476	.1923	.2304	.2632	.2916
	1	.2308	.2788	.3139	.3412	.3632	.3235	.3636	.3933	.4167	.4357	.4058	.4371	.4608	.4799	.4958
	2	.4510	.4697	.4830	.4932	.5015	.5724	.5758	.5795	.5833	.5870	.6641	.6573	.6543	.6533	.6532
	3	.6713	.6515	.6395	.6314	.6257	.7798	.7525	.7347	.7222	.7130	.8486	.8205	.8010	.7866	.7757
	4	.8427	.8030	.7751	.7543	.7381	.9118	.8788	.8534	.8333	.8172	.9479	.9225	.9013	.8836	.8698
	5	.9455	.9121	.8836	.8596	.8392	.9751	.9545	.9347	.9167	.9005	.9876	.9749	.9615	.9483	.9358
	6	.9895	.9758	.9600	.9438	.9281	.9962	.9899	.9816	.9722	.9624	.9985	.9953	.9908	.9852	.9790
	7	1.0000	1.0000	1.0000	1.0000	1.0000	1.0000	1.0000	1.0000	1.0000	1.0000	1.0000	1.0000	1.0000	1.0000	1.0000
8	0	.0529	.0909	.1262	.1579	.1861	.0851	.1282	.1664	.2000	.2295	.1195	.1648	.2041	.2381	.2678
	1	.1833	.2364	.2758	.3065	.3314	.2668	.3147	.3501	.3778	.4003	.3443	.3846	.4147	.4386	.4583
	2	.3733	.4061	.4284	.4452	.4585	.4938	.5105	.5231	.5333	.5419	.5903	.5944	.5990	.6037	.6082
	3	.5806	.5758	.5738	.5732	.5733	.7034	.6845	.6736	.6667	.6621	.7871	.7622	.7465	.7358	.7282
	4	.7620	.7273	.7051	.6896	.6781	.8562	.8205	.7958	.7778	.7640	.9100	.8788	.8554	.8374	.8232
	5	.8910	.8485	.8171	.7931	.7740	.9451	.9138	.8879	.8667	.8491	.9706	.9487	.9288	.9113	.8962
	6	.9636	.9333	.9058	.8818	.8609	.9851	.9681	.9504	.9333	.9176	.9933	.9837	.9724	.9606	.9491
	7	.9932	.9818	.9677	.9527	.9377	.9978	.9930	.9861	.9778	.9698	.9992	.9970	.9934	.9887	.9833
	8	1.0000	1.0000	1.0000	1.0000	1.0000	1.0000	1.0000	1.0000	1.0000	1.0000	1.0000	1.0000	1.0000	1.0000	1.0000
9	0	.0412	.0769	.1113	.1429	.1712	.0681	.1099	.1479	.1818	.2119	.0977	.1429	.1826	.2174	.2480
	1	.1471	.2028	.2450	.2782	.3052	.2214	.2747	.3144	.3455	.3708	.2932	.3407	.3759	.4037	.4265
	2	.3100	.3538	.3835	.4056	.4229	.4257	.4545	.4750	.4909	.5038	.5232	.5385	.5505	.5606	.5694
	3	.5000	.5105	.5182	.5245	.5297	.6300	.6224	.6192	.6182	.6182	.7245	.7063	.6960	.6899	.6860
	4	.6814	.6573	.6433	.6342	.6278	.7951	.7622	.7415	.7273	.7170	.8653	.8322	.8096	.7932	.7810
	5	.8265	.7832	.7545	.7340	.7184	.9051	.8671	.8393	.8182	.8016	.9453	.9161	.8921	.8728	.8570
	6	.9233	.8811	.8484	.8226	.8018	.9651	.9371	.9122	.8909	.8728	.9830	.9650	.9471	.9306	.9158
	7	.9751	.9483	.9222	.8987	.8778	.9908	.9770	.9613	.9455	.9304	.9962	.9890	.9796	.9692	.9586
	8	.9955	.9860	.9734	.9595	.9451	.9987	.9950	.9892	.9818	.9736	.9996	.9980	.9951	.9912	.9864
	9	1.0000	1.0000	1.0000	1.0000	1.0000	1.0000	1.0000	1.0000	1.0000	1.0000	1.0000	1.0000	1.0000	1.0000	1.0000
10	0	.0325	.0659	.0993	.1304	.1588	.0551	.0952	.1328	.1667	.1970	.0807	.1250	.1648	.2000	.2312
	1	.1192	.1758	.2197	.2547	.2833	.1849	.2418	.2845	.3182	.3457	.2507	.3036	.3429	.3739	.3993
	2	.2585	.3107	.3461	.3723	.3929	.3673	.4066	.4338	.4545	.4711	.4632	.4890	.5079	.5230	.5355
	3	.4300	.4545	.4709	.4831	.4928	.5619	.5664	.5712	.5758	.5801	.6632	.6538	.6499	.6485	.6484
	4	.6050	.5944	.5892	.5865	.5850	.7322	.7063	.6913	.6818	.6754	.8163	.7850	.7653	.7519	.7424
	5	.7578	.7203	.6974	.6819	.6706	.8579	.8182	.7917	.7727	.7585	.9143	.8794	.8538	.8346	.8196
	6	.8724	.8252	.7926	.7687	.7503	.9365	.8998	.8711	.8485	.8303	.9668	.9406	.9176	.8982	.8819
	7	.9451	.9051	.8724	.8458	.8239	.9774	.9530	.9298	.9091	.8910	.9899	.9755	.9598	.9445	.9303
	8	.9826	.9590	.9347	.9119	.8913	.9942	.9830	.9691	.9545	.9402	.9978	.9924	.9845	.9753	.9656
	9	.9969	.9890	.9777	.9647	.9511	.9992	.9963	.9914	.9848	.9773	.9998	.9986	.9963	.9929	.9887
	10	1.0000	1.0000	1.0000	1.0000	1.0000	1.0000	1.0000	1.0000	1.0000	1.0000	1.0000	1.0000	1.0000	1.0000	1.0000

n	k	A= 4 B= 6 S=1	S=2	S=3	S=4	S=5	A= 4 B= 8 S=1	S=2	S=3	S=4	S=5	A= 4 B=10 S=1	S=2	S=3	S=4	S=5
15	0	.0119	.0351	.0627	.0909	.1179	.0221	.0526	.0856	.1176	.1476	.0349	.0714	.1082	.1429	.1746
	1	.0474	.0970	.1412	.1789	.2110	.0823	.1404	.1883	.2279	.2611	.1222	.1842	.2331	.2727	.3056
	2	.1130	.1783	.2266	.2638	.2936	.1826	.2487	.2954	.3309	.3591	.2551	.3158	.3580	.3900	.4156
	3	.2076	.2722	.3147	.3456	.3696	.3130	.3661	.4009	.4265	.4465	.4121	.4499	.4757	.4952	.5109
	4	.3245	.3728	.4028	.4242	.4405	.4572	.4835	.5012	.5147	.5256	.5691	.5757	.5824	.5887	.5945
	5	.4532	.4750	.4890	.4993	.5074	.5982	.5942	.5942	.5956	.5976	.7072	.6864	.6763	.6709	.6681
	6	.5818	.5743	.5716	.5708	.5708	.7225	.6935	.6782	.6691	.6633	.8163	.7786	.7567	.7425	.7327
	7	.6999	.6672	.6496	.6386	.6311	.8225	.7786	.7526	.7353	.7231	.8942	.8516	.8235	.8038	.7893
	8	.7998	.7508	.7218	.7023	.6884	.8958	.8483	.8166	.7941	.7773	.9446	.9063	.8774	.8554	.8383
	9	.8776	.8230	.7872	.7619	.7428	.9446	.9025	.8703	.8456	.8262	.9741	.9450	.9194	.8979	.8802
	10	.9327	.8826	.8453	.8168	.7945	.9740	.9422	.9137	.8897	.8697	.9893	.9706	.9506	.9320	.9154
	11	.9673	.9291	.8951	.8668	.8433	.9895	.9693	.9472	.9265	.9078	.9963	.9851	.9725	.9581	.9442
	12	.9873	.9626	.9360	.9112	.8890	.9966	.9861	.9716	.9559	.9404	.9990	.9945	.9867	.9772	.9668
	13	.9963	.9843	.9675	.9493	.9311	.9992	.9951	.9876	.9779	.9672	.9998	.9983	.9949	.9899	.9836
	14	.9994	.9959	.9890	.9797	.9689	.9999	.9990	.9966	.9926	.9875	1.0000	.9997	.9988	.9971	.9946
	15	1.0000	1.0000	1.0000	1.0000	1.0000	1.0000	1.0000	1.0000	1.0000	1.0000	1.0000	1.0000	1.0000	1.0000	1.0000
20	0	.0053	.0217	.0446	.0698	.0950	.0105	.0333	.0616	.0909	.1194	.0175	.0462	.0787	.1111	.1419
	1	.0223	.0613	.1013	.1378	.1702	.0416	.0913	.1373	.1775	.2122	.0657	.1231	.1726	.2145	.2500
	2	.0559	.1149	.1641	.2042	.2372	.0983	.1664	.2186	.2597	.2931	.1474	.2184	.2702	.3103	.3425
	3	.1085	.1793	.2302	.2687	.2991	.1801	.2522	.3013	.3377	.3662	.2565	.3224	.3662	.3987	.4242
	4	.1796	.2513	.2978	.3314	.3571	.2814	.3434	.3828	.4113	.4333	.3812	.4276	.4577	.4800	.4976
	5	.2663	.3281	.3657	.3922	.4122	.3942	.4356	.4616	.4805	.4954	.5090	.5286	.5428	.5542	.5638
	6	.3638	.4071	.4329	.4510	.4647	.5096	.5252	.5364	.5455	.5531	.6287	.6216	.6206	.6218	.6239
	7	.4665	.4862	.4986	.5078	.5150	.6194	.6095	.6065	.6061	.6068	.7329	.7043	.6904	.6828	.6784
	8	.5684	.5632	.5622	.5625	.5633	.7176	.6866	.6712	.6623	.6569	.8175	.7755	.7521	.7375	.7276
	9	.6644	.6366	.6230	.6150	.6098	.8003	.7551	.7301	.7143	.7035	.8820	.8348	.8057	.7861	.7721
	10	.7501	.7050	.6806	.6652	.6545	.8660	.8143	.7829	.7619	.7468	.9281	.8826	.8513	.8289	.8121
	11	.8229	.7671	.7346	.7130	.6976	.9152	.8640	.8296	.8052	.7870	.9589	.9199	.8895	.8661	.8478
	12	.8814	.8221	.7845	.7583	.7390	.9498	.9043	.8701	.8442	.8240	.9782	.9478	.9206	.8980	.8794
	13	.9256	.8696	.8300	.8010	.7787	.9725	.9360	.9044	.8788	.8579	.9894	.9679	.9453	.9248	.9070
	14	.9570	.9091	.8709	.8408	.8169	.9862	.9597	.9329	.9091	.8887	.9953	.9816	.9643	.9469	.9308
	15	.9775	.9407	.9065	.8775	.8533	.9939	.9765	.9556	.9351	.9163	.9982	.9903	.9782	.9646	.9510
	16	.9897	.9647	.9369	.9109	.8878	.9977	.9877	.9730	.9567	.9406	.9994	.9955	.9880	.9782	.9676
	17	.9961	.9816	.9617	.9406	.9203	.9993	.9945	.9855	.9740	.9615	.9998	.9982	.9942	.9881	.9807
	18	.9989	.9924	.9806	.9661	.9505	.9998	.9981	.9937	.9870	.9788	1.0000	.9995	.9978	.9947	.9904
	19	.9998	.9980	.9935	.9864	.9776	1.0000	.9996	.9983	.9957	.9919	1.0000	.9999	.9995	.9985	.9968
	20	1.0000	1.0000	1.0000	1.0000	1.0000	1.0000	1.0000	1.0000	1.0000	1.0000	1.0000	1.0000	1.0000	1.0000	1.0000

25	0	.0027	.0148	.0340	.0566	.0801	.0056	.0230	.0473	.0741	.1010	.0097	.0323	.0608	.0909	.1204
	1	.0118	.0421	.0776	.1121	.1437	.0231	.0640	.1054	.1453	.1800	.0382	.0879	.1350	.1767	.2130
	2	.0305	.0800	.1265	.1665	.2005	.0570	.1188	.1709	.2137	.2493	.0900	.1594	.2139	.2574	.2930
	3	.0613	.1265	.1785	.2197	.2530	.1090	.1833	.2376	.2792	.3123	.1644	.2406	.2935	.3332	.3646
	4	.1052	.1798	.2324	.2717	.3024	.1779	.2543	.3049	.3419	.3706	.2569	.3265	.3714	.4041	.4296
	5	.1619	.2381	.2873	.3225	.3494	.2607	.3289	.3714	.4017	.4250	.3605	.4131	.4462	.4703	.4892
	6	.2300	.3000	.3425	.3721	.3944	.3527	.4046	.4362	.4587	.4761	.4676	.4973	.5169	.5319	.5441
	7	.3070	.3640	.3974	.4204	.4377	.4487	.4792	.4985	.5128	.5242	.5715	.5768	.5829	.5890	.5947
	8	.3898	.4288	.4515	.4674	.4794	.5437	.5512	.5580	.5641	.5696	.6667	.6499	.6438	.6417	.6415
	9	.4752	.4932	.5046	.5130	.5197	.6335	.6192	.6142	.6125	.6124	.7497	.7157	.6993	.6901	.6847
	10	.5597	.5562	.5562	.5573	.5587	.7147	.6822	.6668	.6581	.6529	.8188	.7736	.7493	.7344	.7245
	11	.6404	.6169	.6061	.6001	.5966	.7851	.7395	.7155	.7009	.6911	.8737	.8235	.7939	.7746	.7612
	12	.7147	.6743	.6539	.6415	.6332	.8438	.7905	.7603	.7407	.7272	.9155	.8655	.8332	.8110	.7948
	13	.7808	.7280	.6994	.6813	.6687	.8903	.8352	.8010	.7778	.7611	.9459	.9001	.8674	.8436	.8256
	14	.8374	.7772	.7425	.7196	.7032	.9268	.8736	.8376	.8120	.7929	.9670	.9279	.8967	.8725	.8536
	15	.8841	.8217	.7828	.7562	.7366	.9532	.9057	.8700	.8433	.8227	.9809	.9497	.9214	.8980	.8789
	16	.9211	.8611	.8203	.7910	.7689	.9717	.9319	.8984	.8718	.8504	.9896	.9662	.9419	.9202	.9016
	17	.9491	.8952	.8546	.8240	.8002	.9839	.9528	.9229	.8974	.8762	.9947	.9783	.9584	.9392	.9218
	18	.9692	.9240	.8857	.8551	.8303	.9915	.9688	.9435	.9202	.8998	.9975	.9868	.9715	.9552	.9396
	19	.9827	.9475	.9134	.8840	.8593	.9959	.9805	.9604	.9402	.9214	.9990	.9925	.9814	.9684	.9550
	20	.9912	.9661	.9375	.9108	.8871	.9982	.9888	.9739	.9573	.9408	.9996	.9961	.9887	.9789	.9680
	21	.9961	.9800	.9579	.9351	.9136	.9993	.9942	.9842	.9715	.9579	.9999	.9982	.9938	.9870	.9788
	22	.9986	.9897	.9745	.9568	.9386	.9998	.9974	.9915	.9829	.9727	1.0000	.9993	.9970	.9929	.9874
	23	.9996	.9957	.9871	.9753	.9618	1.0000	.9991	.9963	.9915	.9849	1.0000	.9998	.9989	.9969	.9937
	24	.9999	.9989	.9957	.9901	.9827	1.0000	.9998	.9990	.9972	.9942	1.0000	1.0000	.9997	.9991	.9979
	25	1.0000	1.0000	1.0000	1.0000	1.0000	1.0000	1.0000	1.0000	1.0000	1.0000	1.0000	1.0000	1.0000	1.0000	1.0000

n	k	A= 6 B= 6 S=1	S=2	S=3	S=4	S=5	A= 6 B= 8 S=1	S=2	S=3	S=4	S=5	A= 6 B=10 S=1	S=2	S=3	S=4	S=5
2	0	.2692	.2857	.3000	.3125	.3235	.3429	.3571	.3697	.3810	.3910	.4044	.4167	.4276	.4375	.4464
	1	.7308	.7143	.7000	.6875	.6765	.8000	.7857	.7731	.7619	.7519	.8456	.8333	.8224	.8125	.8036
	2	1.0000	1.0000	1.0000	1.0000	1.0000	1.0000	1.0000	1.0000	1.0000	1.0000	1.0000	1.0000	1.0000	1.0000	1.0000
3	0	.1538	.1786	.2000	.2188	.2353	.2143	.2381	.2588	.2771	.2932	.2696	.2917	.3110	.3281	.3434
	1	.5000	.5000	.5000	.5000	.5000	.6000	.5952	.5916	.5887	.5865	.6740	.6667	.6609	.6563	.6525
	2	.8462	.8214	.8000	.7813	.7647	.9000	.8810	.8639	.8485	.8346	.9314	.9167	.9031	.8906	.8791
	3	1.0000	1.0000	1.0000	1.0000	1.0000	1.0000	1.0000	1.0000	1.0000	1.0000	1.0000	1.0000	1.0000	1.0000	1.0000
4	0	.0923	.1190	.1429	.1641	.1830	.1387	.1667	.1913	.2131	.2326	.1845	.2121	.2364	.2578	.2769
	1	.3385	.3571	.3714	.3828	.3922	.4412	.4524	.4614	.4689	.4752	.5250	.5303	.5349	.5391	.5428
	2	.6615	.6429	.6286	.6172	.6078	.7588	.7381	.7218	.7086	.6977	.8230	.8030	.7868	.7734	.7621
	3	.9077	.8810	.8571	.8359	.8170	.9471	.9286	.9112	.8951	.8802	.9675	.9545	.9419	.9297	.9181
	4	1.0000	1.0000	1.0000	1.0000	1.0000	1.0000	1.0000	1.0000	1.0000	1.0000	1.0000	1.0000	1.0000	1.0000	1.0000
5	0	.0577	.0833	.1071	.1289	.1487	.0924	.1212	.1472	.1705	.1915	.1291	.1591	.1857	.2095	.2308
	1	.2308	.2619	.2857	.3047	.3203	.3235	.3485	.3679	.3836	.3967	.4058	.4242	.4390	.4512	.4616
	2	.5000	.5000	.5000	.5000	.5000	.6176	.6082	.6016	.5967	.5930	.7038	.6894	.6789	.6709	.6647
	3	.7692	.7381	.7143	.6953	.6797	.8529	.8247	.8019	.7832	.7675	.9025	.8788	.8588	.8418	.8271
	4	.9423	.9167	.8929	.8711	.8513	.9706	.9545	.9385	.9231	.9084	.9837	.9735	.9626	.9517	.9409
	5	1.0000	1.0000	1.0000	1.0000	1.0000	1.0000	1.0000	1.0000	1.0000	1.0000	1.0000	1.0000	1.0000	1.0000	1.0000
6	0	.0373	.0606	.0833	.1047	.1246	.0632	.0909	.1167	.1404	.1621	.0922	.1224	.1498	.1746	.1970
	1	.1595	.1970	.2262	.2498	.2693	.2384	.2727	.2994	.3209	.3388	.3136	.3427	.3654	.3840	.3997
	2	.3733	.3918	.4048	.4146	.4223	.4938	.5000	.5049	.5090	.5125	.5903	.5874	.5860	.5854	.5854
	3	.6267	.6082	.5952	.5854	.5777	.7415	.7165	.6983	.6845	.6735	.8173	.7914	.7718	.7563	.7439
	4	.8405	.8030	.7738	.7502	.7307	.9087	.8788	.8538	.8326	.8145	.9450	.9225	.9024	.8845	.8687
	5	.9627	.9394	.9167	.8953	.8754	.9830	.9697	.9555	.9412	.9272	.9915	.9837	.9747	.9551	.9553
	6	1.0000	1.0000	1.0000	1.0000	1.0000	1.0000	1.0000	1.0000	1.0000	1.0000	1.0000	1.0000	1.0000	1.0000	1.0000

7	0	.0249	.0455	.0667	.0873	.1068	.0443	.0699	.0948	.1182	.1400	.0671	.0962	.1233	.1484	.1713
	1	.1120	.1515	.1833	.2095	.2314	.1771	.2168	.2480	.2734	.2947	.2432	.2797	.3083	.3317	.3512
	2	.2783	.3106	.3333	.3505	.3640	.3916	.4126	.4278	.4397	.4493	.4897	.5000	.5082	.5150	.5208
	3	.5000	.5000	.5000	.5000	.5000	.6300	.6166	.6077	.6014	.5967	.7245	.7040	.6898	.6794	.6715
	4	.7217	.6894	.6667	.6495	.6360	.8251	.7914	.7663	.7468	.7312	.8870	.8569	.8332	.8140	.7982
	5	.8880	.8485	.8167	.7905	.7686	.9421	.9138	.8887	.8669	.8478	.9683	.9487	.9300	.9127	.8970
	6	.9751	.9545	.9333	.9127	.8932	.9898	.9790	.9666	.9536	.9404	.9954	.9895	.9821	.9738	.9650
	7	1.0000	1.0000	1.0000	1.0000	1.0000	1.0000	1.0000	1.0000	1.0000	1.0000	1.0000	1.0000	1.0000	1.0000	1.0000
8	0	.0170	.0350	.0545	.0742	.0932	.0316	.0549	.0786	.1013	.1228	.0496	.0769	.1033	.1281	.1512
	1	.0799	.1189	.1515	.1789	.2022	.1328	.1748	.2086	.2365	.2599	.1896	.2308	.2633	.2900	.3124
	2	.2083	.2494	.2788	.3011	.3188	.3099	.3427	.3662	.3843	.3988	.4039	.4266	.4434	.4566	.4676
	3	.3950	.4126	.4242	.4327	.4392	.5279	.5291	.5306	.5321	.5335	.6326	.6224	.6162	.6122	.6095
	4	.6050	.5874	.5758	.5673	.5608	.7322	.7040	.6847	.6706	.6598	.8163	.7855	.7634	.7467	.7336
	5	.7917	.7506	.7212	.6989	.6812	.8808	.8438	.8153	.7926	.7740	.9294	.8998	.8751	.8544	.8369
	6	.9201	.8811	.8485	.8211	.7978	.9625	.9371	.9132	.8916	.8723	.9812	.9650	.9483	.9322	.9170
	7	.9830	.9650	.9455	.9258	.9068	.9937	.9850	.9742	.9624	.9502	.9974	.9930	.9870	.9798	.9719
	8	1.0000	1.0000	1.0000	1.0000	1.0000	1.0000	1.0000	1.0000	1.0000	1.0000	1.0000	1.0000	1.0000	1.0000	1.0000
9	0	.0119	.0275	.0455	.0641	.0824	.0230	.0440	.0662	.0881	.1092	.0372	.0625	.0878	.1121	.1350
	1	.0579	.0949	.1273	.1551	.1791	.1006	.1429	.1778	.2071	.2320	.1487	.1923	.2273	.2563	.2807
	2	.1569	.2028	.2364	.2622	.2829	.2455	.2867	.3164	.3393	.3577	.3325	.3654	.3894	.4080	.4232
	3	.3110	.3427	.3636	.3789	.3906	.4387	.4545	.4657	.4742	.4811	.5468	.5490	.5514	.5538	.5563
	4	.5000	.5000	.5000	.5000	.5000	.6393	.6224	.6117	.6044	.5990	.7398	.7142	.6972	.6851	.6760
	5	.6890	.6573	.6364	.6211	.6094	.8065	.7692	.7431	.7237	.7084	.8776	.8427	.8165	.7961	.7797
	6	.8431	.7972	.7636	.7378	.7171	.9180	.8811	.8514	.8270	.8068	.9553	.9283	.9044	.8836	.8655
	7	.9421	.9051	.8727	.8449	.8209	.9753	.9530	.9309	.9101	.8911	.9886	.9755	.9609	.9460	.9317
	8	.9881	.9725	.9545	.9359	.9176	.9960	.9890	.9797	.9689	.9575	.9985	.9952	.9902	.9840	.9769
	9	1.0000	1.0000	1.0000	1.0000	1.0000	1.0000	1.0000	1.0000	1.0000	1.0000	1.0000	1.0000	1.0000	1.0000	1.0000
10	0	.0085	.0220	.0385	.0561	.0737	.0170	.0357	.0565	.0776	.0981	.0283	.0515	.0756	.0992	.1217
	1	.0426	.0769	.1084	.1362	.1605	.0770	.1181	.1533	.1833	.2091	.1175	.1618	.1981	.2286	.2544
	2	.1192	.1668	.2028	.2310	.2538	.1951	.2418	.2759	.3023	.3236	.2737	.3145	.3441	.3672	.3859
	3	.2449	.2867	.3147	.3351	.3509	.3631	.3916	.4111	.4257	.4372	.4697	.4842	.4948	.5034	.5105
	4	.4100	.4266	.4371	.4445	.4501	.5521	.5490	.5476	.5471	.5471	.6626	.6461	.6361	.6295	.6250
	5	.5900	.5734	.5629	.5555	.5499	.7265	.6958	.6758	.6616	.6509	.8169	.7822	.7582	.7406	.7270
	6	.7551	.7133	.6853	.6649	.6491	.8598	.8182	.7880	.7650	.7468	.9180	.8830	.8553	.8331	.8149
	7	.8808	.8332	.7972	.7690	.7462	.9429	.9081	.8785	.8536	.8325	.9713	.9478	.9254	.9052	.8872
	8	.9574	.9231	.8916	.8638	.8395	.9834	.9643	.9440	.9242	.9057	.9930	.9825	.9697	.9563	.9428
	9	.9915	.9780	.9615	.9439	.9263	.9974	.9918	.9836	.9739	.9633	.9991	.9966	.9925	.9871	.9807
	10	1.0000	1.0000	1.0000	1.0000	1.0000	1.0000	1.0000	1.0000	1.0000	1.0000	1.0000	1.0000	1.0000	1.0000	1.0000

n	k	A= 6 B= 6 S=1	S=2	S=3	S=4	S=5	A= 6 B= 8 S=1	S=2	S=3	S=4	S=5	A= 6 B=10 S=1	S=2	S=3	S=4	S=5
15	0	.0020	.0088	.0196	.0329	.0475	.0046	.0150	.0296	.0454	.0640	.0084	.0227	.0408	.0604	.0803
	1	.0110	.0320	.0564	.0807	.1037	.0232	.0526	.0830	.1116	.1378	.0400	.0766	.1113	.1427	.1707
	2	.0343	.0726	.1078	.1384	.1647	.0667	.1146	.1546	.1878	.2157	.1074	.1603	.2020	.2356	.2635
	3	.0792	.1313	.1716	.2032	.2288	.1420	.1984	.2391	.2702	.2951	.2135	.2670	.3045	.3329	.3555
	4	.1504	.2068	.2451	.2732	.2950	.2491	.2990	.3319	.3559	.3745	.3499	.3871	.4118	.4301	.4445
	5	.2484	.2962	.3260	.3469	.3626	.3800	.4097	.4286	.4422	.4528	.5000	.5103	.5180	.5242	.5295
	6	.3681	.3955	.4118	.4229	.4311	.5212	.5232	.5253	.5273	.5291	.6448	.6277	.6184	.6128	.6093
	7	.5000	.5000	.5000	.5000	.5000	.6573	.6327	.6185	.6093	.6027	.7689	.7322	.7096	.6943	.6831
	8	.6319	.6045	.5882	.5771	.5689	.7753	.7322	.7054	.6867	.6729	.8638	.8193	.7890	.7671	.7504
	9	.7516	.7038	.6740	.6531	.6374	.8670	.8173	.7833	.7583	.7390	.9284	.8870	.8552	.8304	.8106
	10	.8496	.7932	.7549	.7268	.7050	.9305	.8854	.8503	.8226	.8003	.9672	.9358	.9076	.8836	.8632
	11	.9208	.8687	.8284	.7968	.7712	.9690	.9358	.9052	.8787	.8560	.9873	.9678	.9466	.9264	.9079
	12	.9657	.9274	.8922	.8616	.8353	.9889	.9693	.9472	.9255	.9052	.9961	.9864	.9733	.9588	.9442
	13	.9890	.9680	.9436	.9193	.8963	.9971	.9886	.9762	.9619	.9469	.9992	.9957	.9895	.9812	.9718
	14	.9980	.9912	.9804	.9671	.9525	.9996	.9975	.9932	.9870	.9794	.9999	.9992	.9974	.9945	.9905
	15	1.0000	1.0000	1.0000	1.0000	1.0000	1.0000	1.0000	1.0000	1.0000	1.0000	1.0000	1.0000	1.0000	1.0000	1.0000
20	0	.0006	.0043	.0119	.0223	.0345	.0015	.0077	.0182	.0317	.0467	.0031	.0120	.0254	.0416	.0591
	1	.0036	.0162	.0344	.0548	.0754	.0084	.0278	.0519	.0769	.1012	.0158	.0419	.0710	.0996	.1266
	2	.0120	.0377	.0666	.0945	.1199	.0260	.0624	.0983	.1306	.1592	.0461	.0913	.1319	.1669	.1971
	3	.0294	.0698	.1073	.1395	.1670	.0599	.1119	.1549	.1900	.2191	.1000	.1587	.2037	.2393	.2684
	4	.0597	.1130	.1553	.1887	.2157	.1138	.1751	.2194	.2531	.2799	.1793	.2405	.2827	.3142	.3390
	5	.1058	.1668	.2095	.2412	.2658	.1887	.2495	.2894	.3184	.3408	.2807	.3322	.3654	.3895	.4082
	6	.1693	.2300	.2688	.2962	.3169	.2825	.3322	.3630	.3848	.4014	.3970	.4287	.4489	.4637	.4752
	7	.2494	.3012	.3320	.3531	.3688	.3896	.4198	.4381	.4511	.4611	.5183	.5252	.5307	.5354	.5396
	8	.3435	.3783	.3981	.4114	.4211	.5027	.5087	.5130	.5166	.5196	.6347	.6174	.6087	.6038	.6008
	9	.4467	.4590	.4658	.4704	.4737	.6139	.5957	.5861	.5803	.5765	.7383	.7019	.6813	.6679	.6586
	10	.5533	.5410	.5342	.5296	.5263	.7158	.6776	.6560	.6417	.6316	.8237	.7763	.7471	.7272	.7126
	11	.6565	.6217	.6019	.5886	.5789	.8030	.7522	.7212	.7001	.6845	.8891	.8391	.8054	.7811	.7626
	12	.7506	.6988	.6680	.6469	.6312	.8724	.8174	.7809	.7548	.7349	.9354	.8899	.8555	.8292	.8084
	13	.8307	.7700	.7312	.7038	.6831	.9237	.8721	.8341	.8052	.7825	.9655	.9289	.8973	.8713	.8497
	14	.8942	.8332	.7905	.7588	.7342	.9585	.9159	.8801	.8510	.8270	.9835	.9573	.9309	.9071	.8864
	15	.9403	.8870	.8447	.8113	.7843	.9800	.9490	.9185	.8915	.8680	.9930	.9765	.9567	.9368	.9182
	16	.9706	.9302	.8927	.8605	.8330	.9917	.9722	.9492	.9263	.9051	.9975	.9886	.9754	.9603	.9451
	17	.9880	.9623	.9334	.9055	.8801	.9972	.9871	.9720	.9550	.9377	.9993	.9953	.9878	.9779	.9668
	18	.9964	.9838	.9656	.9452	.9246	.9993	.9953	.9876	.9771	.9652	.9999	.9986	.9953	.9900	.9833
	19	.9994	.9957	.9881	.9777	.9655	.9999	.9990	.9965	.9922	.9865	1.0000	.9997	.9989	.9971	.9944
	20	1.0000	1.0000	1.0000	1.0000	1.0000	1.0000	1.0000	1.0000	1.0000	1.0000	1.0000	1.0000	1.0000	1.0000	1.0000

25	0	.0002	.0025	.0079	.0163	.0268	.0006	.0044	.0123	.0234	.0364	.0013	.0071	.0174	.0309	.0462
	1	.0014	.0093	.0232	.0403	.0586	.0035	.0164	.0354	.0571	.0791	.0071	.0253	.0492	.0746	.0996
	2	.0049	.0219	.0452	.0697	.0933	.0114	.0376	.0679	.0975	.1249	.0217	.0566	.0926	.1260	.1559
	3	.0124	.0413	.0733	.1032	.1301	.0276	.0688	.1082	.1427	.1726	.0498	.1010	.1452	.1823	.2135
	4	.0262	.0679	.1068	.1401	.1683	.0551	.1100	.1550	.1914	.2213	.0946	.1574	.2047	.2416	.2713
	5	.0485	.1020	.1453	.1798	.2077	.0964	.1605	.2071	.2425	.2705	.1573	.2237	.2689	.3025	.3288
	6	.0811	.1432	.1880	.2217	.2480	.1526	.2191	.2632	.2952	.3199	.2366	.2973	.3360	.3638	.3853
	7	.1255	.1912	.2344	.2654	.2890	.2229	.2841	.3221	.3488	.3691	.3289	.3756	.4043	.4247	.4405
	8	.1819	.2452	.2839	.3106	.3306	.3051	.3538	.3827	.4028	.4180	.4289	.4556	.4723	.4845	.4941
	9	.2497	.3043	.3358	.3570	.3726	.3958	.4262	.4441	.4567	.4662	.5306	.5348	.5388	.5425	.5459
	10	.3272	.3673	.3895	.4042	.4149	.4904	.4993	.5053	.5099	.5136	.6282	.6108	.6025	.5981	.5956
	11	.4118	.4330	.4444	.4520	.4574	.5842	.5714	.5653	.5620	.5600	.7169	.6817	.6628	.6510	.6430
	12	.5000	.5000	.5000	.5000	.5000	.6727	.6406	.6235	.6127	.6053	.7934	.7460	.7187	.7007	.6880
	13	.5882	.5670	.5556	.5480	.5426	.7525	.7055	.6790	.6616	.6492	.8561	.8028	.7697	.7471	.7304
	14	.6723	.6327	.6105	.5958	.5851	.8208	.7649	.7312	.7083	.6916	.9046	.8515	.8155	.7897	.7702
	15	.7503	.6957	.6642	.6430	.6274	.8765	.8176	.7795	.7526	.7324	.9402	.8919	.8558	.8285	.8071
	16	.8181	.7548	.7161	.6894	.6694	.9195	.8633	.8236	.7941	.7713	.9648	.9244	.8906	.8633	.8411
	17	.8745	.8088	.7656	.7346	.7110	.9508	.9017	.8629	.8326	.8083	.9807	.9496	.9198	.8939	.8720
	18	.9189	.8568	.8120	.7783	.7520	.9721	.9326	.8973	.8677	.8430	.9903	.9682	.9436	.9205	.8998
	19	.9515	.8980	.8547	.8202	.7923	.9856	.9566	.9265	.8992	.8754	.9956	.9813	.9624	.9429	.9244
	20	.9738	.9321	.8932	.8599	.8317	.9933	.9742	.9505	.9269	.9051	.9982	.9899	.9767	.9613	.9457
	21	.9876	.9587	.9267	.8968	.8699	.9974	.9862	.9694	.9506	.9318	.9994	.9952	.9869	.9758	.9636
	22	.9951	.9781	.9548	.9303	.9067	.9991	.9937	.9833	.9699	.9553	.9998	.9981	.9936	.9866	.9780
	23	.9986	.9907	.9768	.9597	.9414	.9998	.9978	.9926	.9848	.9750	1.0000	.9994	.9975	.9940	.9889
	24	.9998	.9975	.9921	.9837	.9732	1.0000	.9995	.9979	.9949	.9904	1.0000	.9999	.9994	.9983	.9963
	25	1.0000	1.0000	1.0000	1.0000	1.0000	1.0000	1.0000	1.0000	1.0000	1.0000	1.0000	1.0000	1.0000	1.0000	1.0000

n	k	A = 8 B = 8					A = 8 B = 10					A = 10 B = 10				
		S = 1	S = 2	S = 3	S = 4	S = 5	S = 1	S = 2	S = 3	S = 4	S = 5	S = 1	S = 2	S = 3	S = 4	S = 5
2	0	.2647	.2778	.2895	.3000	.3095	.3216	.3333	.3439	.3535	.3623	.2619	.2727	.2826	.2917	.3000
	1	.7353	.7222	.7105	.7000	.6905	.7895	.7778	.7672	.7576	.7488	.7381	.7273	.7174	.7083	.7000
	2	1.0000	1.0000	1.0000	1.0000	1.0000	1.0000	1.0000	1.0000	1.0000	1.0000	1.0000	1.0000	1.0000	1.0000	1.0000
3	0	.1471	.1667	.1842	.2000	.2143	.1930	.2121	.2293	.2448	.2588	.1429	.1591	.1739	.1875	.2000
	1	.5000	.5000	.5000	.5000	.5000	.5789	.5758	.5732	.5711	.5694	.5000	.5000	.5000	.5000	.5000
	2	.8529	.8333	.8158	.8000	.7857	.8947	.8788	.8642	.8508	.8385	.8571	.8409	.8261	.8125	.8000
	3	1.0000	1.0000	1.0000	1.0000	1.0000	1.0000	1.0000	1.0000	1.0000	1.0000	1.0000	1.0000	1.0000	1.0000	1.0000
4	0	.0851	.1061	.1253	.1429	.1590	.1195	.1414	.1613	.1795	.1961	.0807	.0979	.1139	.1289	.1429
	1	.3328	.3485	.3611	.3714	.3802	.4135	.4242	.4331	.4406	.4470	.3292	.3427	.3538	.3633	.3714
	2	.6672	.6515	.6389	.6286	.6198	.7444	.7273	.7133	.7016	.6917	.6708	.6573	.6462	.6367	.6286
	3	.9149	.8939	.8747	.8571	.8410	.9449	.9293	.9145	.9005	.8874	.9193	.9021	.8851	.8711	.8571
	4	1.0000	1.0000	1.0000	1.0000	1.0000	1.0000	1.0000	1.0000	1.0000	1.0000	1.0000	1.0000	1.0000	1.0000	1.0000
5	0	.0511	.0707	.0895	.1071	.1237	.0760	.0979	.1183	.1373	.1548	.0471	.0629	.0783	.0931	.1071
	1	.2214	.2475	.2684	.2857	.3003	.2932	.3155	.3334	.3484	.3612	.2153	.2378	.2564	.2721	.2857
	2	.5000	.5000	.5000	.5000	.5000	.5940	.5874	.5825	.5788	.5758	.5000	.5000	.5000	.5000	.5000
	3	.7786	.7525	.7316	.7143	.6997	.8446	.8205	.8005	.7835	.7690	.7847	.7622	.7436	.7279	.7143
	4	.9489	.9293	.9105	.8929	.8763	.9699	.9565	.9430	.9298	.9171	.9529	.9371	.9217	.9069	.8929
	5	1.0000	1.0000	1.0000	1.0000	1.0000	1.0000	1.0000	1.0000	1.0000	1.0000	1.0000	1.0000	1.0000	1.0000	1.0000
6	0	.0316	.0490	.0664	.0833	.0995	.0496	.0699	.0896	.1084	.1260	.0283	.0420	.0560	.0698	.0833
	1	.1484	.1795	.2049	.2262	.2443	.2082	.2378	.2617	.2817	.2988	.1413	.1678	.1902	.2095	.2262
	2	.3673	.3834	.3954	.4048	.4123	.4632	.4709	.4769	.4818	.4860	.3634	.3776	.3886	.3975	.4048
	3	.6327	.6166	.6046	.5952	.5877	.7247	.7040	.6882	.6758	.6656	.6366	.6224	.6114	.6025	.5952
	4	.8516	.8205	.7951	.7738	.7557	.9045	.8788	.8566	.8374	.8206	.8587	.8322	.8098	.7905	.7738
	5	.9684	.9510	.9336	.9167	.9005	.9830	.9720	.9603	.9483	.9364	.9717	.9580	.9440	.9302	.9167
	6	1.0000	1.0000	1.0000	1.0000	1.0000	1.0000	1.0000	1.0000	1.0000	1.0000	1.0000	1.0000	1.0000	1.0000	1.0000

7	0	.0201	.0350	.0508	.0667	.0822	.0331	.0513	.0697	.0877	.1050	.0174	.0288	.0412	.0540	.0667
	1	.1006	.1329	.1601	.1833	.2034	.1487	.1818	.2091	.2322	.2520	.0935	.1206	.1443	.1650	.1833
	2	.2678	.2960	.3170	.3333	.3466	.3570	.3776	.3932	.4056	.4158	.2609	.2858	.3051	.3206	.3333
	3	.5000	.5000	.5000	.5000	.5000	.6049	.5952	.5884	.5834	.5795	.5000	.5000	.5000	.5000	.5000
	4	.7322	.7040	.6830	.6667	.6534	.8146	.7855	.7631	.7450	.7302	.7391	.7142	.6949	.6794	.6667
	5	.8994	.8671	.8399	.8167	.7966	.9405	.9161	.8941	.8744	.8568	.9065	.8794	.8557	.8350	.8167
	6	.9799	.9650	.9492	.9333	.9178	.9901	.9814	.9713	.9606	.9496	.9826	.9712	.9588	.9460	.9333
	7	1.0000	1.0000	1.0000	1.0000	1.0000	1.0000	1.0000	1.0000	1.0000	1.0000	1.0000	1.0000	1.0000	1.0000	1.0000
8	0	.0131	.0256	.0398	.0545	.0693	.0225	.0385	.0554	.0725	.0891	.0110	.0204	.0312	.0427	.0545
	1	.0691	.1002	.1276	.1515	.1725	.1071	.1410	.1698	.1945	.2159	.0625	.0882	.1116	.1326	.1515
	2	.1951	.2308	.2576	.2788	.2960	.2737	.3042	.3271	.3453	.3601	.1865	.2178	.2423	.2622	.2788
	3	.3889	.4048	.4159	.4242	.4308	.4958	.5000	.5033	.5061	.5085	.3849	.3992	.4097	.4178	.4242
	4	.6111	.5952	.5841	.5758	.5692	.7140	.6904	.6735	.6607	.6506	.6151	.6008	.5903	.5822	.5758
	5	.8049	.7692	.7424	.7212	.7040	.8751	.8427	.8168	.7957	.7780	.8135	.7822	.7577	.7378	.7212
	6	.9309	.8998	.8724	.8485	.8275	.9623	.9406	.9198	.9006	.8831	.9375	.9118	.8884	.8674	.8485
	7	.9869	.9744	.9602	.9455	.9307	.9941	.9872	.9787	.9692	.9591	.9890	.9796	.9688	.9573	.9455
	8	1.0000	1.0000	1.0000	1.0000	1.0000	1.0000	1.0000	1.0000	1.0000	1.0000	1.0000	1.0000	1.0000	1.0000	1.0000
9	0	.0087	.0192	.0318	.0455	.0594	.0156	.0294	.0449	.0609	.0768	.0070	.0147	.0241	.0345	.0455
	1	.0481	.0769	.1034	.1273	.1485	.0778	.1109	.1399	.1652	.1875	.0422	.0656	.0879	.1084	.1273
	2	.1426	.1818	.2121	.2364	.2563	.2096	.2466	.2747	.2970	.3154	.1333	.1674	.1948	.2174	.2364
	3	.3001	.3287	.3486	.3636	.3755	.4018	.4194	.4320	.4418	.4496	.2928	.3186	.3374	.3520	.3636
	4	.5000	.5000	.5000	.5000	.5000	.6133	.6008	.5925	.5865	.5820	.5000	.5000	.5000	.5000	.5000
	5	.6999	.6713	.6514	.6364	.6245	.7945	.7620	.7383	.7201	.7055	.7072	.6814	.6626	.6480	.6364
	6	.8574	.8182	.7879	.7636	.7437	.9153	.8830	.8560	.8334	.8142	.8667	.8326	.8052	.7826	.7636
	7	.9519	.9231	.8966	.8727	.8515	.9757	.9570	.9380	.9198	.9027	.9578	.9344	.9121	.8916	.8727
	8	.9913	.9808	.9682	.9545	.9406	.9963	.9910	.9837	.9753	.9662	.9930	.9853	.9759	.9655	.9545
	9	1.0000	1.0000	1.0000	1.0000	1.0000	1.0000	1.0000	1.0000	1.0000	1.0000	1.0000	1.0000	1.0000	1.0000	1.0000
10	0	.0059	.0147	.0259	.0385	.0516	.0110	.0229	.0369	.0519	.0671	.0046	.0108	.0190	.0283	.0385
	1	.0339	.0600	.0851	.1084	.1296	.0571	.0882	.1166	.1420	.1647	.0289	.0495	.0702	.0899	.1084
	2	.1048	.1448	.1767	.2028	.2245	.1608	.2014	.2327	.2580	.2788	.0956	.1299	.1584	.1824	.2028
	3	.2308	.2682	.2947	.3147	.3305	.3235	.3522	.3726	.3882	.4006	.2213	.2549	.2798	.2991	.3147
	4	.4041	.4194	.4296	.4371	.4428	.5193	.5202	.5211	.5222	.5232	.4000	.4141	.4239	.4313	.4371
	5	.5959	.5806	.5704	.5629	.5572	.7072	.6814	.6638	.6508	.6408	.6000	.5859	.5761	.5687	.5629
	6	.7692	.7318	.7053	.6853	.6695	.8527	.8158	.7880	.7663	.7487	.7787	.7451	.7202	.7009	.6853
	7	.8952	.8552	.8233	.7972	.7755	.9422	.9118	.8852	.8622	.8423	.9044	.8701	.8416	.8176	.7972
	8	.9661	.9400	.9149	.8916	.8704	.9841	.9683	.9512	.9342	.9178	.9711	.9505	.9298	.9101	.8916
	9	.9941	.9853	.9741	.9615	.9484	.9977	.9935	.9874	.9799	.9715	.9954	.9892	.9810	.9717	.9615
	10	1.0000	1.0000	1.0000	1.0000	1.0000	1.0000	1.0000	1.0000	1.0000	1.0000	1.0000	1.0000	1.0000	1.0000	1.0000

n	k	A= 8 B= 8					A= 8 B=10					A=10 B=10				
		S=1	S=2	S=3	S=4	S=5	S=1	S=2	S=3	S=4	S=5	S=1	S=2	S=3	S=4	S=5
15	0	.0011	.0048	.0112	.0196	.0294	.0023	.0079	.0164	.0270	.0389	.0007	.0030	.0070	.0127	.0196
	1	.0071	.0207	.0380	.0564	.0747	.0139	.0329	.0544	.0762	.0971	.0051	.0147	.0274	.0416	.0564
	2	.0251	.0536	.0818	.1078	.1312	.0455	.0814	.1140	.1427	.1679	.0198	.0420	.0651	.0873	.1078
	3	.0641	.1069	.1424	.1716	.1960	.1079	.1556	.1927	.2223	.2467	.0547	.0907	.1219	.1487	.1716
	4	.1318	.1816	.2176	.2451	.2669	.2058	.2531	.2860	.3108	.3303	.1194	.1638	.1973	.2237	.2451
	5	.2311	.2755	.3048	.3260	.3422	.3352	.3674	.3886	.4041	.4162	.2190	.2602	.2885	.3096	.3260
	6	.3576	.3839	.4003	.4118	.4204	.4826	.4899	.4950	.4989	.5021	.3502	.3751	.3912	.4029	.4118
	7	.5000	.5000	.5000	.5000	.5000	.6301	.6111	.5995	.5917	.5860	.5000	.5000	.5000	.5000	.5000
	8	.6424	.6161	.5997	.5882	.5796	.7603	.7221	.6973	.6796	.6662	.6498	.6249	.6088	.5971	.5882
	9	.7689	.7245	.6952	.6740	.6578	.8615	.8163	.7842	.7600	.7410	.7810	.7398	.7115	.6904	.6740
	10	.8682	.8184	.7824	.7549	.7331	.9303	.8898	.8572	.8308	.8090	.8806	.8362	.8027	.7763	.7549
	11	.9359	.8931	.8576	.8284	.8040	.9705	.9418	.9146	.8902	.8688	.9453	.9093	.8781	.8513	.8284
	12	.9749	.9464	.9182	.8922	.8688	.9901	.9743	.9558	.9370	.9190	.9802	.9580	.9349	.9127	.8922
	13	.9929	.9793	.9620	.9436	.9253	.9976	.9914	.9820	.9706	.9584	.9949	.9853	.9726	.9584	.9436
	14	.9989	.9952	.9888	.9804	.9706	.9997	.9983	.9955	.9912	.9858	.9993	.9970	.9930	.9873	.9804
	15	1.0000	1.0000	1.0000	1.0000	1.0000	1.0000	1.0000	1.0000	1.0000	1.0000	1.0000	1.0000	1.0000	1.0000	1.0000
20	0	.0003	.0020	.0059	.0119	.0194	.0006	.0034	.0089	.0165	.0259	.0001	.0011	.0033	.0069	.0119
	1	.0019	.0089	.0204	.0344	.0497	.0041	.0148	.0300	.0473	.0653	.0011	.0055	.0131	.0231	.0344
	2	.0072	.0239	.0449	.0666	.0877	.0147	.0384	.0646	.0901	.1140	.0049	.0164	.0320	.0492	.0666
	3	.0200	.0496	.0797	.1073	.1319	.0383	.0769	.1121	.1428	.1694	.0149	.0374	.0618	.0854	.1073
	4	.0449	.0878	.1246	.1553	.1810	.0807	.1314	.1714	.2033	.2295	.0361	.0712	.1033	.1312	.1553
	5	.0865	.1392	.1788	.2095	.2340	.1458	.2013	.2403	.2697	.2929	.0740	.1200	.1564	.1856	.2095
	6	.1479	.2036	.2412	.2688	.2901	.2339	.2840	.3166	.3401	.3583	.1334	.1841	.2202	.2474	.2688
	7	.2299	.2792	.3102	.3320	.3485	.3413	.3758	.3975	.4128	.4246	.2159	.2626	.2931	.3151	.3320
	8	.3297	.3638	.3841	.3981	.4084	.4602	.4725	.4803	.4861	.4907	.3196	.3525	.3730	.3873	.3981
	9	.4418	.4539	.4610	.4658	.4694	.5810	.5691	.5626	.5585	.5558	.4381	.4500	.4572	.4621	.4658
	10	.5582	.5461	.5390	.5342	.5306	.6940	.6612	.6417	.6286	.6191	.5619	.5500	.5428	.5379	.5342
	11	.6703	.6362	.6159	.6019	.5916	.7912	.7449	.7157	.6951	.6797	.6804	.6475	.6270	.6127	.6019
	12	.7701	.7208	.6898	.6680	.6515	.8682	.8174	.7825	.7568	.7370	.7841	.7374	.7069	.6849	.6680
	13	.8521	.7964	.7588	.7312	.7099	.9240	.8768	.8409	.8128	.7903	.8666	.8159	.7798	.7526	.7312
	14	.9135	.8608	.8212	.7905	.7660	.9605	.9228	.8839	.8622	.8390	.9260	.8800	.8436	.8144	.7905
	15	.9551	.9122	.8754	.8447	.8190	.9821	.9559	.9291	.9044	.8823	.9639	.9288	.8967	.8688	.8447
	16	.9800	.9504	.9203	.8927	.8681	.9931	.9777	.9586	.9388	.9198	.9851	.9626	.9382	.9146	.8927
	17	.9928	.9761	.9551	.9334	.9123	.9979	.9905	.9791	.9654	.9509	.9951	.9836	.9680	.9508	.9334
	18	.9981	.9911	.9796	.9656	.9503	.9995	.9970	.9916	.9841	.9749	.9989	.9945	.9859	.9769	.9656
	19	.9997	.9980	.9941	.9881	.9806	.9999	.9994	.9979	.9953	.9915	.9999	.9989	.9957	.9931	.9881
	20	1.0000	1.0000	1.0000	1.0000	1.0000	1.0000	1.0000	1.0000	1.0000	1.0000	1.0000	1.0000	1.0000	1.0000	1.0000

25	0	.0001	.0010	.0035	.0079	.0140	.0002	.0017	.0054	.0112	.0187	.0000	.0005	.0018	.0043	.0079
	1	.0006	.0044	.0123	.0232	.0359	.0014	.0076	.0184	.0322	.0475	.0003	.0024	.0072	.0143	.0232
	2	.0024	.0122	.0274	.0452	.0636	.0054	.0202	.0403	.0520	.0835	.0014	.0074	.0178	.0308	.0452
	3	.0071	.0258	.0493	.0733	.0960	.0149	.0418	.0712	.0992	.1248	.0046	.0174	.0350	.0541	.0733
	4	.0163	.0469	.0782	.1068	.1323	.0334	.0737	.1107	.1428	.1703	.0118	.0343	.0596	.0841	.1068
	5	.0343	.0764	.1138	.1453	.1718	.0646	.1165	.1581	.1915	.2189	.0260	.0598	.0920	.1205	.1453
	6	.0623	.1148	.1558	.1880	.2141	.1111	.1701	.2124	.2445	.2699	.0506	.0953	.1324	.1628	.1880
	7	.1034	.1622	.2036	.2344	.2585	.1743	.2333	.2723	.3006	.3224	.0886	.1413	.1804	.2105	.2344
	8	.1587	.2181	.2565	.2839	.3047	.2532	.3045	.3363	.3588	.3759	.1426	.1978	.2352	.2627	.2839
	9	.2284	.2814	.3136	.3358	.3523	.3450	.3812	.4031	.4183	.4298	.2131	.2638	.2959	.3186	.3358
	10	.3109	.3508	.3739	.3895	.4010	.4449	.4611	.4710	.4781	.4836	.2988	.3378	.3613	.3775	.3895
	11	.4029	.4243	.4364	.4444	.4503	.5470	.5413	.5387	.5374	.5367	.3962	.4174	.4298	.4382	.4444
	12	.5000	.5000	.5000	.5000	.5000	.6454	.6193	.6048	.5954	.5888	.5000	.5000	.5000	.5000	.5000
	13	.5971	.5757	.5636	.5556	.5497	.7349	.6927	.6680	.6515	.6394	.6038	.5826	.5702	.5618	.5556
	14	.6891	.6492	.6261	.6105	.5990	.8116	.7595	.7272	.7048	.6881	.7012	.6622	.6387	.6225	.6105
	15	.7716	.7186	.6864	.6642	.6477	.8734	.8184	.7815	.7549	.7346	.7869	.7362	.7041	.6814	.6642
	16	.8413	.7819	.7435	.7161	.6953	.9202	.8683	.8301	.8012	.7784	.8574	.8022	.7648	.7373	.7161
	17	.8966	.8378	.7964	.7656	.7415	.9533	.9089	.8725	.8432	.8192	.9114	.8587	.8196	.7895	.7656
	18	.9377	.8852	.8442	.8120	.7859	.9749	.9405	.9084	.8805	.8567	.9494	.9047	.8676	.8372	.8120
	19	.9657	.9236	.8862	.8547	.8282	.9878	.9638	.9376	.9129	.8905	.9740	.9402	.9080	.8795	.8547
	20	.9832	.9531	.9218	.8932	.8677	.9948	.9799	.9604	.9401	.9204	.9882	.9657	.9404	.9159	.8932
	21	.9929	.9742	.9507	.9267	.9040	.9981	.9901	.9772	.9620	.9460	.9954	.9826	.9650	.9459	.9267
	22	.9976	.9878	.9726	.9548	.9364	.9994	.9959	.9886	.9786	.9671	.9986	.9926	.9822	.9692	.9548
	23	.9994	.9956	.9877	.9768	.9641	.9999	.9987	.9955	.9902	.9833	.9997	.9976	.9928	.9857	.9768
	24	.9999	.9990	.9965	.9921	.9860	1.0000	.9998	.9989	.9971	.9943	1.0000	.9995	.9982	.9957	.9921
	25	1.0000	1.0000	1.0000	1.0000	1.0000	1.0000	1.0000	1.0000	1.0000	1.0000	1.0000	1.0000	1.0000	1.0000	1.0000

5

Applications

5.1. INTRODUCTION

The contents of this chapter include several disconnected topics. Each concerns how urn models can be used in some more or less (according to individual interest) important field of application.

The fields of application in turn fall into two groups:

1. External applications to specific fields of scientific enquiry.
2. Applications to topics within the general framework of statistical inference and probability theory.

Topics in the first group are discussed in Sections 5.2 through 5.5, and those in the second, in Sections 5.6 through 5.9. In each case the treatment is only introductory but is intended to provide an insight into the nature of the problems associated with the topic under consideration.

Among the fields of application we have omitted we especially mention epidemiology. There is an extensive literature on this topic. We direct the reader's attention to Weiss (1965), Dietz (1966), Downton (1967), Itoh (1973), Abakuks (1974), and Kryscio (1974). A number of applications have also been mentioned incidentally in other chapters. (See, for instance, Example 2.3 in Section 2.3, Examples 3.1 and 3.2 in Section 3.1.1, and a remark after Exercise 3.14 in Section 3.2.1, and Example 3.9a in Section 3.5). Applications to botany, lexicology and numismatics are mentioned by Faucounau (1975).

5.2. APPLICATIONS IN GENETICS

5.2.1. Urn Models in Genetics*

Many models used in the mathematical theory of population genetics are equivalent to urn models and have often been discussed as such. The genes in the population correspond to balls in an urn, and the genetic type of a gene corresponds to the color of a ball. Questions raised by genetic studies often pose new problems in terms of urn models, not otherwise encountered in the urn context. There are two main reasons for this. First, one of the main aims of population genetics theory is the discussion of biological evolution, that is, the changes in the genetic constitution of a population during successive generations. In the context of urn models, this entails consideration of the color composition in a sequence of urns, the composition of each urn in the sequence being arrived at by a stochastic process from the composition of the previous urn. Such processes are not often considered in urn models. Second, in genetics one is seldom able to scrutinize the genetic composition of the entire population and instead takes a sample of genes, which is normally comparatively small in number. In the urn context, this raises questions of the color composition of a sample of balls taken from a very large number in the urn, again a problem not frequently discussed in the context of urns with stochastically varying compositions.

We now describe several genetic models, formulate them as urn models, and discuss the questions raised above in terms of urn model problems.

Consider a population comprising a fixed number m of genes in any generation. Each gene is of one out of two genetic types, A_1 and A_2; each generation will consist of i genes of type A_1 and $(m - i)$ genes of type A_2. Suppose the genetic composition of a daughter generation is derived by binomial sampling from the genes of the parent generation. Then, if in the parent generation there are i genes of type A_1, the probability p_{ij} that in the daughter generation there will be j such genes is

$$p_{ij} = \binom{m}{j}\left(\frac{i}{m}\right)^j\left(1 - \frac{i}{m}\right)^{m-j}. \tag{5.1}$$

While many models have been put forward to describe the genetic evolution of a population, here we consider only the above binomial model.

In terms of urn models, we imagine an urn containing m balls, each ball being either, say, red or green. A second urn is then filled with m balls as follows. Balls are drawn one by one, with replacement, from the first urn. As each ball is drawn, a new ball of the same color as that drawn is placed in the

* This section is based on a draft by W. J. Ewens.

second urn; this process continues until there are m balls in the second urn. Equation (5.1) yields the probability that the second urn will contain j red balls, given that the first urn contained i red balls. In accordance with the evolutionary processes considered in genetics, the first urn is then discarded, a third urn is filled by sampling from the second urn, and so on.

5.2.2. Limiting Cases

Evolutionary studies concern the long-term behavior of such a system. Normally, the mathematical apparatus used for such studies is the theory of Markov chains. However, several important questions can be discussed using only elementary arguments.

We first consider the conditional probability, given that the initial urn contains i red balls, that the urns will eventually contain only red balls. (In evolutionary studies, this is the probability that a certain type of gene will eventually take over in the entire population.) If at any stage all the balls in the urn are of one color (either red or green), they will remain so permanently. Whatever the initial proportion, there is a nonzero probability that this will happen even after one "generation." The probability that it will eventually happen is therefore 1.

Equation (5.1) shows that the expected number of red balls ($m \cdot im^{-1} = i$) does not change from one urn to the next. The limit of the expected number of red balls in urn t (as $t \to \infty$) is

$$0(1 - p) + mp = mp,$$

where p is the required probability. Equating this to the initial number of red balls gives

$$p = im^{-1}. \tag{5.2}$$

The probability that the urns will eventually contain only green balls is $1 - im^{-1}$.

We next consider the rate at which color uniformity is approached. Perhaps the most straightforward approach to this question is to find the mean number of urn replacements needed until uniformity is achieved. No simple explicit answer to this has been found.

EXERCISE 5.1. Show that, if E_i is the expected number of replacements needed to achieve color uniformity when initially there are i red and $(m - i)$ green balls in the first urn, then

$$E_i = 1 + \sum_{j=1}^{m-1} \binom{m}{j} \left(\frac{i}{m}\right)^j \left(1 - \frac{i}{m}\right)^{m-j} E_j.$$

Hence give a formal expression for E_i and evaluate E_i $(i = 1, 2, 3, 4)$ for $m = 5$. (*Answer*: $E_1 = E_4 = 2.58$; $E_2 = E_3 = 531E_1/537 = 2.44$.)

For realistically large values of m, the expression for E_i is completely unwieldy, even using the symmetry relation $E_i = E_{m-i}$.

We consider rather an approach common in the evolutionary study of organisms, where each individual carries two genes, namely (in the urn context), to find the probability F_t that two balls drawn at random from urn t are of the same color. Clearly, $F_t \to 1$ as $t \to \infty$, and the value of F_t gives a measure of color uniformity of the balls in urn t.

To calculate F_t, notice that two balls drawn from urn t will be of the same color if both are copies of the same ball in urn $t - 1$ (probability m^{-1}) or are copies of different balls in urn $t - 1$, which were of the same color. Evidently,

$$F_t = m^{-1} + (1 - m^{-1})F_{t-1}. \tag{5.3}$$

It follows easily from (5.3) (see, for example, Section 1.1.2; Example 1.10) that

$$F_t = 1 - (1 - m^{-1})^{t-1}(1 - F_1). \tag{5.4}$$

The value of F_1 is

$$\left\{\binom{i}{2} + \binom{m-i}{2}\right\}\bigg/\binom{m}{2} = \{i(i-1) + (m-i)(m-i-1)\}/\{m(m-1)\}.$$

Equation (5.4) shows that the loss of color variation is rather slow; in the genetic context, this fact is of great importance in reconciling the Mendelian mechanism of inheritance with the Darwinian theory of natural selection, since it shows that genetic variation tends to be preserved for long time intervals and thus allows ample time for selective forces to act.

Several generalizations of the above model are possible. Suppose, for example, that balls of various colors are allowed. If initially k of the balls are red, (5.2) through (5.4) still stand: note, however, that in (5.4) the calculation of F_1 takes account of the frequencies of all colors initially present.

EXERCISE 5.2. The first urn contains k_j balls of color $\mathscr{C}_j$ $(j = 1, 2, \ldots, s)$. What is the value of F_1?

$$\left[\textit{Answer}: \sum_{j=1}^{s}\binom{k_j}{2}\bigg/\binom{m}{2} = \{m(m-1)\}^{-1}\sum_{j=1}^{s}k_j(k_j - 1).\right]$$

A second generalization allows for the possibility of gene mutation or, in the urn context, of an error in copying the color of any ball from that of its "parent." Suppose, first, that two colors exist, red and green, and that a green ball will be miscopied as red with probability u, and a red ball miscopies as green with probability v. (In genetics, u and v are typically of the order 10^{-5}.)

In this case we are usually interested in the stationary distribution of the number of red balls (i.e., in urn t as $t \to \infty$), and the rate of approach to stationarity. Let μ_t be the mean number of red balls in urn t. Then,

$$\mu_t = \mu_{t-1}(1 - v) + (m - \mu_{t-1})u. \tag{5.5}$$

Solving the difference equation (5.5) we obtain

$$\mu_t = A(1 - v - u)^t + \frac{mu}{u + v}, \tag{5.6}$$

where A is an arbitrary constant defined by μ_1.

Setting $t \to \infty$, we find

$$\lim_{t \to \infty} \mu_t = \mu_\infty = \frac{mu}{u + v}. \tag{5.7}$$

Equation (5.7) gives the mean of the stationary distribution of the number of red balls, and (5.6) shows that the rate of approach to stationarity is geometric with rate $1 - u - v$; for values of u and v arising in genetics, this rate is extremely slow. Similar arguments show that the variance σ_∞^2 of the stationary distribution of the number of red balls is given, to a sufficient approximation, by

$$\sigma_\infty^2 = m^2uv[(u + v)^2(2mu + 2mv + 1)]^{-1}. \tag{5.8}$$

The complete stationary distribution of the number of red balls is not known exactly, although excellent approximations are available (Ewens, 1969, Chapter 6).

Equations (5.7) and (5.8) can be generalized to the case of an arbitrary finite number of colors.

EXERCISE 5.3. Generalize (5.7) to an arbitrary finite number of colors.

We again consider the probability F_t that two balls taken at random from urn t will be of the same color. For $u = v$, generalization of the argument leading to (5.3) yields

$$\begin{aligned} F_t = {} & \{(1 - u)^2 + u^2\}\{m^{-1} + (1 - m^{-1})F_{t-1}\} \\ & + 2u(1 - u)(1 - m^{-1})(1 - F_{t-1}). \end{aligned} \tag{5.9}$$

The solution of the difference equation (5.9) is of form

$$F_t = A\{(1 - m^{-1})(1 - 2u)^2\}^t + \frac{2u(1 - u) + (1 - 2u)^2m^{-1}}{1 - (1 - m^{-1})(1 - 2u)^2}.$$

Letting $t \to \infty$, we find an expression for $F = \lim F_t$:

$$\begin{aligned} F &= \frac{2mu(1-u) + (1-2u)^2}{m\{1-(1-2u)^2\} + (1-2u)^2} \\ &= \frac{1 + 2mu(1-u)(1-2u)^{-2}}{1 + m\{(1-2u)^{-2} - 1\}} \\ &\doteqdot (1 + 2mu)(1 + 4mu)^{-1} \qquad (5.10) \end{aligned}$$

(neglecting terms of second and higher order in u).

An analogous computation is possible for an arbitrary finite number of colors.

EXERCISE 5.4. Obtain the value of $F = \lim_{t\to\infty} F_t$ when there are s different colors and each color has a probability $u/(s-1)$ of being miscopied as any one of the other $(s-1)$ colors.

5.2.3. Mutations

It is also valuable in genetics to proceed to the limiting case which corresponds, in the urn context, to an *infinite* number of colors being possible. A model of such a case is the following. Balls are drawn one by one with replacement from the parent urn. As each ball is drawn, with probability $(1-u)$, a ball of the same color is placed in the "daughter" urn, while, with probability u, a ball of an entirely new color, not previously or currently seen during the history of the process, is placed in the daughter urn. The latter event corresponds to a *mutation*, in which an individual with completely new properties appears.

The properties of this process are rather different from those of the processes just considered. Thus no analogs of (5.7) or (5.8) exist, since *any* particular color *eventually* leaves the system never to reappear. Thus no nontrivial concept of stationarity exists for nominated colors.

However, there does exist a concept of stationarity for the *patterns* or configurations of color numbers, provided no specific color labels are attached to the numbers in the pattern. For example, if $M = 100$*, there will be a stationary probability for the pattern {73, 19, 5, 3} indicating 73 balls of one (unspecified) color, 19 of another, 5 of a third, and 3 of a fourth. While such probabilities are extremely hard to calculate, it is quite simple to find analogs and extensions of limiting results such as (5.10).

As a direct analog of (5.10) consider the probability F_t that two balls taken at random from urn t are of the same color. With the present model, this

* In this subsection the total number of balls in the urn is denoted for notational convenience by M.

requires that both balls be of the same color as their respective parents, and further that both have the same parent or two parents of the same color. Evidently [cf. (5.3) and (5.9)],

$$F_t = (1 - u)^2[M^{-1} + (1 - M^{-1})F_{t-1}],$$

whence

$$F_t = A\{(1 - u)^2(1 - M^{-1})\}^t + \frac{(1 - u)^2 M^{-1}}{1 - (1 - u)^2(1 - M^{-1})} \tag{5.11}$$

for some constant A. Letting $t \to \infty$, we obtain

$$\begin{aligned} F_\infty = F &= \frac{(1 - u)^2}{(1 - u)^2 + M(2u - u^2)} \\ &\doteq (1 + 2Mu)^{-1}, \end{aligned} \tag{5.12}$$

ignoring small-order terms.

Note in passing that (5.11) shows that the rate at which the stationarity of configurations is reached is geometric, with rate $(1 - u)^2(1 - M^{-1})$. This is a rather faster rate than that implied by (5.6) [or the extension of (5.6) for an arbitrary finite number of colors]. This fact is of some importance in evolutionary theory. For a full discussion, see Ewens and Kirby (1975).

From now on we no longer consider time-dependent probabilities (such as F_t) but concentrate solely on limiting stationary probabilities such as that given by (5.12). Thus in the rest of this section we suppose that the urn system has been in progress for sufficient time that stationarity has been effectively achieved. Further, we aim to generalize (5.12) to the case of three or more balls of the same color, and in so doing we are in effect taking steps toward the *sampling theory* of the process, that is, of the properties of the color configuration of a number m of balls, where m is considerably less than the total number M of balls in the urn.

It is convenient to introduce the notation $\theta = 2Mu$, and in place of F to use $F\{2\}$, indicating that it is the probability that *two* balls taken at random from the urn (at stationarity) are of the same color. Equation (5.12) may thus be rewritten

$$F\{2\} = (1 + \theta)^{-1}. \tag{5.13}$$

We now consider the probability $F\{3\}$ that three balls taken at random from the urn are of the same color. This event will occur if each ball is a faithful copy of its parent, and if the one, two, or three parent balls are of the same color. Thus, considering all possible parent combinations and their probabilities,

$$\begin{aligned} F\{3\} = (1 - u)^3[&M^{-2} + 3M^{-1}(1 - M^{-1})F\{2\} \\ &+ (1 - M^{-1})(1 - 2M^{-1})F\{3\}]. \end{aligned} \tag{5.14}$$

Ignoring small-order terms, (5.14) yields

$$F\{3\} \doteqdot 2[(1+\theta)(2+\theta)]^{-1}. \tag{5.15}$$

Continuing in this way, we find that, for sufficiently small integers i,

$$F\{i\} \doteqdot (i-1)![(\theta+1)^{[i-1]}]^{-1} \tag{5.16}$$

to a close approximation.

It is possible now to consider configurations where not all balls are of the same color. The above form of argument does not apply directly unless any color observed is observed on at least two balls (for otherwise it could be a new mutation). Consider then the probability $F\{2, 2\}$ that, if four balls are drawn at random, two will be of one color and two of another. Clearly, all four balls must be of the same color as their parents. Further, all four will derive from two parents (of different colors, each copied twice), from three parents (two of one color, one of another, the singleton copied twice), or from four parents (two of one color, two of another). Recurrence relations analogous to (5.14) show that to a sufficient approximation

$$F\{2,2\} \doteqdot 3\theta[(1+\theta)(2+\theta)(3+\theta)]^{-1}. \tag{5.17}$$

EXERCISE 5.5. Demonstrate (5.17).

We now state (without proof) a general formula for color configuration probabilities. If m ($m \ll M$) balls are drawn at random from the urn, the probability that k different colors, $\mathscr{C}_1, \mathscr{C}_2, \ldots, \mathscr{C}_k$, will be seen with m_1 balls of color $\mathscr{C}_1$, m_2 of color $\mathscr{C}_2, \ldots, m_k$ of $\mathscr{C}_k$ is

$$F\{m_1, \ldots, m_k\} = \frac{m!}{k!\,m_1 m_2 \cdots m_k} \cdot \frac{\theta^k}{\theta^{[m]}}. \tag{5.18}$$

Equation (5.18) was obtained heuristically by Ewens (1972) and proved rigorously by Karlin and McGregor (1972).

From (5.18), the probability that k colors are observed (without regard to frequencies) is

$$P\{k\} = \frac{m!\,\theta^k}{k!\,\theta^{[m]}} \sum \cdots \sum (m_1 m_2 \cdots m_k)^{-1}, \tag{5.19}$$

the summation being over $m_1 + m_2 + \cdots + m_k = m$ $(0 < m_i;\ i = 1, \ldots, k)$.

The right-hand side of (5.19) is of the form $A(k, m)\theta^k/\theta^{[m]}$ $(k = 1, \ldots, m)$. Since

$$\sum_{k=1}^{m} \frac{A(k,m)\theta^k}{\theta^{[m]}} \equiv 1,$$

it follows that

$$A(k, m) = \text{coefficient of } \theta^k \text{ in } \theta^{[m]}$$
$$= |S_m^{(k)}| \qquad [\text{see } (1.25')].$$

Hence

$$P\{k\} = \frac{|S_m^{(k)}|\theta^k}{\theta^{[m]}} \qquad (k = 1, \ldots, m). \tag{5.19'}$$

This is the distribution of K, the number of different colors observed. Note that if the balls were drawn one at a time, *with replacement*, K would have the classical occupancy distribution (3.6′).

Since $\sum_{k=1}^{m} P(k) = 1$, we have $\sum_{k=1}^{m} |S_m^{(k)}|\theta^k = \theta^{[m]}$, and so the moment generating function of K is

$$E[e^{tK}] = \frac{(\theta e^t)^{[m]}}{\theta^{[m]}}. \tag{5.20}$$

We find

$$E[K|m] = (\theta^{[m]})^{-1} \sum_{k=1}^{m} |S_m^{(k)}| k\theta^k = (\theta^{[m]})^{-1}\theta \cdot \frac{d\theta^{[m]}}{d\theta}$$
$$= \theta \frac{d \log \theta^{[m]}}{d\theta}$$
$$= \theta\{\theta^{-1} + (\theta + 1)^{-1} + \cdots + (\theta + m - 1)^{-1}\} \tag{5.21}$$

and

$$\text{var}[K|m] = E[K|m] - \theta^2\{\theta^{-2} + (\theta + 1)^{-2} + \cdots + (\theta + m - 1)^{-2}\}$$
$$= \theta\{(\theta + 1)^{-2} + 2(\theta + 2)^{-2} + \cdots + (m - 1)(\theta + m - 1)^{-2}\}. \tag{5.22}$$

Note that

$$\lim_{\theta \to 0} E[K|m] = 1$$

and

$$\lim_{\theta \to 0} \text{var}[K|m] = 0,$$

as would be expected. Also, $E[K|m]$ increases monotonically with θ, but $\text{var}[K|m]$ is maximized at intermediate values of θ.

EXERCISE 5.6. Evaluate numerically the maximum value of $\text{var}(K|m)$ for $m = 3$ and $m = 4$.

From (5.18) and (5.19′), we find

$$\Pr\left[\bigcap_{j=1}^{k}(M_j = m_j | K = k\right] = \frac{m!}{k!}\left\{|S_m^{(k)}|\prod_{j=1}^{k} m_j\right\}^{-1}. \tag{5.23}$$

Equations (5.18), (5.19), and (5.23) show that M and u enter into the sample probability distributions only through the parameter $\theta\,(= 2Mu)$, and further that K is a sufficient statistic for θ [because (5.23) does not depend on θ]. It follows from standard statistical theory that inference statements made from the sample about the unknown value of θ should be based on the value of K *only*. In particular, estimators of functions of θ should be functions of K only and not depend on $M_1, \ldots, M_K$. This result is of some interest in population genetics theory. The measure $(1 + \theta)^{-1}$ of genetic diversity [see (5.13)] was normally estimated by geneticists from the statistic

$$\frac{M_1^2 + M_2^2 + \cdots + M_K^2}{M^2}.$$

This estimator was known to have poor sampling properties, despite its apparently commonsense appeal. The above observations show the reason for this, namely, that the statistic uses precisely the wrong part of the data vector to estimate $(1 + \theta)^{-1}$. It can be shown that the statistic $G(K)$, defined by

$$G(K) = \frac{\text{coeff } \theta^{K-1} \text{ in } (\theta + 2)(\theta + 3)\cdots(\theta + m - 1)}{\text{coeff } \theta^{K-1} \text{ in } (\theta + 1)(\theta + 2)\cdots(\theta + m - 1)}, \tag{5.24}$$

is the unique minimum variance unbiased estimator of $(1 + \theta)^{-1}$. In practice, $G(K)$ is difficult to compute, and it may be preferable to estimate $(1 + \theta)^{-1}$ by $(1 + \hat{\theta})^{-1}$, where $\hat{\theta}$ is the maximum likelihood estimator of θ. Differentiation in (5.19′) shows that $\hat{\theta}$ is defined implicitly by the equation

$$K = \frac{\hat{\theta}}{\hat{\theta}} + \frac{\hat{\theta}}{\hat{\theta} + 1} + \cdots + \frac{\hat{\theta}}{\hat{\theta} + m - 1}, \tag{5.25}$$

which bears an interesting resemblance to (5.21). Tables of $\hat{\theta}$ for selected values of K and m are given in Ewens (1972, Table III).

We conclude this discussion by considering the probability that on the $(m + 1)$th draw a ball is drawn of a color not seen on draws $1, 2, \ldots, m$. Suppose specifically that the first m balls drawn yield K different colors, with respective numbers $m_1, \ldots, m_K$. Application of (5.18) shows that

$$\Pr[\text{new color on draw } (m + 1) | K, m_1, \ldots, m_K] = \frac{\theta}{\theta + m}. \tag{5.26}$$

Note that this probability is independent of $K, m_1, \ldots, m_K$. In other words, for example, if the first 100 balls drawn are all of different colors, the probability that a new color is seen on draw 101 is the same as if the first 100 balls were all of the same color. It follows that, if we define the event E_i as "new color observed on ith ball drawn," the events $E_2, E_3, \ldots, E_m$ are independent. After some thought, these initially somewhat surprising facts can be reconciled with intuition.

Much of the preceding discussion relates to stationary configurations. In practice (in particular in evolutionary theory) it is important to consider the rate at which stationarity is achieved. This question is fully discussed in Ewens and Kirby (1975) and is not further considered here.

EXERCISE 5.7. Derive (5.25) for the maximum likelihood estimator of θ.

5.3. CAPTURE-RECAPTURE MODELS

5.3.1. Classic Models

The name *capture-recapture* is given to a whole class of methods used to estimate the sizes of natural populations. Many stem from papers by Schnabel (1938) and Chapman (1952), which discuss the estimation of the number of fish f in a lake. The method proposed was essentially:

1. Catch n_0 fish, tag them, and return them to the lake.
2. Catch a new sample of n_1 fish and observe the number, say T_1, of tagged fish.
3. Equate the proportions of tagged fish in the lake before taking the second sample, and in the second sample, giving

$$\frac{n_0}{f} = \frac{T_1}{n_1} \tag{5.27}$$

and leading to the estimator

$$\hat{f} = \frac{n_0 n_1}{T_1} \tag{5.28}$$

for f.

Since $\hat{f}$ can be infinite (if $T_1 = 0$), we seek some small modification in $\hat{f}$. If it is assumed that

1. All sampling is random.
2. Tagging does not affect the chance of being caught in a subsequent sample.

3. There are no changes (by births, deaths, or migration) in the fish population between the first and second samples.

Then T_1 is expected to have a hypergeometric distribution (Section 2.3), with

$$\Pr[T_1 = t_1] = \frac{\binom{n_0}{t_1}\binom{f - n_0}{n_1 - t_1}}{\binom{f}{n_1}}.$$

It can be shown (see, for example, Johnson and Kotz, 1969, p. 145) that

$$E[(T_1 + 1)^{-1}] = \begin{cases} \dfrac{f+1}{(n_0+1)(n_1+1)}\left[1 - \dfrac{(f-n_0)!(f-n_1)!}{(f+1)!(f-n_0-n_1-1)!}\right] & \text{if } f \geq n_0 + n_1 + 1, \\ \dfrac{f+1}{(n_0+1)(n_1+1)} & \text{if } f < n_0 + n_1 + 1, \end{cases}$$

and so

$$\hat{f}^* = (n_0 + 1)(n_1 + 1)(T_1 + 1)^{-1} - 1 \tag{5.29}$$

is a very nearly unbiased estimator of f. When n_0, n_1, and T_1 are large, there is very little difference between $\hat{f}$ and $\hat{f}^*$.

Confidence intervals for f with approximate confidence coefficient $(1 - \alpha' - \alpha'')$ can be constructed by solving for f the equations

$$\sum_{t_1 \geq T_1} \binom{n_0}{t_1}\binom{f - n_0}{n_1 - t_1} = \alpha' \cdot \binom{f}{n_1} \qquad \text{(for the lower limit)},$$

$$\sum_{t_1 \leq T_1} \binom{n_0}{t_1}\binom{f - n_0}{n_1 - t_1} = \alpha'' \cdot \binom{f}{n_1} \qquad \text{(for the upper limit)}.$$

When *no* tagged fish are in the sample of size n_1 it is clearly impossible to estimate f, beyond perhaps getting a lower bound using the first of the two above formulas.

Bell (1974) has suggested, as a purely empirical method, obtaining an "estimate" of f by finding the value which would give a probability of $\frac{1}{2}$ of obtaining no tagged fish—that is, solving for f the equation

$$\binom{f - n_0}{n_1} = \frac{1}{2}\binom{f}{n_1}$$

or equivalently $(f - n_0)^{(n_1)} = \frac{1}{2}f^{(n_1)}$.

Approximating $x^{(r)}$ by $[x - \frac{1}{2}(r - 1)]^r$ gives the approximate value

$$\tilde{f} = n_0\{1 - (\tfrac{1}{2})^{1/n_1}\} + \tfrac{1}{2}(n_1 - 1).$$

Application of calculations by Harris (1975), (who presents a formula omitting the last term in the above expression for $\tilde{f}$) indicates that this is quite a good approximation.

Of course this "estimate" of f can be very inaccurate—it is a 50% lower bound.

There is a natural, and even self-evident, analogy to the above problems in terms of urns and balls. We start with an urn containing f indistinguishable white balls. We take n_0 balls from the urn (without replacement), dye each one black, and then return the black balls to the urn. We then take a random sample (again without replacement) of n_1 balls from the urn [which contained $(f - n_0)$ white and n_0 black balls], and observe the number T_1 of black balls. It is required to estimate f, the number of balls originally in the urn.

There is clearly much oversimplification in this model. It does not allow for natural increase (by birth or perhaps immigration) or decrease (by death or perhaps emigration). It supposes that a genuine random sample of fish is attainable and also (see above) that tagging does not affect the catchability of a fish, relative to the rest of the fish in the lake, on a future occasion.

EXERCISE 5.8. An urn contains an unknown number w of white balls. n_0 black balls are *added* to the urn. A random sample of n_1 balls is then taken (without replacement) from the urn, and the number W_1 of white balls observed. Construct an estimator for w.

EXERCISE 5.9. Solve the problem described in Exercise 5.8, but supposing the sample of n_1 balls is chosen *with replacement*.

Suppose now that, after the first sample of size n_0 has been tagged and returned, we take samples *one at a time* (i.e., $n_1 = n_2 = \cdots = n_k = 1$). Let

$$T_i = \begin{cases} 1 & \text{if the } i\text{th sample is untagged when selected,} \\ 0 & \text{if it is tagged.} \end{cases}$$

Then the likelihood function for the variables $T_1, T_2, \ldots, T_k$ is

$$L = \prod_{i=1}^{k} \left(\frac{f - n_0 - T_1 - \cdots - T_{i-1}}{f} \right)^{T_i} \left(\frac{n_0 + T_1 + \cdots + T_{i-1}}{f} \right)^{1-T_i}. \tag{5.30}$$

Hence

$$\begin{aligned} \ln L = &\sum T_i \ln(f - n_0 - T_1 - \cdots - T_{i-1}) \\ &+ \sum (1 - T_i) \ln(n_0 + T_1 + \cdots + T_{i-1}) \end{aligned}$$

and

$$\frac{\partial \ln L}{\partial f} = \sum_{i=1}^{k} T_i (f - n_0 - T_1 - \cdots - T_{i-1})^{-1} - kf^{-1}.$$

The equation $\partial \ln L/\partial \hat{f} = 0$ can be written

$$\sum_{i=1}^{k} T_i(\hat{f} - n_0 - T_1 - \cdots - T_{i-1})^{-1} = k\hat{f}^{-1}.$$

Equivalently, this can be written

$$\sum_{j=0}^{R-1} (\hat{f} - n_0 - j)^{-1} = k\hat{f}^{-1}, \tag{5.31}$$

where $n_0 + R$ = total number tagged in the whole experiment [cf. Craig (1953), who takes $n_0 = 0$.]

If there is a value of $\hat{f}$ $(\geq n_0 + R)$ satisfying (5.31), it is the maximum likelihood estimator.

EXERCISE 5.10. Show that (5.31) has a unique root $\hat{f} \geq n_0 + R$, provided

$$k \leq (R + n_0) \sum_{j=1}^{R} j^{-1}. \tag{5.32}$$

More generally, if $n_0, n_1, \ldots, n_k$ represent the sizes of samples taken on $(k+1)$ successive occasions, and $U_0, U_1, \ldots, U_k$ the corresponding numbers of individuals not as yet tagged (with of course $U_0 = n_0$), then $U = \sum_{j=0}^{k} U_j$ is a sufficient statistic for the population size f.

The sum $\sum_{j=0}^{k} U_j$ is the number of different individuals included in the $(k+1)$ samples. We can identify this with the number of different individuals on $(k+1)$ committees of sizes $n_0, n_1, \ldots, n_k$ chosen independently from a population of size f. From the right-hand side of (3.111), putting $n_j = w_j$, $n = f$, $m = u$, and changing the range of i from $(1, r)$ to $(0, k)$ we find

$$\Pr[U = u] = \binom{f}{u} \sum_{j=0}^{u} (-1)^{u-j} \binom{u}{j} \prod_{i=0}^{k} \left[\frac{\binom{j}{n_i}}{\binom{f}{n_i}} \right]$$

$$= \binom{f}{u} \Delta^u \prod_{i=0}^{k} \left[\frac{\binom{\mathbf{0}}{n_i}}{\binom{f}{n_i}} \right]$$

$$= \frac{f^{(u)}}{u!} \frac{\Delta^u \prod_{i=0}^{k} \binom{\mathbf{0}}{n_i}}{\prod_{i=0}^{k} \binom{f}{n_i}} \qquad [u \geq \max(n_0, n_1, \ldots, n_k)]. \tag{5.33}$$

We recognize this as a *factorial series distribution* (see Section 2.6.2), which can be obtained from (2.38) by putting $\theta = f$ and $f(\theta) = \prod_{i=0}^{k} \binom{\theta}{n_i}$. Hence, from (2.42), (with $r = 1$)

$$E[U] = f \frac{\left[\prod_{i=0}^{k} \binom{f}{n_i} - \prod_{i=0}^{k} \binom{f-1}{n_i}\right]}{\prod_{i=0}^{k} \binom{f}{n_i}} = f\left[1 - \prod_{i=0}^{k} \left(1 - \frac{n_i}{f}\right)\right].$$

EXERCISE 5.11. Show that the minimum variance unbiased estimator of f is

$$u\left[1 + \frac{\left\{\Delta^{u-1} \prod_{i=0}^{k} \binom{0}{n_i}\right\}}{\left\{\Delta^{u} \prod_{i=0}^{k} \binom{0}{n_i}\right\}}\right] \qquad [u > \max(n_0, n_1, \ldots, n_k]. \tag{5.34}$$

[Note that, if $k = 1$, we obtain the Chapman estimator (5.29) with a small correction, namely, that $\hat{f}^* = \max(n_0, n_1)$ if $U = \max(n_0, n_1)$—which makes it unbiased. Very often n_0 is much greater than n_1 and so $\max(n_0, n_1) = n_0$.]

EXERCISE 5.12 (Schnabel, 1938). Suppose that after the first tagging k successive samples are taken, of sizes $n_1, n_2, \ldots, n_k$. On each occasion, the number $T_1, T_2, \ldots, T_k$ of tagged fish is noted. Except for the last sample, any untagged fish are tagged, and all fish are returned to the lake. The likelihood function is then

$$L = \prod_{i=1}^{k} \binom{n_i}{T_i} \left(\frac{m_i}{f}\right)^{T_i} \left(1 - \frac{m_i}{f}\right)^{n_i - T_i},$$

where $m_i = n_0 + \sum_{j=1}^{i-1} (n_j - T_j)$ = number of tagged fish in the lake when the $(i + 1)$th sample is drawn, if we assume f so large that the hypergeometric distribution can be replaced by the binomial distribution.

The value of f that maximizes L satisfies the equation

$$\sum_{i=1}^{k} \frac{m_i(n_i - T_i)}{f - m_i} = \sum_{i=1}^{k} T_i. \tag{5.35a}$$

[Note that T_i is equal to $(n_i - U_i)$ in the notation used following Exercise 5.10.]

Example 5.1. Table 5.1 shows some data from Seber (1973).

TABLE 5.1

Sample no. (i)	n_i	U_i	$T_i = n_i - U_i$	$\hat{f}_i^*$
0	32	32	0	—
1	54	36	18	95
2	37	6	31	81
3	60	13	47	94
4	41	5	36	101

The last column shows the value of the estimator $\hat{f}_i^*$ [from (5.29)] using the current sample n_i and the total previously tagged $(U_0 + U_1 + \cdots + U_{i-1})$ in place of n_1 and n_0, respectively; that is,

$$\hat{f}_i^* = (U_0 + U_1 + \cdots + U_{i-1} + 1)(n_i + 1)(n_i - U_i + 1)^{-1} - 1.$$

The minimum variance unbiased estimate (5.34) is 94.5. (The unweighted average of the $\hat{f}_i^*$'s is 93.)

The maximum likelihood equation (5.35a) is, for these data,

$$\frac{32 \times 36}{f - 32} + \frac{68 \times 6}{f - 68} + \frac{74 \times 13}{f - 74} + \frac{87 \times 5}{f - 87} = 18 + 31 + 47 + 36$$

or

$$1152(f - 32)^{-1} + 528(f - 68)^{-1} + 962(f - 74)^{-1} + 435(f - 87)^{-1} - 132 = 0.$$

If $f^{(1)}$ is an approximate solution of the equation $h(f) = 0$, then an improved solution $f^{(2)} = f^{(1)} + \delta^{(1)}$ can usually be obtained by equating $h(f^{(1)}) + \delta^{(1)}h'(f^{(1)})$ to zero, that is, taking $\delta^{(1)} = -h(f^{(1)})/h'(f^{(1)})$.

Using $f^{(1)} = 93$ as an initial approximation, we obtain

$$h(93) = 1152(61^{-1}) + 528(25^{-1}) + 962(19^{-1}) + 435(6^{-1}) - 132 = 31.14.$$

The derivative

$$h'(93) = -1152(61^{-2}) - 528(25^{-2}) - 962(19^{-2}) - 435(6^{-2}) = -15.90,$$

whence $\delta^{(1)} = 31.14/15.90 = 1.96$ and $f^{(2)} = 95$. We could repeat the process or just calculate $h(95) = 6.03$; $h(96) = -3.08$, indicating a maximum likelihood estimate of 95 or 96 (recall that f is an integer).

The theory described earlier in this section can be modified to allow for many features of practical interest, though such modifications usually add to the complexity of the analysis. In particular, we may wish to allow for variation in f with time—from births, death, immigration, and emigration. Also, we may wish to allow for the possibility that a tagged fish may be more (or less) easily caught than an untagged fish. (More generally, that inclusion in a previous sample may affect the chance of inclusion in a subsequent sample.)

Suppose, for example, that tagged fish are θ times as likely to be caught as untagged fish, but that all the other assumptions are valid.

The same results, in terms of numbers of tagged fish, would be produced if each chosen tagged (for $\theta < 1$) or untagged (for $\theta > 1$) fish is rejected with probability $|1 - \theta|(1 + \theta)^{-1}$ and replaced by an untagged (or tagged) fish according as $\theta < 1$ or $\theta > 1$. The reader might consider possible application of the randomized occupancy distribution, described in Section 3.4, in this case.

Pollock (1975) has considered a situation in which the effect of tagging disappears after a specified number of subsequent catches.

EXERCISE 5.13. Using the data in Table 5.1, but supposing that the number of fish in the lake varies linearly with i (the sample number), so that the number when the ith sample is taken is $f_0 + ia$, obtain approximate maximum likelihood estimators of f_0 and a.

5.3.2. An Alternative Model (an application in epidemiology).

Wittes et al. (1974) use a model which allows for random variation in the number of balls included in each sample. It is intended to represent the results of combining several independently collected random samples (each without replacement, but with all individuals available for each sample). They suppose that each ball has the same chance of being included in the ith sample, independently of whether or not it is included in one or more other samples, and independently of whether other balls are in the same sample or not. Consequently, the size of the ith sample (N_i) is not prespecified, but it is represented by a binomial random variable with parameters f (the total population size) and p_i (the probability of any one ball being included in the ith sample). As before, we let U denote the total number of different balls in the $(k + 1)$ samples.

The likelihood function is

$$L(f, \{p_i\}) = \left\{\prod_{i=0}^{k} \binom{f}{N_i} p_i^{N_i}(1 - p_i)^{f-N_i}\right\} P(U \mid N_0, N_1, \ldots, N_k),$$

where

$$P(U \mid N_0, N_1, \ldots, N_k) = \text{(probability that } (k+1) \text{ committees of sizes } N_0, N_1, \ldots, N_k \text{ formed randomly from } f \text{ persons will include just } U \text{ persons)}$$

$$= \binom{f}{U} \left\{ \prod_{i=0}^{k} \binom{f}{N_i} \right\}^{-1} \Delta^U \prod_{i=0}^{k} \binom{\mathbf{0}}{N_i}$$

$$(U \geq \max(N_1, \ldots, N_k))$$

(cf. (5.33)). Hence

$$L(f, \{p_i\}) = \binom{f}{U} \prod_{i=0}^{k} (1 - p_i)^{f - n_i} \prod_{i=0}^{k} p_i^{n_i} \Delta^U \prod_{i=0}^{k} \binom{\mathbf{0}}{N_i}. \tag{5.35b}$$

From the conditions $\partial L/\partial \hat{p}_i = 0$ we obtain the maximum likelihood equation

$$\hat{p}_i = \frac{N_i}{\hat{f}} \tag{5.36}$$

and, by equating $L(\hat{f}, \{\hat{p}_i\})$ to $L(\hat{f} - 1, \{p_i\})$, we obtain

$$\prod_{i=0}^{k} (1 - \hat{p}_i) \doteqdot \frac{\hat{f} - U}{\hat{f}}. \tag{5.37}$$

Eliminating the p_i's between (5.36) and (5.37), we obtain (approximately)

$$\prod_{i=0}^{k} (\hat{f} - N_i) \doteqdot \hat{f}^k (\hat{f} - U). \tag{5.38}$$

Since the function

$$g(f) = \prod_{i=0}^{k} (f - N_i) - f^k (f - U) \tag{5.39}$$

is positive when $f = U$ and tends to $-\infty$ as f tends to infinity [provided $\sum_{i=0}^{k} N_i > U > \max(N_0, \ldots, N_k)$], it follows that there is a value of $\hat{f}$, greater than U, that satisfies (5.38). Wittes et al. (1974) (see also Wittes and Sidel, 1968) state that there is only one such value of $\hat{f}$.

The distinction between the model used here and the one described at the beginning of Section 5.3.1 is that, in the former, inferences are made *conditionally on the actually realized sample sizes* $n_0, n_1, \ldots, n_k$. In this section, variation in these sample sizes is allowed for. The difference has little effect on the estimation of f itself, but it can have noticeable effect on the estimated variance of the estimate of f. In general, we prefer to make *inferences* conditional on actual realized sample size, though expected variances are important in the *planning* stages.

Wittes et al. (1974) discuss the likely effect of dependence between the events of a ball belonging to the ith set and to the jth set ($i \neq j$). If there is positive dependence, there will be a greater overlap (i.e., proportion of common elements) between pairs of sets than would be expected on the basis of independence, and conversely if the dependence is negative.

This means that, for given $N_0, N_1, \ldots, N_k$, the value of U tends to be less (greater) if the dependence is positive (negative). This tends to increase (decrease) the right-hand side of (5.38) and so to decrease (increase) the value of $\hat{f}$. That is, positive (negative) dependence tends to lead to under- (over-) estimation of population size.

EXERCISE 5.14. Show that, when $k = 1$ (i.e., two samples), (5.38) leads to

$$\hat{f} = \frac{n_0 n_1}{n_0 + n_1 - U}.$$

EXERCISE 5.15. When the sample size is small compared with f, Wittes has suggested that the equation

$$\prod_{i=0}^{k} (\hat{f} - N_i) = (\hat{f} + 1)^k (\hat{f} - U) \tag{5.38'}$$

is preferable to (5.38). Show that, when $k = 1$, and using (5.38′), we obtain Chapman's estimator (5.29).

Darling and Robbins (1967) have described methods of sequential tagging with a stopping rule designed so that the proportional error in estimation of sample size has a specified value.

We now describe their methods in terms of urn models. In the first method balls are drawn one at a time, their color observed, and they are then returned (if black), or replaced by a black ball (if white). (Of course, all balls are white initially.)

After the Jth white ball has been replaced by a black ball sampling continues (one ball at a time). If a sequence of more than b_j black balls is obtained, sampling ceases forthwith and the population size is estimated as J (which is of course a random variable).

Darling and Robbins consider ways of choosing the constants $b_1, b_2, \ldots$ to best advantage and suggest a stepwise procedure defining b_j (given $b_{j-1}, b_{j-2}, \ldots, b_1$) as the smallest integer satisfying

$$1 - \left(\frac{j}{j+1}\right)^{b_j} \geq \alpha \prod_{i=1}^{j-1} \left\{1 - \left(\frac{i}{j+1}\right)^{b_i}\right\}^{-1}, \tag{5.40}$$

where α is the specified probability that the estimated population size is correct. (Equivalently, all the white balls originally in the urn have been replaced by black balls.)

In order to reduce the needed number of drawings, Darling and Robbins suggested a modification of their method in which sampling stops if the number of drawings when J black balls have been placed in the urn, but before $(J + 1)$ black balls have been placed in the urn, exceeds

$$a_J = \begin{cases} J + 1 & \text{for } J < \exp(1 - c), \\ J \ln J + cJ & \text{for } J \geq \exp(1 - c). \end{cases}$$

The constant c can be chosen arbitrarily. As before, the value of J is then taken as the estimator of population size. Darling and Robbins show that the probability that J equals the population size is asymptotically $\exp(-e^{-c})$. Clearly the sample size cannot exceed a_f.

This type of method is inappropriate unless it is expected that *all* the individuals in the population can be tagged in a reasonable amount of time. Thus it is usually inappropriate when the population is at all large.

A convenient stopping rule for sequential tagging is to stop when the total number of previously tagged individuals chosen in the sequence of samples attains a specified number, say ϕ. Goodman (1953) has shown that, if one individual is chosen at a time (and then tagged—if not already tagged—and replaced), then the distribution of the number of drawings N needed to observe ϕ tagged individuals is such that $N^2 f^{-1}$ is approximately (for large f) distributed as χ^2 with 2ϕ degrees of freedom.

Alternatively, after n_0 individuals (out of the population of size f) have been marked, we can sample (without replacement) until a predetermined number, k say, of marked individuals have been recovered. The distribution of the total number of individuals (Y) selected to achieve this is a negative hypergeometric [Section 2.5, equation (2.31)] with parameters n_0, $f - n_0$, and k.

From (2.33)

$$E[Y] = \frac{k(f + 1)}{n_0 + 1}$$

and

$$\text{var}(Y) = \frac{k(f + 1)(f - n_0)(n_0 - k + 1)}{(n_0 + 1)^2(n_0 + 2)},$$

whence $k^{-1}(n_0 + 1)Y$ is an unbiased estimator of f with variance

$$k^{-1}(n_0 + 2)^{-1}(f + 1)(f - n_0)(n_0 - k + 1).$$

Bunday (1975) points out that if $k \ll n_0 \ll f$, the standard deviation of the estimator is approximately $fk^{-1/2}$. This decreases as k increases, but the amount of sampling increases proportionately to k.

Doctor (1976) has studied the properties of certain stopping rules in Pólya–Eggenberger sampling, for making inferences on the initial proportion of white balls in an urn. By supposing we have an urn containing an unknown number of black balls and add an initial (known) number of white balls, we can make Doctor's results applicable to construction of certain capture-recapture procedures.

5.3.3. Estimation of Population Size without Using Capture-Recapture Methods

If it is possible to ascertain exactly the population in a number of disjoint, well-defined subareas, then the total population can be estimated by multiplying the total observed number by the ratio (total area)/(total of subareas). This is based on the assumption of a random distribution of population among the subareas (with average population proportional to area). Binns (1976) has described a sequential method of measurement intended to avoid unnecessary continuation of sampling once sufficient accuracy has been attained.

The analog in terms of urn models is to suppose that f balls are distributed randomly among m urns (each urn being equally likely to receive each ball). The urns are selected one at a time until the total number of balls in the first t chosen urns is at least $\phi(1 - t/m)$, where ϕ is a constant chosen so as to ensure the required accuracy.

The number of balls in a specified set of t urns is distributed binomially with parameters $f, t/m$. The probability that this will be at least $\phi(1 - t/m)$ is given by the incomplete beta function

$$I_{t/m}\left(\phi\left(1 - \frac{t}{m}\right), f - \phi\left(1 - \frac{t}{m}\right) + 1\right) \qquad \text{[see (2.4)]}.$$

If T is the total number of urns selected under the procedure just described, then

$$\begin{aligned}\Pr[T > t] &= 1 - I_{t/m}\left(\phi\left(1 - \frac{t}{m}\right), f - \phi\left(1 - \frac{t}{m}\right) + 1\right)\\ &= I_{1-t/m}\left(f - \phi\left(1 - \frac{t}{m}\right) + 1, \phi\left(1 - \frac{t}{m}\right)\right).\end{aligned} \tag{5.41}$$

This determines the distribution of T.

The suggested estimator of population size f is

$$\tilde{f} = \frac{m}{T} \times \text{(total number of } \textit{balls} \text{ in the urns chosen when sampling ceases).} \tag{5.42}$$

Binns (1976) has shown that

$$\Pr\left[\frac{\tilde{f}}{f} \le y\right] \doteqdot \Phi\left\{\sqrt{\phi}\left(\sqrt{y} - \frac{1}{\sqrt{y}}\right)\right\}, \tag{5.43}$$

where $\Phi(\cdot)$ is the unit normal cdf defined in (1.43).

EXERCISE 5.16. Show that [assuming (5.43)]

$$\begin{aligned} E[\tilde{f}] &\doteqdot f(1 + \tfrac{1}{2}\phi^{-1}), \\ \operatorname{var}(\tilde{f}) &\doteqdot f^2(\phi^{-1} + \tfrac{5}{4}\phi^{-2}). \end{aligned} \tag{5.44}$$

5.4. APPLICATION OF URN MODELS TO LEARNING PROCESSES

5.4.1. Learning Curves

One of the most intuitively appealing applications of urn models is in the field of mathematical psychology, in particular, in stochastic learning theory.

The process of learning is "notoriously mysterious," and there is a voluminous research literature on this subject. Many current mathematical models of learning originated in the works of Estes (1950) and Bush and Mosteller (1951), although much earlier Thurstone (1919) proposed the notion of a learning function via an urn model and later modified it (Thurstone, 1930). Even though these models do not now play a central role in learning theory, they are indicative of the historical development of the topic. Much of our exposition in Sections 5.4.1–5.4.2 is based on the ideas presented by Restle and Greeno (1970) and Goldberg (1961).

The notion of a learning curve measures the "performance" of an individual as a function of training time or trials. The performance can be (often conveniently, visualized as the probability of a correct response in a multiple-choice problem.

Urn models serve as a useful tool on the elementary level in determining learning curves. This is discussed in the present section. In the following section we use urn models to describe stochastic learning models.

The two elementary mathematical models of learning, involving learning curves, which are most commonly used in practice are: (1) the simple *replacement* model and (2) the simple model of *accumulative* learning. Both can be expressed as urn models (or, equivalently, as a Markov chain model). In our opinion the urn model has the significant advantages of simplicity and visual clarity in the present context. Further, it allows several simpler mathematical derivations.

Moreover, as in other cases, the urn scheme helps to make concrete one's intuitive ideas about learning processes.

Replacement Learning (Linear Operator Model)

In replacement learning some habits (or "stimulus elements") are replaced by others. In the simplest case the total number of entities is constant, and they are substituted one by one. The process of learning is quantitatively summarized by the number of elements of various types and by the numbers and kinds of elements transferred from one type (or state) to another.

These transfers are conveniently visualized by a model consisting of urns and balls, and calculations from such a model represent—to a high degree of approximation—the flow of habits, provided the assumption of random sampling is satisfied.

A subject in a learning experiment is regarded as being associated with an urn S containing black and white balls. On each trial balls are taken from the urn. A black ball is supposed to correspond to a correct response, and a white ball to an error. Let m be the number of balls in the urn. According to the "replacement" idea, m is fixed and *learning is regarded as a process of replacing white balls by black ones.*

In addition to the subject's urn S, we also assume the existence of an experimenter's urn E with an "infinite" (very large) number of balls corresponding to an inexhaustible source of new habits. A proportion $a\ (\leq 1)$ of the balls in urn E is black. This proportion is fixed and, if $a < 1$, then the training process is imperfect. The simple replacement learning process consists in selecting some fixed number k of balls from urn S at each trial and replacing them with an equal number of balls from urn E so that the number of balls in urn S remains at m. The sampling from both urns S and E is supposed to be *random.* Let B_n and W_n be the number of black and white balls in urn S on the nth trial. Clearly,

$$E[B_{n+1} \mid B_n] = B_n(1 - km^{-1}) + ka, \tag{5.45}$$

and so

$$E[B_{n+1}] = E[B_n](1 - km^{-1}) + ka \tag{5.46}$$

(since the balls taken from urn S on the average include kB_n/m black balls and kW_n/m white balls, and the k balls added to urn S consist on the average of ka black balls and $k(1 - a)$ white balls).

The probability of a correct response on trial n is $P_n = B_n/m$ (which is the proportion of black balls in urn S).

Let $\theta = k/m$ be the proportion sampled from urn S. Then,

$$E[P_{n+1}|P_n] = E[m^{-1}B_{n+1}|P_n]. \tag{5.47}$$

From (5.45),

$$E[P_{n+1}|P_n] = (1-\theta)P_n + \theta a.$$

This explains the alternative name, *linear model.*

We also evidently have (taking expected values with respect to P_n)

$$E[P_{n+1}] = E[P_n](1-\theta) + \theta a. \tag{5.48}$$

From (5.48), it follows (see Section 1.1.2) that the general solution to the initial *single linear model* with P_1 (the initial proportion of correct responses) equal to b is

$$E[P_n] = a - (a-b)(1-\theta)^{n-1}. \tag{5.49}$$

Note that, as $n \to \infty$, P_n tends to the asymptotic value a. [Bower (1961) gives some numerical results.]

EXERCISE 5.17. Assume $a = 1$, $b = .5$ (equal probability of error and correct response at the beginning). Show that the expected number of correct responses in n trials is

$$n = \frac{1}{2}\left\{\frac{1-(1-\theta)^n}{\theta}\right\}.$$

[This is approximately $n\{1-(2\theta)^{-1}\}$ when θ is small.]

Accumulation Model

This is essentially the model suggested by Thurstone in 1919 and 1930. Here we have, as before, two urns S and E. However, in the present model *no* balls are removed from urn S. A constant number k of balls is transferred from urn E to urn S at each trial. An accumulation process becomes "stiffer" and more resistant to change as more elements accumulate in urn S. (A replacement process, however, does not "stiffen" with training.) Note that m, the total number of balls in urn S is equal to $B_1 + W_1 + (n-1)k$ at the nth trial. Under these assumptions,

$$E[B_{n+1}] = E[B_n] + ka, \tag{5.50}$$

and

$$E[W_{n+1}] = E[W_n] + k(1-a). \tag{5.51}$$

Thus, if $P_n \equiv B_n/(B_n + W_n)$ is the proportion of black balls in urn S after n trials, then

$$E[P_n] = \frac{B_1 + (n-1)ka}{B_1 + W_1 + (n-1)k}. \tag{5.52}$$

Dividing the numerator and denominator by $B_1 + W_1$ and substituting as before, $b = P_1 = B_1/(B_1 + W_1)$ and $\theta = k/m = k/(B_1 + W_1)$, we have

$$E[P_n] = \frac{b + \theta a(n-1)}{1 + \theta(n-1)} = \frac{b(n-1)^{-1} + \theta a}{(n-1)^{-1} + \theta}.$$

As $n \to \infty$, we have, as in the linear case,

$$\lim_{n\to\infty} E[P_n] = P_\infty = \frac{\theta a}{\theta} = a. \tag{5.53}$$

Hence both learning curves start from $P_1 = b$ and end at $P_\infty = a$, but they follow different routes. [Numerical comparisons of replacement and accumulation curves are given in Restle and Greeno (1970).]

An Urn Model for a Learning Process in Time

Assume now that the sampling occurs very rapidly. Let π_n be the probability of a correct response (i.e., the drawing of a black ball) during the nth unit of time, hence π_n is equal to the proportion of black balls in urn S at that time. Suppose we observe the system by taking m balls from urn S during the nth unit of time (and then replacing them). We assume the *accumulation* model (5.52) to be relevant.

From the previous analysis,

$$E[\pi_n] = E\left[\frac{B_n}{B_n + W_n}\right] = \frac{B_1 + ka(n-1)}{B_1 + W_1 + k(n-1)}.$$

However, now we cannot observe directly the proportion of correct responses, but only the rate c_n of correct response per unit of time. This rate is given by

$$c_n = m\pi_n = \frac{\{mB_1/(B_1 + W_1)\} + \{k/(B_1 + W_1)\}ma(n-1)}{1 + \{k/(B_1 + W_1)\}(n-1)}, \tag{5.54}$$

with the initial rate $c_1 = mB_1/(B_1 + W_1)$ $(=mP_1)$ and the asymptotic rate $c_\infty = ma$.

5.4.2. Mixed (Prediction) Experiments

To investigate more meaningful learning processes, some changes in the experimental procedure have been suggested.

A common procedure is one known as a *reinforced experiment*. A particular case of such an experiment is a *probability learning* or *prediction* experiment.

A typical prediction experiment involves a panel containing two lights, two push buttons, and a "ready" light; this is set out schematically below.

Ready Light

E_1	E_2
A_1	A_2

On each trial the subject must predict which light will come on next, the assumption being that, when the light on the left comes on (event E_1), pushing the button on the left (response A_1) is the correct response and, similarly, when E_2 comes on, A_2 is the correct response.

Schematically, we are here given two experimenter's urns (E_1, E_2), corresponding to two types of trials. The "effect" of a trial corresponds to taking balls from the appropriate urn. The proportion of black balls in urn E_1 is a_1; the proportion of black balls in urn E_2 is a_2. It is possible that different numbers of balls will be taken from the two urns on each trial in which they are chosen. When urn E_i is chosen, we suppose k_i ($i = 1, 2$) balls are taken. In addition, the model involves, as before, the subject urn S, containing (initially) $m = B_1 + W_1$ balls. As before, we consider replacement and accumulation models separately.

Replacement Model

Assume that learning is accomplished by *replacement*. On the trials when urn E_1 is used we have

$$E[B_{n+1} | B_n, E_1] = B_n - P_n k_1 + a_1 k_1 \tag{5.55}$$

and, similarly,

$$E[W_{n+1} | B_n, E_1] = W_n - (1 - P_n)k_1 + (1 - a_1)k_1, \tag{5.56}$$

where, as before, B_n is the number of black balls in urn S on trial n, and $P_n = B_n/(B_n + W_n)$ (the proportion of black balls in urn S) is the probability of a correct response on trial n. P_n is called the *performance measure* at the nth trial. The number of balls in urn S remains constant, at m, throughout the experiment. Hence

$$E[P_{n+1} | B_n, E_1] = \frac{B_n - P_n k_1 + a_1 k_1}{m}.$$

Putting $\theta_i = k_i/m$ and recalling that $B_n + W_n = m$ (a constant), we have

$$E[P_{n+1} | P_n, E_1] = (1 - \theta_1)P_n + \theta_1 a_1. \tag{5.57}$$

On trials when urn E_2 is used we have, analogously,

$$E[B_{n+1} | B_n, E_2] = B_n - P_n k_2 + a_2 k_2$$

and

$$E[W_{n+1} | B_n, E_2] = W_n - (1 - P_n)k_2 + (1 - a_2)k_2.$$

Similarly, at the $(n+1)$th step,

$$E[P_{n+1} \mid P_n, \mathrm{E}_2] = (1-\theta_2)P_n + \theta_2 a_2. \tag{5.58}$$

(Remember that $\theta_2 = k_2/m$.)

Equations (5.57) and (5.58) can be expressed in terms of (linear) *operators* acting on a variable P. If we define O_i—the "type-i trial operator" ($i = 1, 2$) acting on P—by

$$O_i(P) = (1-\theta_i)P + \theta_i a_i, \tag{5.59}$$

then, for example,

$$O_2(O_1(P)) = (1-\theta_1)(1-\theta_2)P + \theta_1 a_1 + \theta_2 a_2 + \theta_1\theta_2 a_1,$$

while

$$O_1(O_2(P)) = (1-\theta_1)(1-\theta_2)P + \theta_1 a_1 + \theta_2 a_2 + \theta_1\theta_2 a_2.$$

These two expressions are not equal in general. They are equal if θ_1 or θ_2 equals 0 or if $a_1 = a_2$. Thus in the replacement model the operators O_1 and O_2 commute if the process has the same asymptotic values ($a_1 = a_2$). (The case $\theta_1 = \theta_2 = 0$ is of no interest, since the contents of urn S remain the same throughout the experiment.)

Accumulation Model

The analysis is very similar to that for the replacement model and is presented in the two following exercises.

EXERCISE 5.18. Show that, in the case of an accumulation model with mixed events, consisting of three urns $\mathrm{E}_1, \mathrm{E}_2$, S, when the balls taken from urn E_1 or E_2 are put into urn S but none are removed from it, the following equation is satisfied after the nth trial if urn E_i is used:

$$E[B_{n+1} \mid B_n, \mathrm{E}_i] = B_n + k_i a_i$$

and

$$E[W_{n+1} \mid W_n, \mathrm{E}_i] = W_n + k_i(1-a_i) \qquad (i = 1, 2),$$

where k_i = number of balls transferred from urn E_i to urn S if urn E_i is used and a_i = proportion of black balls in urn E_i.

EXERCISE 5.19. Show that, after n_1 trials using urn E_1 and n_2 trials using urn E_2 (*irrespective of the order*), the expected probability of correct response on the $(n_1 + n_2 + 1)$th trial will be

$$E[P_{n_1+n_2+1}] = \frac{P_1 + a_1\theta_1 n_1 + a_2\theta_2 n_2}{1 + \theta_1 n_1 + \theta_2 n_2}.$$

(Hence in the accumulation model theory the effects of two kinds of trials are *commutative*.)

Suppose we use urns E_1, E_2 with probabilities π, $1 - \pi$, respectively. The *mean operator is*

$$\bar{O} = \pi O_1 + (1 - \pi)O_2.$$

(Note that O_1 is associated with E_1, and O_2 with E_2.) We have

$$\bar{O}(P_i) = \pi(1 - \theta_1)P_i + \pi\theta_1 a_1 + (1 - \pi)(1 - \theta_2)P_i + (1 - \pi)\theta_2 a_2.$$

If $\theta_1 = \theta_2 = \theta$ (i.e., $k_1 = k_2$), we have

$$\bar{O}(P_i) = (1 - \theta)P_i + \theta\bar{a}, \tag{5.60}$$

with $\bar{a} = \pi a_1 + (1 - \pi)a_2$. To estimate the population average P, we need to find (or estimate experimentally) the proportions f_i of the subjects at various levels P_i ($i = 1, \ldots, g$). Now

$$\sum_{i=1}^{g} f_i[\bar{O}(P_i)] = \sum_{i=1}^{g} f_i[(1 - \theta)P_i + \theta\bar{a}],$$

or (in an obvious notation)

$$\bar{O}(\bar{P}) = (1 - \theta)\bar{P} + \theta\bar{a} \tag{5.61}$$

(since $\sum f_i = 1$). Note the *difference between* (5.60) *and* (5.61). Equation (5.60) is valid for *each i.*

5.4.3. Urn Model Representation of Stochastic Learning Schemes

For many purposes the direct use of learning curves may be misleading. To plot the response variable as a function of the trial number, even for a single individual performance, obscures the fact that the response made at any particular trial may depend on the sequence of responses rewarded or punished during the previous trials.

This leads to descriptions of learning behavior by means of *stochastic processes.* The stochastic process that represents a sequence of discrete learning responses is very conveniently described in terms of a succession of drawings from an urn containing variously colored balls. The basic feature is that at each trial a ball is drawn at random, its color is noted, it is replaced, and the contents of the urn are altered by:

1. Adding a constant number of balls (the Audley–Jonckheere model discussed below).
2. Multiplying the sum total balls of each color by the multiplicative constant specific to that color to determine the number of balls to be added (Luce, 1959).
3. Substituting balls of one color for those of another (Bush and Mosteller, 1955; Estes and Burke, 1950).

The correspondence between urn models and generalized stimulus sampling theory can be established by identifying (1) *balls* with *stimulus* elements, (2) *colors* with *response alternatives*, (3) sum totals of homogeneously colored balls relative to the sum total of balls in the urn with *conditional stimulus-response* probability relations, and (4) the response on each trial with the color of the balls drawn at random.

These models are extensions of the Pólya process (Chapter 4). One variant, the so-called two-response-choice *stochastic learning model* proposed by Audley and Jonckheere (1956) is now discussed in some detail.

We consider a sequence of n trials, in each of which the subject must respond in one of two possible ways. One response is rewarded—positively reinforced—each time it is made. The other is discouraged—negatively reinforced—by some form of punishment administered to the subject. In this way, the $(i + 1)$st response of the subject is likely to be dependent on each of the first i responses.

Audley and Jonckheere (1956) have suggested that the process determining the sequence of responses in such a learning experiment is analogous to the process involved in a representative urn scheme. An urn initially contains b black balls and w white balls, with $m = b + w$. A ball is chosen randomly. The color of the ball determines how many balls of each color are added according to Table 5.2.

TABLE 5.2

Relation between Withdrawn and Added Balls

Color of ball drawn	Number of black balls added	Number of white balls added
Black	a_1	$c_1 - a_1$
White	a_2	$c_2 - a_2$

The drawn ball is replaced along with the added balls, and the procedure is repeated n times. This is recognized as the generalized Pólya–Eggenberger scheme described in Section 4.3, with β_w, β_b, ω_w, ω_b replaced by a_2, a_1, $c_2 - a_2$, $c_1 - a_1$, respectively. The special case $a_1 = 0$, $a_2 = c_1 = c_2 = 1$ is the Ehrenfest model discussed in Section 4.8.2 (with its two urns corresponding to the white and black balls, respectively, in the present model). The drawing of a black ball corresponds to a correct response. The numbers a_1 and $c_1 - a_1$ correspond to the effect of the reward on the subject. For example, if $c_1 = a_1$, the probability of drawing a black ball on the second turn, given that the first drawing was black, is larger than was the probability of drawing the first black ball. This situation corresponds to a rat in a maze being rewarded for making a right turn. Then, if the reward has a large effect on the rat, the probability of it making another right turn will have been increased. The drawing of a white ball corresponds to a subject's incorrect response, and the numbers a_2 and $c_2 - a_2$ correspond to the influence of whatever negative reinforcement is administered.

Note that the case $a_1 = c_1 = c_2 = s$ and $a_2 = 0$ gives us the original *Pólya urn model* (Section 4.2). This, however, does not correspond to a reasonable learning scheme, since in the latter we hope that each trial will in some way increase the probability of success on the next trial, whether or not a correct response is given.

Using this model, Audley and Jonckheere obtained a conditional distribution for the probability of success on the $(n + 1)$th trial, given the results of the previous n trials. (This is of course a generalized Pólya–Eggenberger distribution.)

We define

$$X_i = \begin{cases} 1 & \text{if the } i\text{th drawing is black,} \\ 0 & \text{if the } i\text{th drawing is white.} \end{cases} \tag{5.62}$$

Then,

$$P\{X_{n+1} = 1 \mid X_1 = x_1, \ldots, X_n = x_n\} = \frac{b + a_1 \sum_1^n x_i + a_2\left(n - \sum_1^n x_i\right)}{m + c_1 \sum_1^n x_i + c_2\left(n - \sum_1^n x_i\right)}.$$

Let

$$\rho = \frac{b}{m}, \qquad \alpha = \frac{a_1}{m}, \qquad \beta = \frac{a_2}{m}, \qquad \gamma_i = \frac{c_i}{m} \qquad (i = 1, 2).$$

$$x_{(n)} = \sum_{i=1}^n x_i.$$

Then,

$$\begin{aligned} P\{X_{n+1} = 1 \mid X_1 = x_1, \ldots, X_n = x_n\} &= P\{X_{n+1} = 1 \mid x_{(n)}\} \\ &= \frac{\rho + (\alpha - \beta)x_{(n)} + n\beta}{1 + (\gamma_1 - \gamma_2)x_{(n)} + n\gamma_2}. \end{aligned} \tag{5.63}$$

This conditional distribution has the parameters $\rho, \alpha, \beta, \gamma_i$ $(i = 1, 2)$. As Audley and Jonckheere point out, these are dependent on the individual subject, as well as on the particular reinforcements employed.

Perhaps the most obvious drawback of the model involves the fact that it does not recognize that, in a learning situation, more recent trials may be expected to have a greater influence on the subject's behavior than those in the more distant past. A possible modification which compensates for this is to use weights r^{n-i} $(0 < r \leq 1)$, so that

$$P\{X_{n+1} = 1 | X_i = x_i; i = 1, \ldots, n\} = \frac{\rho + (\alpha - \beta)\sum_1^n r^{n-i}x_i + \beta \sum_1^n r^{n-i}}{1 + (\gamma_1 - \gamma_2)\sum_1^n r^{n-i}x_i + \gamma_2 \sum_1^n r^{n-i}}. \tag{5.64}$$

It is difficult, if not impossible, to visualize a reasonably simple urn model producing this distribution. However, if we discretize the weighting process in a somewhat oversimplified way by supposing that only the most recent h trials have any effect on the subject's behavior, we can write

$$P\{X_{n+1} = 1 | X_1 = x_1, \ldots, X_n = x_n\} = \frac{\rho + (\alpha - \beta) \sum_{i=n-h+1}^{n} x_i + h\beta}{1 + (\gamma_1 - \gamma_2) \sum_{i=n-h+1}^{n} x_i + h\gamma_2}. \tag{5.65}$$

The corresponding urn model is one in which only balls added in the last h steps remain in the urn. Previously added balls are removed after they have been in the urn for h drawings. Each added ball is only a temporary (h-step) visitor.

The expected probability P_{n+1} of drawing a black ball on the $(n + 1)$st trial can be computed by summing

$$P\{X_{n+1} = 1 | X_1 = x_1, \ldots, X_n = x_n\}P\{X_1 = x_1, \ldots, X_n = x_n\}$$

over all 2^n possible sequences $x_1, \ldots, x_n$. However, even with the simplification $c_1 = c_2 = c$ (i.e., $\gamma_1 = \gamma_2 = \gamma$), in the original model (5.63) one obtains a very complicated expression for $E[P_{n+1}]$:

$$E[P_{n+1}] = \left\{\rho + \frac{\beta}{1 + \alpha - \beta}\left[1 + \frac{1 + \gamma}{1 + \alpha - \beta + \gamma} + \frac{(1 + \gamma)(1 + 2\gamma)}{(1 + \alpha - \beta + \gamma)(1 + \alpha - \beta + 2\gamma)} + \cdots + \frac{(1 + \gamma)\cdots(1 + (n - 1)\gamma)}{(1 + \alpha - \beta + \gamma)\cdots(1 + \alpha - \beta + (n - 1)\gamma)}\right]\right\} \times \prod_{j=1}^{n} \left\{\frac{1 + \alpha - \beta + (j - 1)\gamma}{1 + j\gamma}\right\}.$$

Audley and Jonckheere recommend that it be assumed that every subject responds according to the same type of process, but that each subject is governed by his or her own particular set of parameters α, β, γ. It is suggested that the parameters for a given subject be chosen so as to maximize the likelihood of a particular observed sequence of responses. This process consists of solving simultaneous nonlinear equations by some iterative process. Audley and Jonckheere note that, in their experience, they have not come across many cases in which *multiple* maximum likelihood solutions exist.* Alternatively, the individual's parameters can be estimating by minimizing $D_n = \sum_{i=1}^{n} (X_i - E(P_i))^2$, that is, with a least-squares approach. Although this approach sometimes yields probabilities outside [0, 1] it has been found to provide useful results in practice.

A very wide variety of stochastic learning models can be provided by urn models. The model described in Example 2.3 (Section 2.3) has certain similarities with the learning models we have discussed. There are of course inescapable difficulties in interpretation when *different* models can predict similar outcome distributions. (See also Example 4.3.)

5.4.4. Asymptotic Values of Response Probabilities in the Stochastic Learning Model

Wolter and Earl (1969) derived asymptotic values for the response probabilities in the Audley–Jonckheere (1956) learning model discussed in Section 5.4.3.

We have an *urn* initially containing W white balls and $B = (m - W)$ black ones. A trial consists of drawing one ball at random, noting its color, replacing it, and then changing the contents of the urn by adding a specified number of white and black balls. In the Audley–Jonckheere model the number of balls of each color added to the urn is determined by the color of the ball drawn. (See Table 5.2.)

Let P_n represent the probability of drawing a black ball on the $(n + 1)$th trial in a particular sequence and suppose that, on the n trials preceding trial $n + 1$, K black balls and J $(=n - K)$ white balls were drawn. Then, from Table 5.2, we see that

$$P_n = \frac{B + Ka_1 + Ja_2}{m + Kc_1 + Jc_2} = \frac{\rho + K\alpha + J\beta}{1 + K\gamma_1 + J\gamma_2}. \tag{5.66}$$

* The paper was written in 1956. There are by now other examples of such situations in statistical literature, for example, in fitting mixtures of distributions (see, for example, Johnson and Kotz, 1972, Chapter 34) and in estimating factor loadings (Lawley and Maxwell, 1971).

Suppose that a black ball is drawn on trial $n + 1$, then, from (5.66) and Table 5.2, we have

$$P_{n+1} = \frac{\rho + K\alpha + J\beta + \alpha}{1 + K\gamma_1 + J\gamma_2 + \gamma_1}$$
$$= \frac{P_n + \alpha(1 + K\gamma_1 + J\gamma_2)^{-1}}{1 + \gamma_1(1 + K\gamma_1 + J\gamma_2)^{-1}}. \tag{5.67}$$

Similarly, if a white ball is drawn on trial $(n + 1)$, then

$$P_{n+1} = \frac{P_n + \beta(1 + K\gamma_1 + J\gamma_2)^{-1}}{1 + \gamma_2(1 + K\gamma_1 + J\gamma_2)^{-1}}. \tag{5.68}$$

If P_n tends to a limit, say p, as n tends to infinity, then, for large n, $Kn^{-1} \sim p$ and $Jn^{-1} \sim 1 - p$, and, from (5.66),

$$p = \frac{p\alpha + (1 - p)\beta}{p\gamma_1 + (1 - p)\gamma_2}. \tag{5.69}$$

In the special case $\gamma_1 = \gamma_2 = \gamma$ (i.e., $c_1 = c_2 = c$) we obtain

$$p = \frac{\beta}{\gamma - \alpha + \beta}. \tag{5.70}$$

If c is small (but not zero), then P_n approaches p slowly; if c is large, the approach is rapid. If $c = \infty$, then $P_n = p$ for $n \geq 1$. Generally, there is one solution of (5.69) with $0 < p < 1$. As p increases from 0 to 1, the right-hand side of (5.69) varies monotonically from β/γ_2 to α/γ_1, and so the solution gives p between these two values. If it so happens that

$$\frac{\alpha}{\gamma_1} = \frac{\beta}{\gamma_2}, \tag{5.71}$$

then p will equal the common value. Otherwise, p will be a root of the equation

$$(\gamma_1 - \gamma_2)p^2 + (\gamma_2 - \alpha + \beta)p - \beta = 0. \tag{5.72}$$

5.4.5. Reinforcement Models

Arnold (1973)* has provided a more general approach, allowing for variation in the sets of balls added to the subject's urn after each trial. An urn initially contains α_i balls of color $\mathscr{C}_i$ $(i = 1, 2, \ldots, k)$. After each trial a number of balls is added to the urn. These sets of balls are called "reinforcements," and it is supposed that there are p such sets. The number of balls of color i in

* The authors thank Dr. B. C. Arnold for providing his rough notes for this section.

set j is α_{ij}; the total number of balls in set j is $\sum_{i=1}^{k} \alpha_{ij}$. On the nth trial it is possible to

1. Select a ball at random from the reinforcement set added after the $(n-g)$th trial $(g = 1, \ldots, n-1)$, or
2. Select a ball at random from all the balls in the urn.

We denote the probabilities in case 1 by δ_g $(g = 1, \ldots, n-1)$ and the probability in case 2 by $\lambda_n = 1 - (\delta_1 + \delta_2 + \cdots + \delta_{n-1})$. Note that the δ's do not depend on n (the number of the trial), and that the probability of drawing from a particular reinforcement set depends on the number of trials since it was introduced. Clearly (since $\delta_1 + \cdots + \delta_{n-1} < 1$ and each δ is nonnegative), $\sum_{h=1}^{\infty} \delta_h$ must converge and, as $n \to \infty$, $\lambda_n \to \lambda = 1 - \sum_{h=1}^{\infty} \delta_h$. The ball that is drawn is returned to the urn.

To complete the model we need a rule to decide which reinforcement set to use after any given trial. Denoting by the random variable Y_h the set chosen after the hth trial, we may suppose

$$\Pr[Y_h = j] = \pi_j \qquad (j = 1, 2, \ldots, p), \tag{5.73}$$

with $Y_1, Y_2, \ldots$ mutually independent. This is called *noncontingent reinforcement* (because π_j does not depend on earlier results—either of drawings or of selection of reinforcement sets).

Generally, let X_n take the value i if a ball of color $\mathscr{C}_i$ is chosen on the nth trial. Then,

$$\Pr\left[X_n = i \,\middle|\, \bigcap_{j=1}^{n-1} (Y_j = a_j)\right] = \lambda_n \left\{ \frac{\alpha_i + \sum_{h=1}^{n-1} \alpha_{ia_h}}{\sum_{g=1}^{k} \left(\alpha_g + \sum_{h=1}^{n-1} \alpha_{ga_h}\right)} \right\} + \sum_{h=1}^{n-1} \left\{ \frac{\delta_{n-h} \alpha_{ia_h}}{\sum_{g=1}^{k} \alpha_{ga_h}} \right\}. \tag{5.74}$$

Taking expected values with respect to $Y_1, Y_2, \ldots, Y_{n-1}$, assuming model (5.74), we obtain

$$\Pr[X_n = i] = \lambda_n E[W_n] + \sum_{h=1}^{n-1} \delta_{n-h} E[Z_h],$$

where

$$W_n = \left(\alpha_i + \sum_{h=1}^{n-1} \alpha_{iY_h}\right) \left\{ \sum_{g=1}^{k} \left(\alpha_g + \sum_{h=1}^{n-1} \alpha_{gY_h}\right) \right\}^{-1},$$

$$Z_h = \alpha_{iY_h} \left(\sum_{g=1}^{k} \alpha_{gY_h} \right)^{-1}.$$

[*Note*: W_n = proportion of balls with color $\mathscr{C}_i$ just before nth trial; Z_h = proportion of balls with color $\mathscr{C}_i$ in hth reinforcement set.]

Now Z_h takes the value $\alpha_j(\sum_{g=1}^{k}\alpha_{gj})^{-1}$, with probability π_j, hence

$$E[Z_h] = \sum_{j=1}^{p}\left\{\pi_j\alpha_{ij}\left(\sum_{g=1}^{k}\alpha_{gj}\right)^{-1}\right\}. \tag{5.75}$$

In the numerator of W_n we note that [if (5.73) is valid] the random variables α_{iY_h} ($h = 1, 2, \ldots, n-1$) are independent. By the strong law of large numbers,

$$(n-1)^{-1}\text{ (numerator of } W_n) \to E[\alpha_{iY_h}] = \sum_{j=1}^{p}\pi_j\alpha_{ij} \tag{5.76}$$

with probability 1 as $n \to \infty$.

By an extension of the argument, as $n \to \infty$,

$$W_n \to \left(\sum_{j=1}^{p}\pi_j\alpha_{ij}\right)\left(\sum_{g=1}^{k}\sum_{j=1}^{p}\pi_j\alpha_{gj}\right)^{-1} \tag{5.77}$$

with probability 1, and this is also $\lim_{n\to\infty} E[W_n]$. Hence

$$\lim_{n\to\infty}\Pr[X = i] = \frac{\lambda\sum_{j=1}^{p}\pi_j\alpha_{ij}}{\sum_{g=1}^{k}\sum_{j=1}^{p}\pi_j\alpha_{gj}} + (1-\lambda)\sum_{j=1}^{p}\left\{\frac{\pi_j\alpha_{ij}}{\sum_{g=1}^{k}\pi_j\alpha_{gj}}\right\}, \tag{5.78}$$

where $\lambda = 1 - \sum_{h=1}^{\infty}\delta_h$.

A generalization of the model of Audley and Jonckheere (5.63) to more than two colors corresponds to $\delta_h = 0$ for all h, and so $\lambda = 1$ and

$$\lim_{n\to\infty}\Pr[X = i] = \left(\sum_{j=1}^{p}\pi_j\alpha_{ij}\right)\left(\sum_{g=1}^{k}\sum_{j=1}^{p}\pi_j\alpha_{gj}\right)^{-1} \quad \text{(for } i = 1, 2, \ldots, k). \tag{5.79}$$

If further, the number of balls in each reinforcement set is the same, say α, so that

$$\sum_{g=1}^{k}\alpha_{gj} = \alpha \qquad \text{(for all } j),$$

then

$$\lim_{n\to\infty}\Pr[X_n = i] = \alpha^{-1}\sum_{j=1}^{p}\pi_j\alpha_{ij}. \tag{5.80}$$

EXERCISE 5.20. Investigate the special cases:

(i) $\pi_j = k^{-1}$ $(j = 1, \ldots, k)$.
(ii) $\pi_1 = \theta, \pi_j = (1-\theta)(k-1)^{-1}$ $(j = 2, \ldots, k)$.

5.4.6. Further Modifications of the Audley-Jonckheere Model

There are numerous modifications of the basic (typical) model described above. The Audley–Jonckheere model is *subject-controlled*, in that the numbers of black and white balls added depend only on the subject's responses. From the mathematical point of view the following *experimenter-controlled* model is the simplest. It is assumed that, instead of drawing a ball to decide the number of balls of each color to add to urn S, we add a_1 black and $(c_1 - a_1)$ white balls with probability π on any given trial regardless of the response made. With probability $1 - \pi$ we add a_2 black and $(c_2 - a_2)$ white balls. Hence, as the number of trials increases, the proportion of black balls in the urn tends to

$$P_\infty = \frac{\pi\alpha + (1 - \pi)\beta}{\gamma} = \frac{\pi a_1 + (1 - \pi)a_2}{c}, \tag{5.81}$$

if $c_1 = c_2 = c$.

This model can be extended by supposing that a_j black and $(c_j - a_j)$ white balls are added with probability π_j $(j = 1, \ldots, s)$ and $\sum \pi_j = 1$.

EXERCISE 5.21. Show that, if $s = 2$ and $c_1 \neq c_2$, then

$$P_\infty = \frac{\pi a_1 + (1 - \pi)a_2}{\pi c_1 + (1 - \pi)c_2}.$$

EXERCISE 5.22. Show that, for general s,

$$P_\infty = \frac{\sum_{j=1}^{s} \pi_j a_j}{\sum_{j=1}^{s} \pi_j c_j}.$$

Further modification can take into account the possibility of partial control by both subject and experimenter. The experimenter adds black and white balls as just described, and *also* adds black and white balls according to Table 5.2.

EXERCISE 5.23. Find P_∞ for this case, when $s = 2$ and $c_1 \neq c_2$.

The ideas described may be elaborated to form inexhaustible series of models of varying complexity. The following is just one example, presented for purposes of illustration.

We could construct a model of "chain learning," by supposing that the subject's urn S in Section 5.4.1 is used as an experimenter's urn for a *second* subject, for whom we have an urn S′. We suppose further that, after each stage of experimentation with urns E and S, there is an experiment with urns

S and S′, in which k' balls are selected at random, *with replacement*, from urn S and balls of the same colors are added to urn S′. This can be regarded as an extension of the accumulation model in Section 5.4.1.

After the nth stage, the proportion of black balls in urn S is P_{n+1}, and so the number of balls added to urn S′ at the nth stage (say Y_n) is distributed binomially with parameters k', P_{n+1}. Hence, conditionally on $P_2, P_3, \ldots, P_n$, the expected value of the proportion P'_n of black balls in urn S′ at the nth stage is

$$E[P'_n | P_2, P_3, \ldots, P_n] = \frac{B'_1 + k'(P_2 + P_3 + \cdots + P_n)}{B'_1 + W'_1 + k'(n-1)},$$

where B'_1, W'_1 are the numbers of black and white balls, respectively, in urn S′ at the beginning of the experiment.

It follows that [remembering (5.52)]

$$\begin{aligned} E[P'_n] &= \frac{B'_1 + k'\{E[P_2] + E[P_3] + \cdots + E[P_n]\}}{B'_1 + W'_1 + k'(n-1)} \\ &= \{B'_1 + W'_1 + k'(n-1)\}^{-1} \\ &\quad \times \left[B'_1 + k' \sum_{j=2}^{n} \{B_1 + W_1 + k(j-1)\}^{-1}\{B_1 + ka(j-1)\} \right]. \end{aligned}$$

It is easy to show that $\lim_{n\to\infty} E[P'_n] = a$, thought the convergence is less rapid than for $E[P_n]$ (cf. (5.53)).

A variety of other models is described by Estes (1972) in a useful and interesting survey article which includes 85 references to related work.

5.5. MISCELLANEOUS APPLICATIONS

5.5.1. Military Applications

In this section we describe a few applications of results described earlier in this book to situations of a military nature. (See also Exercise 3.11.)

Occupancy distributions (Chapter 3) naturally apply to situations in which we are concerned with hitting as many as possible out of a set of m targets. Identifying the targets with urns and supposing that n missiles (identified with balls) are aimed at the targets in such a way that each target has an equal chance of being hit, the numbers of different targets that are hit, or escape being hit, have the classic occupancy distribution described in Section 3.1.

If some missiles may miss the targets altogether—say with probability $1 - p$—then the randomized occupancy distribution described in Section 3.3 applies. (The probability $1 - p$ is the probability of "falling through" in the terminology of that section.) Harkness (1970) describes another situation in which this distribution may be expected to arise: "In 'air-battle' theory (or predator-prey models) one identifies the m cells (*urns*) with 'bombers' 'missiles' (or more generally, with 'offensive forces') and the n balls with 'fighters' or 'anti-missiles' (or more generally, with 'defensive forces')... p is a 'kill' parameter corresponding to the probability that a fighter is able to intercept and destroy a bomber."

The number of occupied urns then corresponds to the number of bombers destroyed. If the randomized occupancy distribution is used, it implies, among other things, an assumption that each fighter will engage only one bomber. It is of course possible to allow for multiple engagements by means of appropriate modification of the model. This in general leads to other distributions, though one might hope under some circumstances to be able to use the randomized occupancy distribution, with an appropriately modified (increased) value of n, as a useful approximation.

The situation discussed in Section 3.2.1, where the urns (targets) are classified according to the probability that a ball (missile) will hit a given target, corresponds to a situation in which targets are, for example, of different sizes and so cannot be assumed to have equal chances of being hit. The problem, also studied in Section 3.2.1, of the distribution of hits on previously *unhit* targets is of interest in that it contributes to assessment of (1) the worthwhileness of continuing further attack when a certain number of hits have already been achieved, and (2) The advantage of restricting attack to targets not already hit (with a view to balancing this against probable increased difficulty and cost associated with such restriction).

Sprott (1957) gives a useful pioneering account of some of these problems.

EXERCISE 5.24. Show that, if the randomized occupancy distribution (3.62) applies, then the distribution of the number of shots S needed to hit exactly m' targets out of m is

$$\Pr[S = s] = p\binom{m-1}{m'-1}\Delta^{m'-1}\{1 - p + (m'-1)m^{-1}p - m^{-1}p\mathbf{0}\}^{s-1}.$$

EXERCISE 5.25. Show that the expected value of S (in Exercise 5.24) is

$$\left(\frac{m}{p}\right)\sum_{j=0}^{m'-1}(m-j)^{-1}.$$

The distributions of $M_2, M_3, \ldots$ (the number of targets hit by more than one missile each) are relevant in cases where it is expected that more than one hit is needed to destroy a target. If this is so, the discussion of the advantages of restricting future attacks will need to take into account *how many* hits have already been scored on each target, and also how many more are estimated to be needed to achieve destruction.

Urn models can be constructed to represent more complicated situations. For example, if it is supposed that a hit on one target can produce serious damage on another target, one could suppose each urn to contain a number of cards, each card stating which other (neighboring) urns are to be regarded as also having received a ball (or possibly, having received a "fraction of a ball"). When a ball is assigned to an urn, a card is chosen at random from the urn, and the instructions on the card are followed for assigning additional balls (or fractions of balls) to the specified urns.

EXERCISE 5.26. Under the conditions of Exercise 5.24, find the distribution of the number of shots needed to hit each of m' targets out of m at least twice.

5.5.2. Urn Representation of Some Filing Systems (Burville and Kingman, 1973; Burville, 1974; Kingman, 1975)

In terms of urn models we can imagine m urns in a row, each containing one ball; when ball b_i is requested, we start by looking in the left-hand urn (urn 1) and proceed to the right until we find the urn (say urn J,) that contains b_i. The "delay" is then $J - 1$.

We then put the ball in urn $J - 1$ into urn J, that in urn $J - 2$ into urn $J - 1$, and so on. This leaves urn 1 empty, and b_i is put into urn 1. The process is repeated many times.

The probability that b_i is in an urn to the right of the urn containing b_g is the probability that, of b_i and b_g, the last to be requested was b_g. If the process is repeated sufficiently often, this probability tends to $p_g/(p_i + p_g)$, where $p_i = \Pr[b_i \text{ chosen}]$. Since each urn to the left of b_i contributes 1 to the delay, the limiting value of $E[\text{delay} | b_i \text{ chosen}]$ is

$$\sum_{g \neq i}^{m} \frac{p_g}{p_i + p_g}.$$

Since $\Pr[b_i \text{ chosen}] = p_i$, the limiting value of $E[\text{delay}]$ is

$$\sum_{i=1}^{m} p_i \cdot E[\text{delay} | b_i \text{ chosen}] = \sum_{i \neq g}^{m} \sum^{m} \frac{p_i p_g}{p_i + p_g} = \delta_1. \tag{5.82}$$

Since $p_i p_g \leq \frac{1}{4}(p_i + p_g)^2$,

$$E[\text{delay}] \leq \tfrac{1}{4} \sum_{i \neq g}^{m} \sum^{m} (p_i + p_g) = \tfrac{1}{2}(m - 1).$$

If there have been *any* previous requests for either b_i or b_g, the conditional probability that the last one was for b_g is $p_g/(p_i + p_g)$. The probability that there have been no such requests at all, in a series of n requests, is $(1 - p_i - p_g)^n$.

The expected delay is then

$$\sum^{m}\sum^{m}_{i \neq g} p_i \left[\Pr[b_g \text{ initially above } b_i](1 - p_i - p_g)^n + \{1 - (1 - p_i - p_g)^n\} \frac{p_g}{p_i + p_g} \right].$$

If we suppose that the balls are assigned to the urns initially at random, then $\Pr[b_g \text{ initially above } b_i] = \frac{1}{2}$, and the expected delay is

$$\sum^{m}\sum^{m}_{i \neq g} p_i \left[\tfrac{1}{2}(1 - p_i - p_g)^n + \{1 - (1 - p_i - p_g)^n\} \frac{p_g}{p_i + p_g} \right].$$

Note that, as $n \to \infty$, this tends to (5.82) which corresponds to the stationary state.

If the balls are assigned to the urns so that the corresponding probabilities

$$p_{[1]} \geq p_{[2]} \geq \cdots \geq p_{[m]}$$

decrease from left to right, then the expected delay is

$$\delta_2 = \sum_{g=1}^{m} (g - 1) p_{[g]} = \sum^{m-1}\sum^{m}_{i<g} p_{[g]},$$

because there is a delay of $j - 1$ when b_j is chosen.

Now

$$\delta_1 = \sum^{m-1}\sum^{m}_{i \neq g} \frac{p_{[i]} p_{[g]}}{p_{[i]} + p_{[g]}} = 2 \sum^{m-1}\sum^{m}_{i<g} \frac{p_{[i]} p_{[g]}}{p_{[i]} + p_{[g]}}$$

(since the expression is symmetric in the p_i's).

If $i < g$, then, since $p_{[i]} \geq p_{[g]}$,

$$\frac{p_{[g]}}{2} \leq \frac{p_{[i]} p_{[g]}}{p_{[i]} + p_{[g]}} \leq p_{[g]},$$

so

$$\sum\sum^{m}_{i<g} p_{[g]} \leq \delta_1 \leq 2 \sum\sum^{m}_{i<g} p_{[g]},$$

and finally the following inequality is established:

$$\delta_2 \leq \delta_1 \leq 2\delta_2. \tag{5.83}$$

Thus, if the rank order of popularity of the balls is known, they should be filed in descending order of popularity, but, if not, the system described at the beginning of this section at most doubles the average search time.

5.5.3. A Biological Application

The following discussion is based on Cohen (1966). A *biotope*, a place inhabited by living organisms, such as an island or a desert oasis, is studied in hopes of being able to make predictions regarding relative frequencies of species. For the purpose of constructing the model, the biotope is partitioned into subniches. A subniche can be a certain category of types of places within the biotope, such as wooded areas or swamplands. It can be a certain way in which a group of animals utilizes the environment, such as the amount of vegetation serving as food. Cohen (1966) and others (e.g., Bailey, 1972) have discussed the difficulty in precisely defining a subniche and other unavoidable basic inaccuracies of the model which follow. Despite its faults, it has been found useful in some applications.

Let the biotope be represented by a collection of m urns. Each urn corresponds to a disjoint subniche. The life process is represented by a game involving s ($\leq m$) players who take turns throwing balls at the collection of urns. Each "player" represents a different species and each "ball" represents a member of that species. Every time a player throws a ball, it lands at random in one of the m urns. Each urn is capable of holding an unlimited number of balls, and each player has an unlimited number of balls available to him. The play proceeds according to the following rules:

1. Each player must throw at least one ball.
2. Each urn must eventually contain at least one ball.
3. No two players may "inhabit" the same number of urns (although they may share certain urns).
4. Each player throws the minimum number of balls necessary to satisfy rules 1 through 3.

Rules 1 and 2 ensure that there are no "phantom" species or subniches. Rule 3 is (according to Cohen) a strong form of the principle of competitive exclusion, known in biology as *Gause's axiom.*

Note that rules 3 and 4 imply that the jth player must eventually "inhabit" just j urns.

For convenience and definiteness we label the players by numbers, $r = 1, 2, \ldots, s$. We denote the number of balls tossed by the rth player by $N(m - r)$. This is also the number of balls necessary to *occupy* r urns [i.e., to have exactly $(m - r)$ empty urns]. The total number of balls thrown $(\sum_{i=1}^{r} N(m - i))$ is denoted by T.

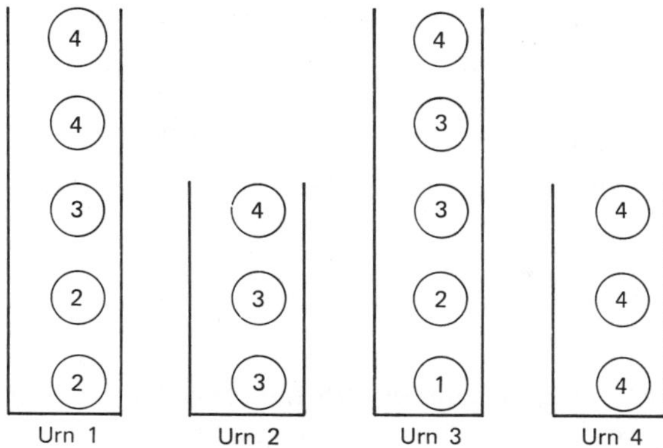

To help clear up difficulties, a sample game is represented above. Here $m = s = 4$, and player 1 played first, player 2 second, and so on (the order of the play is arbitrarily determined by the "referee," nature.) Player 1's ball landed in the third urn. He did not need to throw any more balls. Player 2's first ball landed in urn 1, and so did his second; therefore he had to throw a third ball in order to inhabit two urns and live up to his "name." His third ball fell in urn 3, and he did not need to throw any more balls as he now occupied two urns. Similarly, player 3 finished when his fifth ball landed in urn 1, and player 4 when his seventh ball landed in urn 2. In this example

$$\frac{N(m-1)}{T} = \frac{1}{16} \qquad \frac{N(m-2)}{T} = \frac{3}{16}, \qquad \frac{N(m-3)}{T} = \frac{5}{16}, \qquad \frac{N(m-4)}{T} = \frac{7}{16}.$$

A point of some interest is the amount of time (i.e., number of balls) used by the jth player in fulfilling the requirements. This may be approximated by

$$\frac{E[N(m-j)]}{\sum_{i=1}^{s} E[N(m-i)]}.$$

From Exercise 3.34, this ratio is

$$\left\{\sum_{h=1}^{j} (m-h+1)^{-1}\right\}\left\{\sum_{i=1}^{s}\sum_{h=1}^{i} (m-h+1)^{-1}\right\}^{-1}$$
$$= \left\{\sum_{h=1}^{j} (m-h+1)^{-1}\right\}\left\{\sum_{h=1}^{s} (s-h+1)(m-h+1)^{-1}\right\}^{-1}. \tag{5.84}$$

If $s = m$, we have

$$\sum_{i=1}^{m} E[N(m-i)] = m\sum_{h=1}^{m} (m-h+1)(m-h+1)^{-1} = m^2,$$

and (5.84) becomes

$$m^{-2}\sum_{h=1}^{j}(m-h+1)^{-1}. \tag{5.85}$$

The actual proportion of time occupied by the jth player is

$$\frac{N(m-j)}{\sum_{i=1}^{s}N(m-i)},$$

and (5.85) is only an approximation to the expected value of this statistic. Bailey (1972) gives an exact formula for this expected value for the case $s = m$. It is quite complicated. If a corrective term

$$-m^{-3}\sum_{h=1}^{j}(h-1)(m-h+1)^{-2}$$

is added to (5.85), quite a good approximation is obtained.

The numbers of balls in the different urns represent the relative abundance of the corresponding species in the different subniches.

Cohen (1966) also describes an economic interpretation of a somewhat more elaborate version of this model in which "players" are firms and "balls" are amounts of investment capital.

A biological application of the B–E occupancy distribution due to Hill (1970) is discussed in Chapter 6 (Section 6.2.4); examples of applications of related occupancy models in chemistry and bacteriology are given in Tukey (1949).

EXERCISE 5.27. Prove that

$$\Pr[N(m-k)=x+1]=\frac{k-1}{m}\Pr[N(m-k)=x]$$
$$+\frac{m-k+1}{m}\Pr[N(m-k+1)=x].$$

[*Hint*: (i) Express $\Pr[N(m-k)=x+1]$ in terms of

$$\Pr[N(m-k+1)=x-j] \qquad (j=0,1,\ldots,x-(k-1)).$$

(ii) Do the same with x replaced by $(x-1)$.]

EXERCISE 5.28. Show that the pgf

$$p_k(u)=\sum_{x=0}^{\infty}\Pr[N(m-k)=x]u^x$$

is equal to

$$\prod_{i=1}^{k} \frac{(m - i + 1)u}{m - (i - 1)u}.$$

[*Hint*: Prove and use the recursive formula

$$P_k(u) = \frac{(m - k + 1)u}{m - (k - 1)u} P_{k-1}(u)$$

and the initial condition $P_0(u) = 1$.]

EXERCISE 5.29. Verify that $\Pr[N(m - k) = x] = S_{x-1}^{(k-1)} m^{(k)}/m^x$, where S is a Stirling number of the second kind (see Section 1.1.2).

The result given by Exercise 5.29 allows us to write

$$P_k(u) = \sum_{x=k}^{\infty} \left(S_{x-1}^{(k-1)} \frac{m^{(k)}}{m^x} \right) u^x.$$

5.5.4. An Application in Computer Theory

An interesting example of the use of an urn model to represent a problem in computer theory has been provided by Fagin (1975), extending the work of Denning and Schwartz (1972), which we discussed in Example 3.1.

Suppose we have n balls, labeled $1, \ldots, n$, and we have a method of choosing these numbers (and the corresponding balls) with probabilities $p_1^{(n)}, p_2^{(n)}, \ldots, p_n^{(n)}$ ($\sum_{j=1}^{n} p_j^{(n)} = 1$). (This may itself be done using a suitable urn system.) There is an urn that can contain a proportion β_n of the balls (i.e., $n\beta_n$ balls in all). We commence by choosing $n\beta_n$ of the balls at random with the appropriate probabilities and putting them in the urn. Then numbers (1 through n) continue to be chosen at random. If the corresponding ball is not in the urn, it is placed in the urn and the ball that has been in the urn longest (without its number being selected) is removed from the urn. Note that, if a ball is already in the urn when its number is selected, there is no transfer of balls into or out of the urn, but the ball becomes the most recently selected and so cannot be removed until at least $n\beta_n$ further selections have been made. The urn thus always contains the most recently selected $n\beta_n$ balls.

We consider the sum of the selection probabilities for the numbers on the balls in the urn—called the *weight* of the urn—after n selections. This is clearly a random variable, and Fagin (1975) shows that its expected value is approximately

$$1 - M(S^{-1}(\beta_n)), \tag{5.86}$$

where

$$M(x) = \sum_{i=1}^{n} p_i^{(n)}(1 - p_i^{(n)})^x$$

and

$$S(y) = n - \sum_{i=1}^{n} (1 - p_i^{(n)})^y.$$

The weight is equal to the probability that a chosen number will be on a ball already in the urn, and that no transfer will be necessary. Regarding the urn as a computer's memory and the balls as pages of either data or instructions, the weight is the probability that a chosen page is already available in the memory. The expected value of the weight is called the *least recently used hit ratio.*

By taking $p_i^{(n)} = F(in^{-1}) - F((i-1)n^{-1})$, where $F(x)$ is a cumulative distribution function with $F(0) = 0$, $F(1) = 1$, and letting n tend to infinity, Fagin (1975) obtains the asymptotic results:

$$\text{Expected value} \doteqdot 1 - h(g^{-1}(\beta)),$$

where

$$h(x) = \int_0^1 f(y)\exp\{-xf(y)\}\,dy$$

and

$$g(x) = 1 - \int_0^1 \exp\{-xf(y)\}\,dy,$$

with

$$f(x) = \frac{dF(x)}{dx}.$$

EXERCISE 5.30. Obtain an approximate expression for the expected value by choosing

$$F(x) = x^{1-\theta} \qquad (0 \le x \le 1, 0 < \theta < 1).$$

Show that, for this choice of $F(x)$,

$$p_i^{(n)} \doteqdot n^{-1}(1-\theta)(in^{-1})^{-\theta}.$$

(This is proportional to $i^{-\theta}$. It is called *Zipf's law.*)

5.5.5. Solution of Difference Equations

Urn models can sometimes be used to solve difference equations. As an example we consider the quantities $P(k, l)$ [used by Barlow et al. (1972, p. 134)] in connection with testing hypotheses on ordered means, which satisfy the equation

$$P(l, k) = \frac{1}{k} P(l - 1, k - 1) + \left(1 - \frac{1}{k}\right) P(l, k - 1). \tag{5.87}$$

We can imagine an urn containing k balls. A ball is chosen to be added to the urn in such a way that the probability it is white is $(k + 1)^{-1}$ and the probability that it is black is $1 - (k + 1)^{-1}$.

Suppose we start with an empty urn ($k = 0$). Let $P(l, k)$ denote the probability that the urn contains l white balls when it contains k balls in all. Then $P(l, k)$ satisfies the recurrence relation (5.87).

Also, $P(l, k)$ is the coefficient of t^l in the expansion of

$$t(\tfrac{1}{2} + \tfrac{1}{2}t)(\tfrac{2}{3} + \tfrac{1}{3}t) \cdots \left(\frac{k-1}{k} + \frac{1}{k}t\right) = \prod_{j=1}^{k} \left\{\frac{j - 1 + t}{j}\right\} = \frac{1}{k!} t^{[k]}.$$

Hence [using (1.25′)]

$$P(l, k) = \frac{|S_k^{(l)}|}{k!}. \tag{5.88}$$

[It should be noted that in Barlow et al. (1972) the difference equation is derived from the solutions.]

EXERCISE 5.31. (Chacko, 1963). Show that

$$\sum_{l=1}^{k} (-1)^l P(l, k) = 0.$$

EXERCISE 5.32. Solve the difference equation (5.87) with initial values $P(0, 0) = 1$, $P(1, 0) = 1$.

5.6. SAMPLING SYSTEMS

5.6.1. Sampling Heterogeneous Populations

When sampling human, animal, or plant populations, it is essential to take account and, if possible, advantage of their nonhomogeneous nature. For example, human populations can be partitioned according to geographical locations (state, county, city, block, dwelling unit) within which the variation may be expected to be less than it is overall.

For practical purposes it is often convenient to select first not the individuals who will ultimately be measured, but some larger unit within which individual sampling will be applied. This is called *multistage sampling*. Often there is more than one stage of selection of units, subunits, and so on, before the ultimate selection of individuals occurs. For simplicity, however, we discuss only *one* intermediate stage.

We can imagine one large urn (the population) containing μ smaller urns (the units) marked with numbers 1 to μ. The first stage of sampling randomly selects m of these smaller urns, and the second (and final, in this case) stage selects, at random, n_i individuals from the ith urn if it is chosen. It is possible, and indeed usual, for the smaller urns to contain different numbers, $\nu_1, \nu_2, \ldots, \nu_\mu$, of balls. While simple random sampling may be used to select the m urns from among the μ available, it is also possible (and often desirable) to select them at random, but with the probability of selection proportional to the numbers of balls in the urns. This is called *probability proportional to size* (PPS) *sampling*. Sampling within each urn is supposed random; the ratio n_i/ν_i is the *sampling fraction* for the ith urn.

We suppose the value of a character observed for an individual to be inscribed on the corresponding ball. Denote by $x_{i1}, x_{i2}, \ldots, x_{i\nu_i}$ the numbers on the ν_i balls in the ith of the smaller urns.

The sample is of total size $N = n_{A_1} + n_{A_2} + \cdots + n_{A_m}$, where $A_1, A_2, \ldots, A_m$ are the numbers marked on the chosen urns. Note that, if the n_i's are fixed in advance, then N will be a random variable.

Suppose we wish to estimate the total of the numbers on all the balls in the large urn, that is,

$$\xi = \sum_{i=1}^{\mu} \sum_{j=1}^{\nu_i} x_{ij} = \sum_{i=1}^{\mu} \nu_i \bar{x}_i, \tag{5.89}$$

where $\bar{x}_i = \nu_i^{-1} \sum_{j=1}^{\nu_i} x_{ij}$.

We denote the numbers on the balls chosen in the sample by Y_{gh} ($g = A_1, A_2, \ldots, A_m; h = 1, 2, \ldots, n_g$). *Conditionally* on the urns chosen being those marked $A_1, A_2, \ldots, A_m$ (in that order), we have

$$\Pr[Y_{gh} = x_{gj}] = \nu_g^{-1} \qquad (g = A_1, A_2, \ldots, A_m; h = 1, \ldots, n_g; j = 1, \ldots, \nu_g) \tag{5.90}$$

$$\Pr[(Y_{gh} = x_{gj}) \cap (Y_{gh'} = x_{gj'})] = \nu_g^{-1}(\nu_g - 1)^{-1} \qquad (h \neq h', j \neq j'), \tag{5.91}$$

whence, putting $Y_{g\cdot} = \sum_{h=1}^{n_g} Y_{gh}$ and $\bar{Y}_{g\cdot} = n_g^{-1} Y_{g\cdot}$,

$$E[Y_{g\cdot}] = n_g E[Y_{gh}] = n_g \nu_g^{-1} \sum_{j=1}^{\nu_g} x_{gj} = n_g \bar{x}_g, \tag{5.92}$$

$$E[\bar{Y}_{g\cdot}] = \bar{x}_g, \tag{5.93}$$

and

$$\begin{aligned}\operatorname{var}(Y_{g\cdot}) &= n_g \operatorname{var}(Y_{gh}) + n_g(n_g - 1)\operatorname{cov}(Y_{gh}, Y_{gh'}) \qquad (h \neq h')\\ &= \frac{n_g}{v_g}\sum_{j=1}^{v_g}(x_{gj} - \bar{x}_g)^2 + \frac{n_g(n_g - 1)}{v_g(v_g - 1)}\sum_{j\neq j'}^{v_g}\sum^{v_g}(x_{gj} - \bar{x}_g)(x_{gj'} - \bar{x}_g).\end{aligned}$$

Since $\sum_{j=1}^{v_g}(x_{gj} - \bar{x}_g) = 0$, we have $\{\sum_{j=1}^{v_g}(x_{gj} - \bar{x}_g)\}^2 = 0$, whence

$$\sum_{j\neq j'}^{v_g}\sum^{v_g}(x_{gj} - \bar{x}_g)(x_{gj'} - \bar{x}_g) = -\sum_{j=1}^{v_g}(x_{gj} - \bar{x}_g)^2,$$

and so

$$\begin{aligned}\operatorname{var}(Y_{g\cdot}) &= \left\{\frac{n_g}{v_g} - \frac{n_g(n_g - 1)}{v_g(v_g - 1)}\right\}\sum_{j=1}^{v_g}(x_{gj} - \bar{x}_g)^2\\ &= \left(1 - \frac{n_g}{v_g}\right)n_g\left[\frac{1}{v_g - 1}\sum_{j=1}^{v_g}(x_{gj} - \bar{x}_g)^2\right]\end{aligned} \tag{5.94}$$

and

$$\operatorname{var}(\bar{Y}_{g\cdot}) = \left(1 - \frac{n_g}{v_g}\right)n_g^{-1}\left[\frac{1}{v_g - 1}\sum_{j=1}^{v_g}(x_{gj} - \bar{x}_g)^2\right]. \tag{5.95}$$

Formulas (5.94) and (5.95) can be simplified in appearance by introducing the symbols

$$\sigma_g^2 = (v_g - 1)^{-1}\sum_{j=1}^{v_g}(x_{gj} - \bar{x}_g)^2 \qquad (g = 1, \ldots, \mu). \tag{5.96}$$

The quantity σ_g^2 can be thought of as the variance within the gth of the smaller urns.

Note that formulas (5.90) through (5.96) are all conditional on the urns $A_1, A_2, \ldots, A_m$ being chosen. Subject to the same conditions, the $Y_{g\cdot}$'s (and so the $\bar{Y}_{g\cdot}$'s) are uncorrelated. Hence subject to these conditions, for any constants $b_1, b_2, \ldots, b_m$, we have

$$\operatorname{var}\left(\sum_{g=A_1}^{A_m} b_g\bar{Y}_{g\cdot}\right) = \sum_{g=A_1}^{A_m} b_g^2 \operatorname{var}(\bar{Y}_{g\cdot}) = \sum_{g=A_1}^{A_m} b_g^2\left(1 - \frac{n_g^2}{v_g}\right)\frac{\sigma_g^2}{n_g} \tag{5.97}$$

and also of course

$$E\left[\sum_{g=A_1}^{A_m} b_g\bar{Y}_{g\cdot}\right] = \sum_{g=A_1}^{A_m} b_g\bar{x}_g. \tag{5.98}$$

It is now convenient to define $\overline{Y}_{g\cdot} = 0$ if the gth smaller run is not included in the sample. Then, from (5.93) and (5.95), we have *unconditionally*

$$E\left[\sum_{g=1}^{\mu} b_g \overline{Y}_{g\cdot}\right] = \sum_{g=1}^{\mu} \Pr[g\text{th urn is in sample}] b_g \bar{x}_g$$

$$= \sum_{g=1}^{\mu} P_g b_g \bar{x}_g \tag{5.99}$$

and

$$\operatorname{var}\left(\sum_{g=1}^{\mu} b_g \overline{Y}_{g\cdot}\right) = \sum_{g=1}^{\mu} P_g b_g^2 \left(1 - \frac{n_g}{v_g}\right)\frac{\sigma_g^2}{n_g} + \operatorname{var}\left(\sum_{g=A_1}^{A_m} b_g \bar{x}_g\right)$$

$$= \sum_{g=1}^{\mu} P_g b_g^2 \left(1 - \frac{n_g}{v_g}\right)\frac{\sigma_g^2}{n_g} + \sum_{g=1}^{\mu} P_g b_g^2 \bar{x}_g^2 + \sum_{g \neq g'}^{\mu}\sum^{\mu} P_{g,g'} b_g b_{g'} \bar{x}_g \bar{x}_{g'}$$

$$- \left(\sum_{g=1}^{\mu} P_g b_g \bar{x}_g\right)^2$$

$$= \sum_{g=0}^{\mu} P_g b_g^2 \left(- \frac{n_g}{v_g}\right)\frac{\sigma_g^2}{n_g} + \sum_{g=1}^{\mu} P_g (1 - P_g) b_g^2 \bar{x}^2$$

$$+ \sum_{g \neq g'}^{\mu}\sum^{\mu} (P_{g,g'} - P_g P_{g'}) b_g b_{g'} \bar{x}_g \bar{x}_{g'}, \tag{5.100}$$

where $P_g = \Pr[g\text{th urn is in the sample}]$ and $P_{g,g'} = \Pr[g\text{th and } g'\text{th urns are both in the sample}]$.

In order that $\sum_{g=1}^{\mu} b_g \overline{Y}_{g\cdot}$ [which is equal to the sum of the products (b_g times the gth urn sample mean) over the urns included in the sample] shall be an unbiased estimator of $\xi = \sum_{i=1}^{\mu} v_i \bar{x}_i$ we must have

$$P_g b_g = v_g;$$

that is, we must take

$$b_g = \frac{v_g}{P_g}. \tag{5.101}$$

If PPS is used in choosing the urns, then v_g/P_g does not depend on g and so the b_g's are all the same and $b_1 \overline{Y}_{1\cdot} + \cdots + b_\mu \overline{Y}_{\mu\cdot}$ is just a multiple of the sum of the $\overline{Y}_{g\cdot}$'s.

For any given sampling scheme for the urns (i.e., values of P_g and $P_{g,g'}$) the variance given by (5.100) is minimized by minimizing

$$\sum_{g=1}^{\mu} P_g b_g^2 \left(1 - \frac{n_g}{v_g}\right)\frac{\sigma_g^2}{n_g} = \sum_{g=1}^{v} P_g b_g^2 \sigma_g^2 (n_g^{-1} - v_g^{-1})$$

with respect to the n_g's, subject to certain restrictions on the n_g's. This is clearly equivalent to minimizing $\sum_{g=1}^{\nu} P_g b_g^2 \sigma_g^2 n_g^{-1}$. Strictly speaking, we should take into account the fact that the n_g's must be integers, but useful approximate solutions can be obtained by regarding the n_g's as continuously varying quantities.

Suppose, for example that the sum n of the n_g's is fixed, so that

$$n = n_1 + n_2 + \cdots + n_\mu.$$

(Note that this does not necessarily imply a fixed total *sample size* N.)

For any set of μ positive numbers $w_1, w_2, \ldots, w_\mu$, the sum $w_1 n_1^{-1} + w_2 n_2^{-1} + \cdots + w_\mu n_\mu^{-1}$ is minimized, subject to $n = n_1 + n_2 + \cdots + n_\mu$, by taking n_g proportional to $\sqrt{w_g}$, so that

$$n_g = \frac{n\sqrt{w_g}}{\sum_{g=1}^{\mu} \sqrt{w_g}}. \tag{5.102}$$

This can be seen from the identity

$$\begin{aligned}\sum_{g=1}^{\mu} n_g^{-1} w_g &= \sum_{g=1}^{\mu} n_g\left(\frac{\sqrt{w_g}}{n_g}\right)^2 \\ &= \sum_{g=1}^{\mu} n_g(n_g^{-1}\sqrt{w_g} - \sqrt{w})^2 + n(\sqrt{w})^2,\end{aligned} \tag{5.103}$$

where

$$\sqrt{w} = n^{-1} \sum_{g=1}^{\mu} n_g(n_g^{-1}\sqrt{w_g}\,) = n^{-1} \sum_{g=1}^{\mu} \sqrt{w_g}.$$

[The last term in (5.103) does not depend on individual n_g values, and the other term on the right-hand side is minimized (and equal to zero) by making $n_g^{-1}\sqrt{w_g}$ constant.] Applying (5.103) with $w_g = P_g b_g^2 \sigma_g^2$, we see that the variance is minimized (approximately) by taking the n_g's as nearly as possible proportional to $b_g \sigma_g \sqrt{P_g}$, that is, to $v_g \sigma_g / \sqrt{P_g}$ if b_g satisfies (5.101).

If we wish to keep fixed the value D of a linear function of the n_g's—$\sum_{g=1}^{n} d_g n_g$—with positive coefficients d_g, we can utilize (5.102) by noting that

$$\sum_{g=1}^{\mu} n_g^{-1} w_g = \sum_{g=1}^{\mu} (d_g n_g)^{-1} (d_g w_g).$$

We can minimize $\sum_{g=1}^{\mu} n_g^{-1} w_g$, subject to $\sum_{g=1}^{n} d_g n_g = D$, by taking

$$d_g n_g \propto \sqrt{d_g w_g},$$

that is,

$$n_g \propto \sqrt{\frac{w_g}{d_g}}. \tag{5.104}$$

Example 5.2. Suppose we wish to control the total *expected* sample size at, say τ. This means we require

$$\sum_{g=1}^{n} P_g n_g = \tau. \tag{5.105}$$

Hence (putting $d_g = P_g$ and $w_g = P_g b_g^2 \sigma_g^2$) we see that the variance of

$$\sum_{g=1}^{\mu} b_g \bar{Y}_{g\cdot}$$

[see (5.100)] is minimized when

$$n_g \propto b_g \sigma_g. \tag{5.106}$$

[Note that this proportionality does not depend on P_g. The actual value of n_g is of course $\tau b_g \sigma_g (\sum_{j=1}^{\mu} P_j b_j \sigma_j)^{-1}$.]

The reader is again reminded that the n_g's actually have to be integers.

EXERCISE 5.33. Find a formula for the n_g's that will minimize the variance of the unbiased estimator of ξ, subject to the condition that the *expected cost* of the sample has a specified value, say C, given that

(i) The cost of obtaining a ball from the gth small urn and observing the number on it is c_g ($g = 1, \ldots, \mu$).

(ii) All other costs can be ignored.

EXERCISE 5.34. If *all* the smaller urns are always selected (so that $P_g = P_{g,g'} = 1$) we will have a situation corresponding to *stratified sampling*, in which the population is divided into μ strata and n_g individuals are chosen from the gth stratum ($g = 1, 2, \ldots, \mu$). By a suitable choice of strata it is often possible to reduce considerably the variance of the unbiased estimator of ξ,

$$\sum_{g=1}^{\mu} v_g \bar{Y}_{g\cdot} \qquad \text{(because } P_g = 1\text{)}.$$

From (5.100), with $b_g = v_g$, $P_g = P_{g,g'} = 1$, we have

$$\operatorname{var}\left(\sum_{g=1}^{\mu} v_g \bar{Y}_{g\cdot}\right) = \sum_{g=1}^{\mu} v_g^2(n_g^{-1} - v_g^{-1})\sigma_g^2. \tag{5.107}$$

Clearly, the smaller the σ_g^2's, the smaller this variance. It is therefore (for this purpose) desirable to arrange that the values of the measured character for individuals in the same stratum are as nearly alike as possible. Equivalently, we try to make the differences among different strata as large as possible.

EXERCISE 5.35. Sometimes we do not select all the small urns but, for each of those chosen, the numbers on every ball contained in them are observed. [This corresponds to a form of *cluster sampling*, in which the sampling is applied to naturally occurring groups (households, for example) and all members of the chosen groups are observed.]

Obtain formulas for estimating ξ in this case and investigate which kinds of groupings are most favorable to obtaining as accurate estimators of ξ as possible for a specified amount of effort (e.g., in terms of total cost).

5.6.2. Sampling Account Numbers

Jones (1959) (cf. Section 3.2.1) has considered the selection of a random sample from a set of numbers (e.g., customers' account numbers). In practice this may be done by generating (either from a list or with a computer) random numbers within a range including all the numbers from among which the sample is to be chosen. The list of random numbers so generated is then cleared of duplicates and of numbers not included in the set of account numbers. For any given number n of generated random numbers, the number V of usable numbers is itself a random variable.

We can apply the results of Section 3.4.2 directly to this problem. We regard the μ different numbers in the range from which the random numbers are chosen as urns, and the subset of τ different actual account numbers as class I urns. Then V is the number of occupied class I urns after random assignment of n balls among all the μ urns. We have a very simple case of (3.86), with $k = 1$, $b_1 = \tau$, $p_1 = \mu^{-1}$, $p_0 = 1 - \tau\mu^{-1}$, and find

$$\Pr[V = v] = \binom{\tau}{v}\Delta^v(1 - \tau\mu^{-1} + \mu^{-1}\mathbf{0})^n \tag{5.108}$$

[cf. (3.25)]. This formula is given by Jones (1959) [as equation (4.5)], but his method of derivation is different. (See also Glasser, 1963, equation 1.)

A related problem, discussed by Glasser (1963), considers the distribution of the number N of random numbers needed to be chosen in order to obtain a specified number v of different account numbers.

Clearly, since the last random number chosen must raise the number of chosen account numbers from $(v - 1)$ to v,

$$\begin{aligned}
\Pr[N = n|v] &= \Pr[V = v|n] - \Pr[V = v|n-1] \\
&= \binom{\tau}{v}\Delta^v[(1 - \tau\mu^{-1} + \mu^{-1}\mathbf{0})^n - (1 - \tau\mu^{-1} + \mu^{-1}\mathbf{0})^{n-1}] \\
&= \binom{\tau}{v}\Delta^v(\mu^{-1}\mathbf{0} - \tau\mu^{-1})(1 - \tau\mu^{-1} + \mu^{-1}\mathbf{0})^{n-1} \\
&= \frac{v - \tau}{\mu}\binom{\tau}{v}\Delta^{v-1}(1 - \tau\mu^{-1} + \mu^{-1}\mathbf{0})^{n-1}
\end{aligned}$$

[see Exercise 1.10(v)]

$$= \frac{\tau}{\mu}\binom{\tau - 1}{v - 1}\Delta^{v-1}(1 - \tau\mu^{-1} + \mu^{-1}\mathbf{0})^{n-1}. \tag{5.109}$$

EXERCISE 5.36. Obtain formulas for the expected values and variances of V and N in (5.108) and (5.109), respectively.

5.6.3. Randomized Response Methods

There are cases in which an "urn" system (or something very similar physically) is actually used to determine the questions asked in a survey sample enquiry. The randomized response technique suggested by Warner (1965), and further developed by Greenberg et al. (1969) and Liu and Chow (1976) is a noteworthy example. This technique is used when a "sensitive" question (e.g., Have you ever had an abortion? Do you smoke marijuana?) is to be asked, in an attempt to encourage correct replies by putting the question to a randomly chosen subset (unknown to the enquirers) of the respondents. The selection of the subset is in fact done by the respondents themselves, who are asked to perform an "urn experiment" the results of which determine the question(s) they are to answer.

In the original formulation (Warner, 1965) the respondent chooses (in effect) a ball from an urn containing g green and r red balls. It is arranged that, if a green ball is chosen the respondent will answer ("yes" or "no") to the statement, "I am a member of X." If a red ball is chosen, the statement is, "I am not a member of X." (X of course will be the particular sensitive point at issue.)

The enquirer does not know the colors of the balls chosen by the respondents, but the probability of choosing a green ball $[P = g(g + r)^{-1}]$ is known. If out of n respondents Y answer "yes" then, assuming (1) randomness of the drawing of balls, and (2) accuracy of answers, Y will have a binomial distribution with parameters n, $P\omega + (1 - P)(1 - \omega)$, where ω denotes the proportion of the population who are members of X.

An unbiased estimator of ω is provided by

$$(Yn^{-1} - 1 + P)(2P - 1)^{-1}.$$

Its variance is

$$\begin{aligned}\{n(2P - 1)\}^{-2}\operatorname{var}(Y) &= n^{-1}(2P - 1)^{-2}\{P\omega + (1 - P)(1 - \omega)\} \\ &\quad \times \{1 - P\omega - (1 - P)(1 - \omega)\} \\ &= n^{-1}(2P - 1)^{-2}\{(2P - 1)\omega + 1 - P\} \\ &\quad \times \{(2P - 1)(1 - \omega) + 1 - P\} \\ &= \frac{\omega(1 - \omega)}{n} + \frac{P(1 - P)}{n(2P - 1)^2}.\end{aligned}$$

Note that, if regular random sampling (i.e., with just one statement) could have been used, with assumption (2) still valid, the proportion of "yes" answers in a sample of size n would have been $n^{-1}\omega(1 - \omega)$.

The additional variance

$$n^{-1}(2P - 1)^{-2}P(1 - P)$$

is large when P is near $\frac{1}{2}$. It will be zero if $P = 0$ or 1, but then of course we would be back to regular random sampling, and it would be known which statement was being answered. It is necessary that P be sufficiently far from 0 or 1 to reassure respondents that it is not possible to guess which statement is being answered, with much confidence.

In an attempt to improve the procedure, Greenberg et al. (1969) have suggested that, if a red ball is chosen, an *unrelated* (and innocuous) statement, "I am a member of X^*"—for example "Is your Social Security number even?" —be substituted for "I am not a member of X." If the proportion ω^* of the population belonging to X^* is known, this is of considerable assistance in the analysis.

EXERCISE 5.37. Using the notation of this section show that

$$P^{-1}\{n^{-1}Y - \omega^*(1 - P)\}$$

is an unbiased estimator of ω and its variance is

$$(nP^2)^{-1}\{\omega P + \omega^*(1 - P)\}\{1 - \omega P - \omega^*(1 - P)\}.$$

EXERCISE 5.38. When ω^* is not known, two independent samples of sizes n_1, n_2 are taken with P having different values P_1, P_2. Denoting the numbers of "yes" answers in the two samples by Y_1, Y_2, respectively, show that

$$(P_1 - P_2)^{-1}\{n_1^{-1}Y_1(1 - P_2) - n_2^{-1}Y_2(1 - P_1)\}$$

is an unbiased estimator of ω. Obtain an expression for its variance.

Liu and Chow (1976) have described a further application of the model, applicable when the answer to the sensitive question (e.g., How many times have you cheated on examinations in the last 2 years?) is numerical (0, 1, 2, . . .). In this case the red balls are numbered $0, 1, 2, \ldots, k$ in specified proportions $p_0, p_1, \ldots, p_k$. (k should be chosen at least as large as the expected greatest value of the answer to the sensitive question.) If a green ball is chosen, the sensitive question is answered, but if a red ball is chosen, the number on the ball is recorded. Denoting by ω_j the proportion of the population for which the correct answer to the sensitive question is j, and by Y_j the number of j answers out of n in all, an unbiased estimator of ω_j is

$$\hat{\omega}_j = P^{-1}\{Y_j n^{-1} - p_j(1 - P)\}.$$

The variances and covariances of the $\hat{\omega}_j$'s can be obtained from the joint distribution of $Y_0, Y_1, \ldots, Y_k$, which is multinomial with parameters n; $\{\omega_j P + p_j(1 - P)\}$ $(j = 0, 1, \ldots, k)$.

5.7. TESTS OF EMPTY BOXES

5.7.1. The Test and Its Power

The easily-applied test of "empty boxes"—originally proposed by David (1950)—can be described as follows.

We are testing the hypothesis H_0 that an independent sample $X_1, \ldots, X_n$ is taken from a given continuous population with known cdf $F(\cdot)$. We choose the points $z_0 = -\infty < z_1 < \cdots < z_{m-1} < z_m = \infty$ in such a manner that $F(z_k) - F(z_{k-1}) = 1/m$ $(k = 1, \ldots, m)$. The test is based on a statistic M_0 which is equal to the number of intervals (z_{k-1}, z_k) containing none of the observed values X_i. The test is of the form: H_0 is accepted if $M_0 \leq C$ and rejected if $M_0 > C$, where the constant C is chosen in such a manner that the error of the first kind (also known as the *level* of the test), that is, the probability of rejecting H_0 when it is correct, equals, say γ.

Under the null hypothesis [that $F(\cdot)$ is indeed the population cumulative distribution] the distribution of M_0 is the same as that of the number of empty urns when n balls are thrown independently into m urns, assuming that the probability of any particular ball falling into any particular urn is $1/m$. The observations $X_1, \ldots, X_n$ correspond to the throws of the balls, and the intervals correspond to the urns. This distribution is in fact the classical occupancy distribution (see Section 3.1.1).

Chapter 6 contains a detailed study of limit distributions of M_0 and related variables $M_1, M_2, \ldots$. For completeness we anticipatorily quote several results needed for a discussion of some basic properties of the test.

To study the distribution of the test statistic M_0 for large values of m and n we utilize the generalized generating function (cf. Section 1.4.3).

$$\Phi(x; z) = \sum_{n=0}^{\infty} \sum_{k=0}^{m} \left(\frac{mz}{n!}\right)^n x^k \Pr[M_0 = k].$$

From (3.17b), with $j = 0$, we have

$$\Phi(x; z) = (e^z + x - 1)^m.$$

Utilizing this generating function one can obtain asymptotic formulas for $E[M_0]$ and $\text{var}[M_0]$, when $m, n \to \infty$ and the average number of balls per urn, $\alpha = n/m$, is bounded. These are

$$m^{-1}E[M_0] \sim p_0 \tag{5.110a}$$

and

$$m^{-1}\,\text{var}(M_0) \sim \sigma_0^2, \tag{5.110b}$$

where

$$p_0 = e^{-\alpha}, \qquad \sigma_0^2 = e^{-\alpha}\{1 - (1 + \alpha)e^{-\alpha}\}. \tag{5.110c}$$

Rényi (1962) (cf. Section 6.2.2) has shown that the random variable M_0 is approximately normal with parameters mp_0, $\sqrt{m}\sigma_0$. This enables us to compute approximately the constant C in the test of empty boxes when the level γ is given. Denoting by u_ε the upper $100\varepsilon\%$ point of the normal distribution, determined by

$$\Phi(u_\varepsilon) = 1 - \varepsilon, \tag{5.111}$$

we observe that under H_0, for m and n sufficiently large and with n/m fixed,

$$\Pr\{M_0 \geq mp_0 + u_\gamma \sqrt{m}\sigma_0 | H_0\} \approx \gamma. \tag{5.112}$$

An approximate value for C is $mp_0 + u_\gamma\sigma_0\sqrt{m}$. Alternative conditions for the validity of the normal approximation are $n^2/m \to \infty$ or $(1/m)e^{n/m} \to 0$ (cf. Rényi, 1962; Békéssy, 1963; Weiss, 1958).

Example 5.3. From (3.13),

$$E[M_0] = m(1 - m^{-1})^n$$
$$\text{var}[M_0] = m[(m - 1)(1 - 2m^{-1})^n + (1 - m^{-1})^n - m(1 - m^{-1})^{2n}].$$

Using the assumption that M_0 is approximately normally distributed with these moments, compare the exact values $\Pr[M_0 = m_0]$ with the values of $\Pr[m_0 - \frac{1}{2} < M_0 < m_0 + \frac{1}{2}]$ calculated under the approximate distribution, for $m = 20$, $n = 20$.

David (1950) obtained the following results:

m_0	Exact	Approximate	m_0	Exact	Approximate
3	.0029	.0044	8	.2365	.2355
4	.0216	.0240	9	.1215	.1231
5	.0874	.0883	10	.0378	.0394
6	.2031	.1997	11	.0070	.0077
7	.2811	.2768	12	.0007	.0010

To determine the power of the empty boxes test we utilize the generalized version of the occupancy problem [cf. (3.5)]. We suppose we have m urns into which n balls are thrown independently in such a manner that the probability that any given ball falls into urn j is p_j $(j = 1, 2, \ldots, m)$, with $p_j \geq 0$; $\sum_{j=1}^{m} p_j = 1$. Laurent (1963) discusses this problem.

We shall see in Chapter 6 that, in this model, the number of empty urns M_0 is also asymptotically normally distributed with parameters $E[M_0]$, $\sqrt{\text{var}[M_0]}$, provided $n, m \to \infty$, and all mp_j $(j = 1, \ldots, m)$ and n/m are (uniformly) bounded (Chistyakov, 1964; Holst, 1971). From this limit theorem it follows that in the case where the probabilities p_j, on the *alternative* hypothesis H_1 are of the form

$$p_j = m^{-1}(1 + b_j n^{-1/4}) \qquad (j = 1, \ldots, m)$$

and

$$\lim_{m \to \infty} m^{-1} \sum_{j=1}^{m} b_j^2 = b^2 > 0 \tag{5.113}$$

(the case of *closely competing* hypotheses), then the error of the first kind γ and that of the second ω (the latter equals, by definition, the probability that H_0 is accepted when H_1 is correct) are related asymptotically as $m, n \to \infty$ in the following way:

$$u_\gamma + u_\omega = \frac{b^2 p_0 \alpha^2}{2\sigma_0}, \tag{5.114}$$

where u_γ and u_ω are the corresponding percentiles of the normal distribution given by (5.111) and p_0 and σ_0 are as given by (5.110c). This result was obtained by Chistyakov (1964). A similar result was derived by Okamoto (1952) under more stringent conditions on the behavior of n and m (cf. Section 6.2.3).

David (1950) suggests that nearly optimum power may be obtained by taking $m = n$, that is, making the expected number of observations in each urn equal to 1. (She further suggests that, when the sample size n exceeds 30, it is probably better to use a χ^2 test of goodness of fit, rather than the empty-boxes test.)

For n, m small ($2 \le m \le 10, 5 \le n \le 50$) Csörgö and Guttman (1962), taking $\alpha = .01$ and $.05$, constructed a table of values of the constant C based on the exact formula for the distribution of M_0.

The tables give values of a such that

$$\Pr[M_0 \ge a] \le \alpha$$

and

$$\Pr[M_0 \ge a - 1] > \alpha,$$

where $\alpha = .01$ or $\alpha = .05$. These are the (approximate) upper 1% and 5% points of the distribution of M_0. Also, the exact values of $\Pr[M_0 \ge a]$ are presented. An extract from their tables is given in Table 5.3.

TABLE 5.3

Approximate Upper 1% Points of M_0

$m \backslash n$	10	15	20	25	30
2	1	1	1	1	1
	(.0020)	(.0001)	(.0000)	(.0000)	(.0000)
3	2	1	1	1	1
	(.0001)	(.0069)	(.0009)	(.0001)	(.0000)
4	2	2	2	2	2
	(.0059)	(.0002)	(.0000)	(.0030)	(.0007)
5	3	2	2	2	1
	(.0010)	(.0047)	(.0004)	(.0000)	(.0062)
6	4	3	2	2	2
	(.0003)	(.0006)	(.0045)	(.0006)	(.0001)

NOTE: Exact values of $\Pr[M_0 \ge \text{tabulated value}]$ are given in parentheses.

5.7.2. A Two-Sample Empty Box Test

A two-sample empty box test has also been studied by Csörgö and Guttman (1962, 1964). The observed values in the first sample themselves are regarded as constituting a subdivision of the real line into cells. Using this subdivision, we observe the number of these cells that do not contain *any* of the values observed in the second sample. More precisely, suppose we have a random sample of n_1 independent observations taken from a population with an unknown continuous cdf $F(\cdot)$. Let $X_{(1)} < X_{(2)} < \cdots < X_{(n_1)}$ be their order statistics. Let a second sample of n_2 observations taken from a population with cdf $G(\cdot)$ (also continuous but otherwise unknown).

Define cells $I_1, \ldots, I_{n_1+1}$ by

$$I_i = (X_{(i-1)}, X_i] \qquad (i = 1, \ldots, n_1 + 1; X_{(0)} = -\infty; X_{(n_1+1)} = +\infty).$$

Let $R_1, \ldots, R_{n_1+1}$ denote the numbers of observations of the second sample that lie in $I_1, \ldots, I_{n_1+1}$. Let S_0 be the number of empty cells [i.e., the number of I_i's ($i = 1, \ldots, n_1 + 1$) for which $R_i = 0$]. The hypothesis $F \equiv G$ implies that all $(n_1 + n_2)$ possible orderings of the two samples are equally likely. The number of ways of choosing the m_0 cells that are to be empty is $\binom{n_1+1}{m_0}$. We must fill the remaining $(n_1 + 1 - m_0)$ cells with n_2 observations in such a way that none of these remaining cells are empty. This can be done in $\binom{n_2 - 1}{n_1 - m_0}$ ways (cf. Table 1.2 with $n = n_2$; $m = n_1 + 1 - m_0$). Hence, in this case.

$$\Pr[M_0 = m_0] = \frac{\binom{n_1+1}{m_0}\binom{n_2-1}{n_1-m_0}}{\binom{n_1+n_2}{n_1}}, \tag{5.115}$$

where the range of M_0 is

$$m_0 = [k, k+1, \ldots, n_1] \qquad \text{and} \qquad k = \max[0, n_1 + 1 - n_2].$$

This is a hypergeometric distribution (see (2.2.4)).

Tables in Csörgö-Guttman (1962) are given which yield b values such that $\Pr[M_0 \geq b] \leq \alpha$ and $\Pr[M_0 \geq b - 1] > \alpha$, where $\alpha = .01$ or $.05$. The tables also give $\Pr[M_0 \geq b]$; that is, they present the exact levels. Table 5.4 is an extract from them.

In this model,

$$E[M_0] = \frac{n_1^{[2]}}{n_1 + n_2}, \tag{5.116a}$$

$$\operatorname{var}(M_0) = \frac{1}{n_1 + n_2} \cdot \frac{n_1^{[2]} n_2^{(2)}}{(n_1 + n_2)^{(2)}} \tag{5.116b}$$

[cf. (2.27)].

TABLE 5.4

Approximate Upper 1% Points of M_0 in a Two-Sample Empty-Cell Test

$n_1 \backslash n_2$	10	15	20	25	30
5	5 (.0020)	5 (.0004)	4 (.0055)	4 (.0026)	4 (.0014)
6	6 (.0009)	5 (.0055)	5 (.0018)	5 (.0007)	4 (.0076)
7	7 (.0004)	6 (.0024)	6 (.0006)	5 (.0047)	5 (.0023)
8	8 (.0076)	7 (.0011)	6 (.0048)	6 (.0017)	6 (.0007)
9	9 (.0044)	7 (.0088)	7 (.0021)	6 (.0088)	6 (.0039)

NOTE: Exact value of $\Pr[M_0 \geq \text{tabulated value}]$ are given in parentheses.

If $n_2 = \rho n_1 + O(1)$ $(\rho > 0)$, then

$$E[M_0] = n_1\left[\left(\frac{1}{1+\rho}\right) + O\left(\frac{1}{n_1}\right)\right] \to \frac{1}{1+\rho} \tag{5.116c}$$

$$\operatorname{var}(M_0) = n_1\left(\frac{\rho^2}{(1+\rho)^3} + O\left(\frac{1}{n_1}\right)\right) \to 0, \tag{5.116d}$$

as $n_1, n_2 \to \infty$. Hence, from Chebyshev's inequality [cf. equation (1.130b)], $M_0/(n+1) \to (1+\rho)^{-1}$ (in probability) if $F \equiv G$.

To test the hypothesis $F \equiv G$ at an (approximate) $100\alpha\%$ level, Csörgö and Guttman propose the test:

$$\begin{cases} \text{Reject if } M_0 \geq b, \\ \text{Accept otherwise,} \end{cases}$$

where b is obtained from the tables.

Example 5.4. As an illustration suppose that two sets of sample values are:

Sample 1: 202, 210, 225, 237, 250.
Sample 2: 197, 198, 205, 211, 215, 222, 226, 229, 230, 231.

We take the "cells" $I_1 = (-\infty, 202]$; $I_2 = (202, 210]$; $I_3 = (210, 225]$; $I_4 = (225, 247]$; $I_5 = (237, 250]$; $I_6 = (250, \infty)$, based on the values in the first sample. The number of cells unoccupied by values in the second sample is two (I_5 and I_6). We have $n_1 = 5, n_2 = 11$. From Table 5.4 we see that there is no significant difference between the samples.

5.7.3. Generalizations of the Test

The test constructed in Section 5.7.1 is based on the statistic M_0 under the "equiprobability" assumption. (Its power depends on the distribution of M_0 when the cell probabilities are not all equal.) It seems natural to utilize the random variables M_r (representing the number of urns containing r balls) in constructing other statistical tests of the type described. Various statistical tests based on the statistics $M_0, \ldots, M_r$ have been developed (see, for example, Viktorova and Chistyakov, (1966). The joint distribution of the random variables $M_{r_1}, M_{r_2}, \ldots, M_{r_s}$ has the following generalized generating function:

$$\Phi_{r_1, r_2, \ldots, r_s}(x_1, \ldots, x_s; z) = \sum_{n=0}^{\infty} \frac{(m^n)}{n!} z^n E[x_1^{M_{r1}} x_2^{M_{r2}} \cdots x_s^{M_{rs}}],$$

which is equal to

$$\left\{e^z + \sum_{k=1}^{s} \frac{z^{r_k}}{r_k!}(x_k - 1)\right\}^m \tag{5.117}$$

[see (3.82)].

Utilizing this generating function, Sevast'yanov and Chistyakov (1964) obtained the following asymptotic (for $n, m \to \infty$) expressions for the expectations $E[M_r]$ and covariances $\operatorname{cov}(M_r, M_s)$ under the assumption that $\alpha = n/m$ is bounded (cf. Section 6.2.2):

$$\left.\begin{aligned} E[M_r] &\sim mp_r, \\ \operatorname{cov}(M_r, M_s) &\sim m\sigma_{rs}, \\ \text{where} \quad & \\ \sigma_{rs} &= -p_r p_s\left\{1 + \frac{(\alpha - r)(\alpha - s)}{\alpha}\right\} \quad (r \neq s), \\ \sigma_{rr} &= p_r\left[1 - p_r\left\{1 + \frac{(\alpha - r)^2}{\alpha}\right\}\right], \\ \text{and} \quad & \\ p_r &= \frac{\alpha^r}{r!} e^{-\alpha}. \end{aligned}\right\} \tag{5.118}$$

It was also shown in the same paper that the random vector $(M_{r_1}, M_{r_2}, \ldots, M_{r_s})$ is asymptotically normal with parameters given by the asymptotic expressions for $E[M_r]$ and $\operatorname{cov}(M_r, M_s)$ given in (5.118). The asymptotic result for the equiprobable scheme is valid as $n, m \to \infty$, provided $\alpha = n/m$ is bounded (cf. Section 6.2.2). Later, Chistyakov and Viktorova (1965) proved a similar theorem in the case of unequal cell probabilities a_j, provided $a_j m$ is bounded for all j.

Moreover, if the probabilities p_j in the alternative hypothesis H_1 satisfy the closely competing hypothesis conditions (5.113), then the asymptotic expressions for $E_{H_1}[M_r]$ and $\operatorname{cov}_{H_1}(M_r, M_s)$ are as follows:

$$m^{-1}E_{H_1}[M_r] \simeq p'_r$$

$$m^{-1}\operatorname{cov}_{H_1}(M_r, M_s) \simeq \sigma'_{rs},$$

where

$$p'_r = p_r + \frac{b^2(m)p_r}{2\sqrt{m}}[(\alpha - r)^2 - r] + O\left(\frac{1}{m^{3/4}}\right), \tag{5.119}$$

and

$$\sigma'_{rs} = \sigma_{rs} + O\left(\frac{1}{\sqrt{m}}\right)$$

(provided n/m is bounded from above and below: $0 < \alpha_0 \leq n/m \leq \alpha_1 < \infty$).

It was shown in a further paper by Viktorova and Chistyakov (1966) that a good test with power asymptotically the same as that of the likelihood ratio (Neyman-Pearson) test is based on a linear statistic of form

$$T_r = C_{0r}M_0 + C_{1r}M_1 + \cdots + C_{kr}M_r$$

having the critical region $T_r \leq C$. The two types of errors γ and ω are related in this case by the expression

$$u_\gamma + u_\omega = \frac{b^2 \sum_{k=0}^{r} C_{kr}p_k[(\alpha - k)^2 - k]}{2\sqrt{\sum_{k=0}^{r}\sum_{l=0}^{r} C_{kr}C_{lr}\sigma_{kl}}} \tag{5.120}$$

[cf. (5.114)]. [More precisely, for a given error of the first kind γ, prescribed for all m and n, the error of the second kind ω, with $n = \alpha m$, is determined by (5.120).]

Moreover, there exist, for $\alpha = n/m$, values of C_{kr} denoted $C^*_{kr}(\alpha)$ such that, for a fixed error of the first kind γ, the error of the second kind ω is minimal in the class of linear tests. In other words the power $1 - \omega$ is maximal. These coefficients were obtained in Viktorova and Chistyakov's paper (1966) and are given by

$$C_{kr}^* = 1 - \frac{2k}{\alpha} + \frac{k(k-1)}{\alpha^2} + \frac{\alpha - r}{\alpha}\theta_r + \frac{\theta_r[\alpha(1+\theta_r) - k][\theta_r(\alpha - r) + ((\alpha - r)^2 + r)/\alpha]}{\alpha[1 - \theta_r(\alpha - r - 1 + \alpha\theta_r)]},$$

where

$$\theta_r = \frac{\alpha^r/r!}{\sum_{k=r+1}^{\infty} (\alpha^k/k!)}.$$

Since

$$\lim_{\alpha\to\infty} C_{kr}^* = 1 \qquad (k = 0, \ldots, r),$$

$$\lim_{\alpha\to 0} \alpha^2 C_{kr}^* = 2\binom{r-k+2}{2} \qquad (k = 0, \ldots, r), \tag{5.121}$$

we have approximately optimal tests based on the statistics (see Exercise 5.39)

$$T_r = \binom{r+2}{2}M_0 + \binom{r+1}{2}M_1 + \cdots + 3M_{r-1} + M_r \qquad \text{(for small } \alpha)$$

and

$$T_r = \sum_{i=0}^{r} M_i \qquad \text{(for large } \alpha).$$

EXERCISE 5.39. Show that $\sum_{k,l}^{r} C_{kr}C_{lr}\sigma_{kl}$ appearing in the denominator of the expression for $u_\gamma + u_\omega$ is equal to

$$\sum_{k,l=0}^{r} C_{kr}C_{lr}\sigma_{kl} = \sum_{k=0}^{r} C_{kr}^2 p_k - \left(\sum_{k=0}^{r} C_{kr}p_k\right)^2 - \alpha\left[\sum_{k=0}^{r} C_{kr}p_k - \sum_{k=0}^{r-1} C_{k+1,r}p_k\right]^2.$$

This formula is more suitable for computational purposes.

EXERCISE 5.40. Show that

$$\lim_{\alpha\to 0} \alpha^2 C_{kr}^* = 2\binom{r-k+2}{2}.$$

5.8. TOLERANCE REGIONS

Suppose we have m balls with numbers $a_1 < a_2 < \cdots < a_m$ inscribed on them, one number to each ball. We select (without replacement) a sample of n balls at random from the m available. Denote the numbers on the chosen balls, in ascending order, as $X_{1,n} < X_{2,n} < \cdots < X_{n,n}$. The X's are random variables; they are the *order statistics* of the sample. (See also Section 4.2.)

Now let r, s be integers with $1 \le r < s \le n$. What is the distribution of the number, say G, of a's between $X_{r,n}$ and $X_{s,n}$? Or, in another form, how is the *proportion* G/m of the a's included between $X_{r,n}$ and $X_{s,n}$ distributed?

The event $G = g$ is the union of events $(X_{r,n} = a_u) \cap (X_{s,n} = a_{u+g+1})$ for $u = r, r+1, \ldots, m-g-1-n+s = t$, say. Note that there are $(r-1)$ X's and $(u-1)$ a's less than a_u, $(s-r-1)$ X's and g a's between a_u and a_{u+g+1}, and $(n-s)$ X's and $(m-u-g-1)$ a's greater than a_{u+g+1}. Hence

$$\Pr[G = g] = \sum_{u=r}^{t} \Pr[(X_{r,n} = a_u) \cap (X_{s,n} = a_{u+g+1})]$$

$$= \binom{m}{n}^{-1} \sum_{u=r}^{t} \binom{u-1}{r-1}\binom{g}{s-r-1}\binom{m-g-1-u}{n-s}$$

$$= \binom{m}{n}^{-1} \binom{g}{s-r-1} \sum_{u=r}^{t} \binom{u-1}{r-1}\binom{m-g-2-(u-1)}{n-s}$$

$$= \binom{m}{n}^{-1} \binom{g}{s-r-1}\binom{m-g-1}{n-s+r} \quad \text{(using Exercise 1.12).} \tag{5.122}$$

This formula is valid for $g = s-r-1, s-r, \ldots, m-r-n+s$.

It is important to note that (5.122) does not depend on the a's. It shows that the distribution of the proportion of the population included between the rth and sth order statistics of a random sample (without replacement) does not depend on the population distribution (provided there are no ties in the population).

The distribution of G/m corresponding to (5.122) is sometimes called the *tolerance distribution.* (Care should be taken not to confuse this with the same term used in bioassay for the distribution of threshold values of effective drug dosage.)

If we imagine m to increase without limit with the a's also varying in such a way that

$$\lim_{m\to\infty} a_{[mf]} = \xi_f \qquad [\text{with } F(\xi_f) = f] \tag{5.123}$$

for all f $(0 < f < 1)$, were $F(\cdot)$ is a continuous cdf, we then approach a situation in which $X_{1,n}, \ldots, X_{n,n}$ are order statistics in random samples of size n from a population in which X has the continuous distribution defined by the cdf $F(\cdot)$.

The distribution of the proportion of the population between $X_{r,n}$ and $X_{s,n}$, that is, of

$$Y_{r,s} = F(X_{s,n}) - F(X_{r,n}) = F_s - F_r,$$

can be derived by direct analysis, starting from the joint density function of F_r and F_s

$$p_{F_r,F_s}(f_r, f_s) = \frac{n!}{(r-1)!(s-r-1)!(n-s)!} f_r^{r-1}(f_s - f_r)^{s-r-1}(1-f_s)^{n-s}$$

$$(0 \le f_r \le f_s \le 1).$$

The result obtained is

$$p_{Y_{r,s}}(y) = \frac{n!}{(r-1)!(s-r-1)!(n-s)!} y^{s-r-1} \int_0^{1-y} f_r^{r-1}(1-y-f_r)^{n-s}\, df_r$$

$$= \frac{n!}{(s-r-1)!(n-s+r)!} y^{s-r-1}(1-y)^{n-s+r} \qquad (0 \le y \le 1).$$

(5.124)

$Y_{r,s}$ has a standard beta distribution with parameters $s - r$, $n - s + r + 1$ [cf. (2.20)].

From (5.124) we can calculate the probability that the population proportion included in the *tolerance interval* $X_{r,n}$ to $X_{s,n}$ lies between any two specified values.

Example 5.5. How large a sample is needed for the probability that more than 75% of the population is included between the second greatest and second least observation to exceed 90%?

We want to have $\Pr[Y_{2,n-1} > .75] > .90$. Since $Y_{2,n-1}$ has a standard beta distribution with parameters $(n-3, 4)$ we have

$$\Pr[Y_{2,n-1} > .75] = 1 - \Pr[Y_{2,n-1} \le .75]$$
$$= 1 - I_{.75}(n-3, 4) = I_{.25}(4, n-3),$$

where $I_y(\cdot,\cdot)$ is the incomplete beta function ratio [see (1.54)]. We require $I_{.25}(4, n-3) > .90$.

From tables of the incomplete beta function ratio (Pearson (1968)) we find

$$I_{.25}(4, 21) = .885, \qquad I_{.25}(4, 22) = .904.$$

The least value of n is therefore $22 + 3 = 25$.

EXERCISE 5.41. Show that $E[Y_{r,s}] = (s - r)/(n + 1)$

(i) By using the distribution of $Y_{r,s}$.

(ii) By calculating the probability that a newly selected X, X_{n+1} falls between $X_{r,n}$ and $X_{s,n}$.

The interested reader will find an extensive discussion of tolerance intervals and their uses in Guttman (1970).

5.9. MARKOV CHAINS

Markov chains provide one of the most well known and popular tools in applications of probability theory to real-world models involving uncertainty. In this section, *which is mainly of didactic and pedagogical interest*, we show how urn model concepts can be utilized to elucidate the presentation of this theory.

Much of the work in Chapters 3 and 4 could have been presented in terms of Markov chains. Also, in Section 5.4 we utilized urn models *instead* of Markov chains to describe basic concepts of stochastic learning theory. We now follow a line of approach formulated by Feller (1957).

Suppose we have m urns numbered (1), (2), ..., (m), respectively. These urns each contain balls numbered 1, 2, ..., m. The proportion of balls numbered j in the urn numbered (i) is p_{ij}. Of course,

$$p_{i1} + p_{i2} + \cdots + p_{im} = \sum_{j=1}^{m} p_{ij} = 1.$$

Starting with an arbitrarily chosen urn—say (a_1)—we draw a ball at random, note the number on it—say A_2—and return it to urn (a_1). We then take the next ball from urn (A_2), note its number—A_3—replace it in (A_2), and so on.

The sequence of values $a_1, A_2, A_3, \ldots$ form a *Markov chain*.* In such a sequence, the distribution of each variate is defined by the value of the immediately preceding variate in the chain. In fact,

$$\Pr[A_{h+1} = j \mid A_h = i] = p_{ij}. \tag{5.125}$$

The quantities p_{ij} are known as *transition probabilities*.

If the current urn (i) is regarded as representing the current "state" of a system, then p_{ij} represents the probability of transition from state (i) to state (j). It is convenient to think of such transitions as occurring one after another with a fixed time in between successive transitions.

* After A. A. Markov (1856–1922).

It is often of interest to consider transition probabilities after k (>1) transitions. Denoting these by ${}_k p_{ij}$, we have

$$ {}_k p_{ij} = \sum p_{ia_1} p_{a_1 a_2} \dots p_{a_{k-1} j}, \tag{5.126}$$

where the summation $\sum$ is over all paths $i, a_1, a_2, \dots, a_{k-1}, j$ of k steps from (i) to (j).

Formula (5.126) can be succinctly expressed using matrix notation. Defining the $m \times m$ *transition matrix*

$$\mathbf{P} = (p_{ij}) = \begin{pmatrix} p_{11} & p_{12} & \cdots & p_{1m} \\ p_{21} & p_{22} & \cdots & p_{2m} \\ \vdots & \vdots & \vdots & \vdots \\ p_{m1} & p_{m2} & \cdots & p_{mm} \end{pmatrix}, \tag{5.127}$$

we see that, since

$$ {}_2 p_{ij} = \sum_{a=1}^{m} p_{ia} p_{aj},$$

we have

$$ {}_2\mathbf{P} = ({}_2 p_{ij}) = \mathbf{P}^2 \tag{5.128}$$

and, similarly, for any positive integer k,

$$ {}_k\mathbf{P} = ({}_k p_{ij}) = \mathbf{P}^k. \tag{5.129}$$

There are a number of special kinds of Markov chains which have special names. If for a certain subset $\mathscr{S}$ of urns (indicated by being painted blue, say) the transition probabilities to any urn *outside* the subset are all zero, then once having sampled from any urn in $\mathscr{S}$, all subsequent samplings will be within $\mathscr{S}$. This set acts like a "trap" and does not let the sampler ever escape from it.

Such a set of states is called *closed*. If $\mathscr{S}$ contains only one state, that state is called an *absorbing state*.

Considering, for the sake of simplicity, the case where the first m_1 urns (states) constitute a closed set $\mathscr{S}_1$, the next m_2 urns a closed set $\mathscr{S}_2$, and so on for k closed sets in all $\mathscr{S}_1, \mathscr{S}_2, \dots, \mathscr{S}_k$ containing $m_1 + m_2 + \cdots + m_k = m' \le m$ of the m urns, we see that the transition matrix $\mathbf{P}$ can be partitioned

$$\mathbf{P} = \begin{pmatrix} \mathbf{P}_1 & \mathbf{0} & \cdots & \mathbf{0} & \mathbf{0} \\ \mathbf{0} & \mathbf{P}_2 & \cdots & \mathbf{0} & \mathbf{0} \\ \vdots & \vdots & \ddots & \vdots & \vdots \\ \mathbf{0} & \mathbf{0} & & \mathbf{P}_k & \mathbf{0} \\ \hdashline & & \mathbf{P}_{(R)} & & \end{pmatrix}, \tag{5.130}$$

where $\mathbf{P}_j$ is a square $(m_j \times m_j)$ matrix of transition probabilities within $\mathscr{S}_j$, the **0**'s in the last column in (5.130) each occupy $(m - m')$ columns of **P**, and $\mathbf{P}_{(R)}$ is a $(m - m') \times m$ matrix of transition probabilities from states not belonging to any of the closed sets $\mathscr{S}_1, \mathscr{S}_2, \ldots, \mathscr{S}_k$.

A Markov chain containing two or more closed sets ($k \geq 2$) is said to be *decomposable* or *reducible*. The last term refers to the fact that one can regard each $\mathscr{S}_j$ as if it were a single state. (This is the so-called *solidarity result*.)

Readers familiar with other concepts associated with Markov chains (such as irreducible chains, transient states, periodic chains, etc.) will have no difficulty in formulating them in terms of urns along the lines described.

EXERCISE 5.42. Show that, if a Markov chain is not decomposable, then there is a nonzero probability of reaching any state from any other state in at most $(m - 1)$ transitions.

Another kind of grouping of the urns arises when all the urns in the same group are indistinguishable from one another. This means that we can only observe the group in which a state is, rather than the state itself. In such circumstances we cannot expect to be able to obtain information on certain aspects of the Markov chain. Situations of this kind are described by Burke and Rosenblatt (1958) and Restle and Greeno (1970, Chapter 10).

5.10. A DECISION THEORY APPLICATION

Urn models can be used with advantage in the representation of problems in decision theory. In this section we describe in detail one such representation, due to Shepp (1969) and Boyce (1973), by way of illustration.

An (m, p) urn is defined as one containing m balls of value -1 (minus balls) and p balls of value $+1$ (plus balls), and one is allowed to draw balls randomly without replacement until one decides to stop. The *score* is the sum of the values of the chosen balls. The *value of an* (m, p) *urn* is defined as $V(m, p) = \max(0, E(m, p))$, where $E(m, p)$ is the expected value of drawing from an (m, p) urn. [If $E(m, p) > 0$, we draw; hence $V(m, p) = E(m, p) > 0$; but if $E(m, p) \leq 0$, $V(m, p) = 0$.]

The player can draw any number from zero to $n = (m + p)$ of the balls randomly, one at a time without replacement. He can stop at any time, even before drawing any balls at all. The objective is to draw in such a way as to maximize the *expected score* $V(m, p)$ on stopping.

Observe that for an (m, p) urn we draw a minus ball with probability $m/(m + p)$, which gives us an observed -1 and the opportunity to draw from an $(m - 1, p)$ urn; while drawing a plus ball with probability $p/(m + p)$

gives us $+1$ and the opportunity to draw from an $(m, p-1)$ urn. Thus the total expected value $E(m, p)$ of drawing from an (m, p) urn is

$$E(m, p) = \frac{m}{n}[-1 + V(m-1, p)] + \frac{p}{n}[+1 + V(m, p-1)], \quad (5.131)$$

where

$$V(m, p) = \max\{0, E(m, p)\}.$$

The following properties of $V(m, p)$ [stated in the form of theorems proved by Boyce (1973)] give additional insight to the structure of the "optimal" drawing policy.

Theorem 5.1. If $E(m, p) \geq 0$, then

(i) $V(m, p+1) \geq V(m, p) + (n+1)^{-1}$ (adding a plus ball *never* hurts).
(ii) $V(m, p+1) \leq V(m, p) + 1$.

Proof. By induction on $n = m + p$. The theorem is clearly true for $n = 1$; we thus assume its validity for n. Applying the recursive relation (5.131) to $E(m, p+1)$ and noting that $V(m, p) = \max\{0, E(m, p)\} \geq E(m, p)$, we have

$$V(m, p+1) \geq \frac{m}{n+1}(-1 + V(m-1, p+1)) + \frac{p+1}{n+1}[+1 + V(m, p-1)]. \quad (5.132)$$

When $E(m, p) \geq 0$, $V(m, p) = E(m, p)$ and from (5.131) we obtain

$$(n+1)V(m, p) = m[-1 + V(m-1, p)] + p[1 + V(m, p-1)] + V(m, p). \quad (5.133)$$

Subtracting (5.133) from $(n+1) \times (5.132)$, we obtain (when $E(m, p) \geq 0$)

$$\begin{aligned}(n+1)\{V(m, p+1) - V(m, p)\} &= m\{V(m-1, p+1) - V(m-1, p)\} \\ &\quad + p\{V(m, p) - V(m, p-1)\} + 1.\end{aligned}$$

By the induction hypothesis, the quantities in braces on the right-hand side are nonnegative; hence

$$(n+1)\{V(m, p+1) - V(m, p)\} \geq 1,$$

and part (i) is proved.

The proof of part (ii) is very similar. We may assume that $V(m, p+1) = E(m, p+1)$, and an argument similar to that in part (i) yields

$$(n+1)\{V(m, p+1) - V(m, p)\} \leq m + p + 1 = n + 1.$$

Other properties are discussed in Exercises 5.41 through 5.45. For computational purposes the following interesting result established by Boyce (1973) is useful.

Theorem 5.2. $D(m, p) = \binom{m+p}{m} \times V(m, p)$ is an integer.

The proof is, as before, by induction on m and p. Since both $V(0, 1)$ and $V(1, 0)$ are integers (the first being 1 and the second 0), the case $n = 1$ is trivial.

Assuming the validity of the theorem for $(n - 1)$, we multiply (5.132) and (5.133) by $\binom{m+p}{m}$ to obtain

$$D(m, p) = \max\left\{0, \binom{m+p}{m}\frac{m}{n}[-1 + V(m-1, p)] + \binom{m+p}{m}\frac{p}{n}[1 + V(m, p-1)]\right\},$$

but since

$$\binom{m+p}{m}\frac{m}{n} = \binom{m+p-1}{m-1} \quad \text{and} \quad \binom{m+p}{m}\frac{p}{n} = \binom{m+p-1}{m}$$

we obtain

$$D(m, p) = \max\left\{0, \binom{m+p-1}{m} - \binom{m+p-1}{m-1}\right\} + D(m-1, p) + D(m, p-1).$$

By the induction hypothesis this yields that $D(m, p)$ is an integer.

To simplify these expressions we introduce the notation

$$A(m, p) = \binom{m+p-1}{m} - \binom{m+p-1}{m-1}.$$

In view of the Pascal triangle relation (Section 1.1.3)

$$\binom{m+p}{m} = \binom{m+p-1}{m} + \binom{m+p-1}{m-1}$$

we have

$$A(m, p) = A(m-1, p) + A(m, p-1),$$

with the initial values $A(1, 0) = -1$, $A(0, 1) = 1$.

Now we are ready to present the computational algorithm for evaluation of $V(m, p)$.

Set $A(m, p) = D(m, p) = 0$, whenever m or p is negative, and set $A(0, 0) = A(1, 0) = -1$ and correspondingly $A(0, 1) = 1$. Compute recursively $A(m, p)$ and $D(m, p)$, using the relationships

$$A(m, p) = A(m-1, p) + A(m, p-1),$$

$$D(m, p) = \max\{0, A(m, p) + D(m-1, p) + D(m, p-1)\},$$

and

$$V(m, p) = \frac{\binom{m+p}{m}}{D(m, p)}.$$

We present in Tables 5.5 and 5.6, for convenience, values of $A(m, p)$ and $D(m, p)$ for $m, p < 10$ reproduced from Boyce (1973).

We present in Table 5.7 some values of $V(m, p)$. Observe from this table that when $m > p$ it is possible to increase the number of balls, maintaining the proportions of minus and plus balls but decreasing the *values* of the urn. [For example, $V(3, 2) = 0.20$, but $V(6, 4) = 0.07$.]

EXERCISE 5.43. Consider the urn (2, 1).

(i) Show that there is no method of drawing that yields a positive value. (So not drawing is an optimal strategy for this urn.)

TABLE 5.5

Values of $A(m, p)$

$p \backslash m$	0	1	2	3	4	5	6	7	8	9
9	1	8	35	110	275	572	1001	1430	1430	0
8	1	7	27	75	165	297	429	429	0	−1430
7	1	6	20	48	90	132	132	0	−429	−1430
6	1	5	14	28	42	42	0	−132	−429	−1001
5	1	4	9	14	14	0	−42	−132	−297	−572
4	1	3	5	5	0	−14	−42	−90	−165	−275
3	1	2	2	0	−5	−14	−28	−48	−75	−110
2	1	1	0	−2	−5	−9	−14	−20	−27	−35
1	1	0	−1	−2	−3	−4	−5	−6	−7	−8
0	0	−1	−1	−1	−1	−1	−1	−1	−1	−1

TABLE 5.6

Values of $D(m, p)$

$p \backslash m$	0	1	2	3	4	5	6	7	8	9
9	9	81	396	1388	3885	9165	18760	33796	53683	74131
8	8	64	280	882	2222	4708	8594	13606	18457	20448
7	7	49	189	527	1175	2189	3457	4583	4851	3421
6	6	36	120	290	558	882	1136	1126	697	0
5	5	25	70	142	226	282	254	122	0	0
4	4	16	36	58	70	56	14	0	0	0
3	3	9	15	17	12	0	0	0	0	0
2	2	4	4	2	0	0	0	0	0	0
1	1	1	0	0	0	0	0	0	0	0
0	0	0	0	0	0	0	0	0	0	0

(ii) Show that $E(2, 1) = 0$. (So an expected value of zero may be obtained by drawing as well as by not drawing.)

(iii) Verify that drawing until you obtain a plus is also an optimal strategy for the (2, 1) urn.

(An urn is called *neutral* if drawing from it may be included in or omitted from an "optimal" strategy as one chooses. Numerical calculations up to $p = 100$ show that there is no neutral urn other than a (2, 1) urn, but the conjecture that the (2, 1) urn is the only neutral urn is still open.)

TABLE 5.7

Values of $V(m, p)$

$p \backslash m$	0	1	2	3	4	5	6	7	8	9
9	9	8.10	7.20	6.31	5.43	4.58	3.75	2.95	2.11	1.53
8	8	7.11	6.22	5.35	4.49	3.66	2.86	2.11	1.43	0.84
7	7	6.13	5.25	4.39	3.56	2.76	2.01	1.34	0.75	0.30
6	6	5.14	4.29	3.45	2.66	1.91	1.23	0.66	0.23	0
5	5	4.17	3.33	2.54	1.79	1.12	0.55	0.15	0	0
4	4	3.20	2.40	1.66	1.00	0.44	0.07	0	0	0
3	3	2.25	1.50	0.85	0.34	0	0	0	0	0
2	2	1.33	0.67	0.20	0	0	0	0	0	0
1	1	0.50	0	0	0	0	0	0	0	0
0	0	0	0	0	0	0	0	0	0	0

EXERCISE 5.44. Prove assertion (ii) in Theorem 5.1.

EXERCISE 5.45 Prove the following analog of Theorem 5.1. If $E(m+1, p) \geq 0$ then:

(i) $V(m, p) \geq V(m+1, p) + (n+1)^{-1}$.
(ii) $V(m, p) \leq V(m+1, p) + 1$.

[Part (i) means that adding a minus ball *never* helps.]

EXERCISE 5.46. Show that for each p there is a maximum m for which $V(m, p) > 0$. [Shepp (1969) showed that this maximum m is approximately $p + \alpha\sqrt{2p}$, where $\alpha \doteqdot .84$ is the unique real root of the equation

$$\frac{\alpha}{1-\alpha^2} = \frac{\Phi(\alpha)}{\phi(\alpha)},$$

where $\Phi(\cdot)$ and $\phi(\cdot)$ are the standard normal integral and density defined in (1.43). Additional computational details are given by Boyce (1973).]

EXERCISE 5.47. Show that under an optimal play the ball drawn last is always a plus. [*Hint*: Observe that just before a minus ball is drawn in an optimal game, $m > 0$ and $E(m, p) \geq 0$, but $V(m-1, p) > V(m, p)$.]

The subject of "optimal stopping rules" is of considerable importance. Urn models and related concepts play a basic role in studies on this topic, as can be seen from a survey article by Gottinger (1975). The particular model we have discussed in this section is a prototype for a class of random urn models with applications to financial and marketing problems (Boyce, 1970).

References

Abakuks, A. (1974) A note on supercritical carrier-borne epidemics, *Biometrika*, **61**, 271–275.

Arnold, B. C. (1973) Response distributions for a generalized urn scheme under noncontingent reinforcement, *J. Math. Psychol.* **10**, 232–239.

Audley, R. J. (1957) A stochastic description of the learning behaviour of an individual subject, *Q. J. Exp. Psychol.*, **9**, 12–20.

Audley, R. J. and Jonckheere, A. R. (1956) The statistical analysis of the learning process, II, *Brit. J. Stat. Psychol.* **9**, 87–94.

Bailey, R. C. (1972) *A Montage of Diversity*, Ph.D. thesis, Emory University.

Barlow, R. E., Bartholomew, D. J., Bremner, J. M., and Brunk, H. D. (1972) *Statistical Inference under Order Restrictions*, New York: John Wiley and Sons.

Békéssy, A. (1963) On the classical occupancy problem, I, *Magy. Tud. Akad. Mat. Kutato Int. Kozl.*, **8**, 59–71.

Bell, G. (1974) Population estimates from recapture studies in which no recaptures have been made, *Nature*, **248**, 616.

Berg, S. (1974) Factorial series distributions, with applications to capture-recapture problems, *Scand. J. Stat.*, **1**, 145–152.

Berg, S. (1976) *Recurrence relation for the UMVU estimate in a multiple-recapture census*, Technical Report, No. 10, University of Lund.

Bernoulli, D. (1769–1770) Disquisitiones analyticae de novo problemate conjecturale, *Novi Comment. Acad. Sci. Imp. Petropolitanae*, **14**, par. 1, 3–25.

Binns, M. R. (1976) A sequential counting procedure for estimating the total number of randomly distributed individuals, *J. Am. Stat. Assoc.*, **71**, 74–79.

Boyce, W. M. (1970) Stopping rules for selling bonds, *Bell J. Econ. Manage. Sci.*, **1**, 27–53.

Boyce, W. M. (1973) On a simple optimal stopping problem, *Discrete Math.*, **5**, 297–312.

Bunday, B. D. (1975) Methods for population estimation by random sampling, *Powder Technol.*, 283–286.

Burke, C. J. and Rosenblatt, M. (1958) A Markovian function of a Markov chain, *Ann. Math. Stat.*, **29**, 1112–1122.

Burville, P. J. (1974) HEAPS: A concept in optimization, *J. Inst. Math. Appl.*, **13**, 263–278.

Burville, P. J. and Kingman, J. F. C. (1973) On a model for storage and search, *J. Appl. Prob.*, **10**, 697–701.

Bush, R. R. (1960) A survey of mathematical learning theory, in *Developments in Mathematical Psychology*, R. D. Luce (Ed.), New York: Free Press, pp. 120–165.

Bush, R. R. and Mosteller, F. (1951) A mathematical model for simple learning, *Psychol. Rev.*, **58**, 313–323.

Bush, R. R. and Mosteller, F. (1955) *Stochastic Models for Learning*, New York: John Wiley and Sons.

Chacko, V. J. (1963) Testing homogeneity against ordered alternatives, *Ann. Math. Stat.*, **34**, 945–956.

Chapman, D. G. (1952) Inverse multiple and sequential censuses, *Biometrics*, **8**, 286–306.

Chistyakov, V. P. (1964) On the calculation of the power of the test of empty boxes, *Theory Prob. Appl.*, **9**, 648–653 (English translation).

Chistyakov, V. P. and Viktorova, I. I. (1965) Asymptotic normality in a problem of balls when the probabilities of falling into different boxes are different, *Theory Prob. Appl.* **10**, 149–154 (English translation).

Cohen, J. E. (1966) *A model of simple competition*, Cambridge, Mass.: Harvard University Press.

Consul, P. C. and Mittal, S. P. (1975) A new urn model with predetermined strategy, *Biom. Z.*, **17**, Heft 2, 67–75.

Coombs, C. H., Davies, R. H., and Tversky, A. (1970) *Mathematical Psychology, An Elementary Introduction*, Chap. 9, Englewood Cliffs, N. J.: Prentice Hall.

Craig, C. C. (1953) On utilization of marked specimens in estimating the population of flying insects, *Biometrika*, **40**, 170–176.

Csörgö, M. and Guttman, I. (1962) On the empty cell test, *Technometrics*, **4**, 235–247.

Csörgö, M. and Guttman, I. (1964) On the consistency of the two-sample empty cell test, *Bull. Can. Math. Soc.*, **7**, 57–64.

Darling, D. A. and Robbins, H. (1967) Finding the size of a finite population, *Ann. Math. Stat.*, **38**, 1392–1397.

David, F. N. (1950) Two combinatorial tests of whether a sample has come from a given population, *Biometrika*, **37**, 97–110.

David, F. N. (1972a) Measurements of diversity, in *Proceedings of the 6th Berkeley Symposium on Mathematical Statistics and Probability Theory*, Vol. 1, Berkeley: University of California Press, pp. 631–648.

David, F. N. (1972b) Measurements of diversity: Multiple cell contents, in *Proceedings of the 6th Berkeley Symposium on Mathematical Statistics and Probability Theory*, Vol. 4, Berkeley: University of California Press, pp. 109–136.

David, F. N. and Barton, D. E. (1962) *Combinatorial Chance*, London: Griffin.

Decomps, B. and Kastler, A. (1963) Repartition de N particules entre g cellules; Loi de fluctuations, *C. R. Acad. Sci. (Paris)*, **256**(5), 1087–1089.

Denning, P. J. and Schwartz, S. C. (1972) Properties of the working-set model, *Commun. ACM*, **15**, 191–198.

Dietz, K. (1966) On the model of Weiss for the spread of epidemics by carriers, *J. Appl. Prob.*, **3**, 375–382.

Doctor, P. (1976) *Some Sequential Inference Problems*, Ph.D. thesis, Iowa State University.

Downton, F. (1967) Epidemics with carriers: A note on a paper by Dietz, *J. Appl. Prob.*, **4**, 264–270.

Estes, W. K. (1950) Towards a statistical theory of learning, *Psychol. Rev.*, **57**, 94–107.

Estes, W. K. (1959) The statistical approach to learning theory, in *Psychology: a Study of a Science*, Vol. II: *General Systematic Formulations, Learning and Special Processes*, S. Koch (Ed.), New York: McGraw-Hill.

Estes, W. K. (1972) Research and theory on learning probabilities, *J. Am. Stat. Assoc.*, **67**, 81–102.

Estes, W. K. and Burke, C. J. (1953) A theory of stimulus variations in learning, *Psychol. Rev.*, **60**, 276–286.

Ewens, W. J. (1969) *Population Genetics*, London: Methuen.

Ewens, W. J. (1972) The sampling theory of selectivity neutral alleles, *Theoret. Pop. Biol.*, **3**, 87–112.

Ewens, W. J. and Kirby, K. (1975) The eigenvalues of the neutral alleles process, *Theoret. Pop. Biol.*, **7**, 212–220.

Fagin, R. (1975) *Asymptotic miss ratios over independent references*, IBM Research Report Rc5415, Yorktown Heights, N.Y. (See also *Not. Am. Math. Soc.*, Nov. 1975, **22**(7), A-715).

Faucounau, J. (1975) Note sur une loi de probabilité discrète méconnue, *Math. et Sci. Humaines*, **52**, 55–68.

Feller, W. (1957) *An Introduction to Probability Theory and Its Applications*, Vol. I. 2nd ed., New York: John Wiley and Sons.

Glasser, G. J. (1963) Random numbers, sample selection and occupancy problems, *J. Roy. Stat. Soc. Ser. A*, **126**, 115–119.

Goldberg, S. (1961) *Introduction to difference equations*, New York: Wiley.

Goodman, L. A. (1953) Sequential sampling tagging for population size problems, *Ann. Math. Stat.*, **24**, 56–69.

Gottinger, H. W. (1975) *Sequential Analysis and Optimal Stopping*, Working Paper No. 31, Institute of Mathematical Economics, University of Bielefeld, Rheda, Germany.

Greenberg, B. G., Abdel-Latif, A. A-E., Simmons, W. R., and Horvitz, D. G. (1969) The unrelated question randomized response model: Theoretical framework, *J. Am. Stat. Assoc.*, **64**, 520–539.

Greeno, J. G. (1974) Representation of learning as discrete transition in a finite state space, in *Contemporary Developments in Mathematical Psychology*, D. H. Krantz *et al.* (Eds.), Vol. 1, San Francisco: W. H. Freeman, pp. 1–43.

Guttman, I. (1970) *Statistical Tolerance Regions: Classical and Bayesian*, Darien, Conn.: Hafner.

Harkness, W. L. (1970) The classical occupancy problem revisited, in *Random Counts in Physical Sciences*, G. P. Patil (Ed.), University Park, Pa.: Penn State University Press, pp. 107–126.

Harris, C. C. (1975) Estimating small populations by sampling, *Powder Technol.*, **12**, 85–91.

Hill, B. M. (1970) Zipf's law and prior distributions for the composition of a population, *J. Am. Stat. Assoc.*, **65**, 1220–1232.

Holst, L. (1971) Limit theorems for some occupancy problems, *Ann. Math. Stat.*, **42**, 1671–1680.

Itoh, Y. (1973) On a ruin problem with interaction, *Ann. Inst. Stat. Math.*, *Tokyo*, **25**, 635–641.

Johnson, N. L. and Kotz, S. (1969) *Distributions in Statistics: Discrete Distributions*, New York: John Wiley and Sons.

Johnson, N. L. and Kotz, S. (1972) *Distributions in Statistics: Continuous Multivariate Distributions*, New York: John Wiley and Sons.

Jones, H. L. (1959) How many of a group of random numbers will be useable in selecting a particular sample, *J. Am. Stat. Assoc.*, **54**, 102–122.

Karlin, S. and McGregor, J. (1972) Addendum to a paper of W. Ewens, *Theoret. Pop. Biol.*, **3**, 113–116.

Kingman, J. F. C. (1975) *Random Discrete Distributions*, *J. Roy. Stat. Soc.*, *Ser. B*, **37**(1), 1–22.

Kitabatake, S. (1958) A remark on a non-parametric test, *Math. Jap.*, **5**, 45–49.

Kryscio, R. J. (1974) On the extended simple stochastic model, *Biometrika*, **61**, 200–202.

Laurent, A. G. (1963) Probability distributions, factorial moments, empty cell tests, in *Classical and Contagious Discrete Distributions*, G. P. Patil (Ed.), Oxford: Pergamon Press.

Lawley, D. N. and Maxwell, A. E. (1971) *Factor Analysis as a Statistical Method*, 2nd ed., New York: American Elsevier.

Liu, P. T. and Chow, L. P. (1976) A new discrete quantitative randomized response model, *J. Am. Stat. Assoc.*, **71**, 72–73.

Luce, R. D. (1959) *Individual Choice Behavior: A Theoretical Analysis*, New York: John Wiley and Sons.

Mertz, D. B. and Davies, R. B. (1968) Cannibalism of the pupal stage by adult flour beetles, *Biometrics*, **24**, 247–275.

Mielke, P. W. and Siddiqui, M. M. (1965) A combinatorial test for independence of dichotomous responses, *J. Am. Stat. Assoc.*, **60**, 437–441.

Millward, R. B. and Wickens, T. D. (1974) Concept-identification models, in *Contemporary Developments in Mathematical Psychology*, Vol. 1, D. H. Krantz *et al.* (Eds.), San-Francisco: W. H. Freeman, 45–100.

Norman, M. F. (1972) *Markov Processes and Learning Models*, New York: Academic Press.

Okamoto, M. (1952) On a non-parametric test, *Osaka Math. J.*, **4**, 77–85.

Pearson, K. (Ed.) (1968) *Tables of the Incomplete Beta-Function* (2nd edition) Cambridge University Press.

Pollock, K. H. (1975) A *K*-sample tag-recapture model allowing for unequal survival and catchability, *Biometrika*, **62**, 577–583.

Rapoport, A. (1951) The probability distribution of distinct hits on closely packed targets, *Bull. Math. Biophys.*, **13**, 133–138.

Rashevsky, N. (1955) Note on a combinatorial problem in mathematical biology, *Bull. Math. Biophys.*, **19**, 163–170.

Rényi, A. (1962) Three new proofs and generalizations of a theorem of I. Weiss, *Magy. Tud. Akad. Mat. Kutató Int. Közl.*, **7**, 1–2, 203–214.

Restle, F. and Greeno, J. G. (1970) *Introduction to Mathematical Psychology*, Reading, Mass: Addison Wesley.

Schnabel, Z. E. (1938) The estimation of the total fish population of a lake, *Am. Math. Mon.*, **45**, 348–352.

Seber, G. A. F. (1973) *The Estimation of Animal Abundance and Related Parameters*, London: Griffin.

Sevast'yanov, B. A. (1969) The test of empty boxes and its generalizations, *Tbilis. Gos. Univ., Inst. Appl. Math.*, **2**, 229–233 (in Russian).

Sevast'yanov, B. A. and Chistyakov, V. P. (1964) Asymptotic normality in the classical ball problem, *Theory Prob. Appl.*, **9**, 198–211 (English translation). Also Letter to the Editors, *ibid.*, 513–514.

Shepp, L. A. (1969) Explicit solutions to some problems of optimal stopping, *Ann. Math. Stat.*, **40**, 993–1010.

Spence, K. W. (1956) *Behavior Theory and Conditioning*, New Haven: Yale University Press.

Sprott, D. A. (1957) Probability distributions associated with direct hits on targets, *Bull. Math. Biophys.*, **19**, 163–170.

Sternberg, S. (1963) *Stochastic Learning Theory* in *Handbook of Mathematical Psychology*, Vol. II, R. D. Luce, R. R. Bush, and E. Galanter (Eds.), New York: John Wiley and Sons, pp. 1–120.

Stevens, W. L. (1937) Significance of grouping, *Ann. Eugenics, London*, **8**, 57–69.

Thionet, P. (1971) Quelques problèmes de tirages sans remise dans l'urne, in *Studi di Probabilità, Statistica e Ricerca Operativa in Onore di Giuseppe Pompilj*, Gubbio: Oderisi, pp. 394–412.

Thurstone, L. L. (1919) *The Learning Curve Equation*, *Psychol. Rev., Monogr.*, **26**(3).

Thurstone, L. L. (1930) The learning function, *J. Gen. Psychol.*, **3**, 469–491.

Tukey, J. W. (1949) Moments of random group size distributions, *Ann. Math. Stat.*, **20**, 523–529.

Viktorova, I. I. and Chistyakov, V. P. (1966) Some generalizations of the test of empty boxes, *Theory Prob. Appl.*, **11**, 270–276 (English translation).

Warner, S. L. (1965) Randomized response: A survey technique for eliminating evasive answer bias, *J. Am. Stat. Assoc.*, **60**, 63–69.

Weiss, G. H. (1965) On the spread of epidemic carriers, *Biometrics*, **21**, 481–490.

Weiss, I. (1958) Limiting distributions in some occupancy problems, *Ann. Math. Stat.*, **29**, 878–884.

Wittes, J. T., Colton, T., and Sidel, V. W. (1974) Capture-recapture methods for assessing the completeness of case ascertainment when using multiple information sources, *J. Chron. Dis.*, **27**, 25–36.

Wittes, J. T. and Sidel, V. W. (1968) A generalization of the simple capture-recapture model with applications to epidemiological research, *J. Chron. Dis.*, **21**, 287–301.

Wolter, D. G. and Earl, R. W. (1969) The asymptotic values of response probabilities in the Audley–Jonckheere learning model, *Psychometrika*, **34**, 203–214.

Wolter, D. G. and Earl, R. W. (1972) The asymptotic distributions of response probabilities in the Audley–Jonckheere learning model, *Psychometrika*, **37**, 167–177.

Woodroofe, M. and Hill, B. M. (1972) On Zipf's law, *J. Appl. Prob.*, **12**, 425–434.

6

Limit Distributions for Urn Models

6.1. INTRODUCTION AND SUMMARY

The material in this chapter is more complicated (mathematically) than that in any of the earlier chapters. Although most of the results can be stated in a fairly simple manner, proofs are often rather complicated. In particular, cumbersome (though usually straightforward) algebraic manipulations are needed. We give some details of these proofs because we believe that more mathematically inclined readers will find them of interest.

Although we have tried to give the simplest of the available proofs, some of them do make use of concepts (e.g., Fourier and Laplace transforms, Rolle's theorem, absolute and uniform continuity, entire functions, contour integration, and saddle-point methods) we have not defined in earlier chapters. We felt it would add unduly to the length of the book, while not adding materially to the essential topics of the book, were we to provide such definitions.

The necessary mathematical tools can be found in many standard textbooks on real and complex analysis. A possible exception is the group of *saddle-point techniques.* This subject is lucidly treated in Sveshnikov and Tikhonov (1971) and Daniels (1954) gives some interesting applications.

In the field of limit distributions for occupancy and Pólya distributions (especially the former) there is available a great number of basic results with variants of varying importance and mathematical generality. A recent brief survey by Kolchin and Chistyakov (1974) lists over 150 major references in this field.

Therefore it was decided to present only selected results. The main criteria for inclusion are importance and relevance of a particular result in the general framework of this field of investigation, but another aim is exposition of a wide range of useful techniques and indication of the spirit of the subject sufficient to assist and encourage a determined reader to plunge into the lion's den of available literature on this topic. Even if this aim fails, we hope to arouse interest and curiosity in regard to the present direction of research and possible applications.

There are several relatively simple results on limit distributions, which are so well-known that we do not devote special attention to their derivations. For convenience, we list some of them here.

We suppose that n balls are randomly assigned to m urns, and that each ball is equally likely to fall in any one of the urns. M_r denotes the number of urns containing r balls after the assignments are completed.

1. (Exercise 3.4)

$$E[M_r] = m\binom{n}{r}\left(\frac{1}{m}\right)^r\left(1 - \frac{1}{m}\right)^{n-r} \qquad (r = 0, 1, \ldots, n).$$

If $n, m \to \infty$, with $nm^{-1} \to \lambda < \infty$, then

$$\lim_{n\to\infty} E[m^{-1}M_r] = \frac{\lambda^r}{r!}e^{-\lambda}.$$

For fixed n and r, with $m \to \infty$,

$$\lim_{m\to\infty} E[M_0] = \infty, \qquad \lim_{m\to\infty} E[M_1] = n,$$

and

$$\lim_{m\to\infty} E[M_r] = 0 \qquad (\text{for } r \geq 2).$$

2. [From (3.6)]

$$\Pr[M_0 = k] = \binom{m}{k}\Delta^{m-k}\left(\frac{\mathbf{0}}{m}\right)^n = (-1)^{m-k}\binom{m}{k}\Delta^{m-k}\left(1 - \frac{k + \mathbf{0}}{m}\right)^n.$$

If $n, m \to \infty$, with $me^{-n/m} \to \lambda$, then

$$\lim_{n\to\infty} \Pr[M_0 = k] = \frac{\lambda^k}{k!}e^{-\lambda}$$

(von Mises, 1939, 1964; Feller, 1957).

3. (Geiringer, 1938) If $n, m \to \infty$, with $nm^{-1} \to \lambda < \infty$, then the limit standardized distribution of M_r is unit normal, and

$$E[M_r] = m\binom{n}{r}\left(\frac{1}{m}\right)^r\left(1 - \frac{1}{m}\right)^{n-r} \sim \frac{\lambda^r}{r!}e^{-\lambda}$$

$$\frac{\operatorname{var}(M_r)}{E[M_r]} \sim 1 - \frac{\lambda^r e^{-\lambda}}{r!}\{1 + \lambda^{-1}(r - \lambda)^2\}.$$

4. If $n, m \to \infty$, with $nm^{-1} \to \lambda$, then the B–E (Bose–Einstein) expected proportion of urns (with r balls) given by

$$\frac{\binom{m+n-r-2}{n-r}}{\binom{m+n-1}{n}}$$

tends to $\lambda^r(1 + \lambda)^{-r-1}$ (Feller, 1957).

5. The B–E probability that a group of k *prescribed* urns contains a total of exactly j balls is

$$\frac{\binom{k+j-1}{j}\binom{m-k+n-j-1}{n-j}}{\binom{m+j-1}{j}}.$$

As $n, m \to \infty$, with $nm^{-1} \to \lambda$, this tends to

$$\binom{k+j-1}{j}\frac{\lambda^j}{(1+\lambda)^{k+j}}.$$

Note that the expressions in results 1 through 5, which represent *limits of expected values*, are of the same mathematical form as those for probabilities in Poisson, geometric, and negative binomial *distributions*.

6. From the randomized occupancy distribution [(3.62) with $p = m'/m$ and k replaced by m', and also $x = m'$], we see that the probability that each of a specified set of m' urns (out of the m urns), is occupied is

$$\frac{1}{m'^n}\Delta^{m'}\left\{\left(1 - \frac{m'}{m}\right)m' + \frac{m'}{m}\mathbf{0}\right\}^n = \sum_{h=0}^{m'}(-1)^h\binom{m'}{h}\left(1 - \frac{h}{m}\right)^n.$$

As $n, m \to \infty$, with $n/m \to \lambda$, this tends to $(1 - e^{-\lambda})^{m'}$.

The reader is urged to verify the limits in results 1 through 6 as exercises.

6.2. OCCUPANCY DISTRIBUTIONS

There is an almost bewildering array of results on limit distributions of occupancy distributions. We therefore start this section with a telescopic survey of a majority of the more notable results.

6.2.1. Telescopic Survey

Limit distribution theorems on occupancy distributions can be divided into four main categories according to the method(s) used in proof. We first note a well-tried method which is, unfortunately, *not* usually available. This is the representation of random variables as sums of independent random variables. For example, if we use indicator random variables Z_i to represent the empty ($Z_i = 0$) or nonempty ($Z_i = 1$) state of the ith urn, then $Z_1, Z_2, \ldots, Z_m$ are *not* usually independent.

The methods most completely used, with references to classic examples of their use, are:

1. Method of moments (Sevast'yanov, 1972; Harris and Park, 1971a; Okamoto, 1952; Weiss, 1949).
2. Reduction to conditional distributions of independent random variables (Kolchin, 1968; Dwass, 1969). (This is especially useful for multivariate extensions.)
3. Saddle-point methods (Kolchin, 1966, 1967; Sevast'yanov and Chistyakov, 1964).

In the following we suppose, as in Chapter 3, that there are m urns and n balls thrown at random, so that for each ball the probability of falling into the jth urn is p_j ($j = 1, 2, \ldots, m$; $p_1 + p_2 + \cdots + p_m = 1$).

Results on the Distribution of M_0 (Number of Empty Urns)

We first assume $p_1 = p_2 = \cdots = p_m = m^{-1}$ (the *equiprobable* case) and consider limiting cases as $m \to \infty$ and $n \to \infty$.

1. If $\frac{1}{2}n^2m^{-1} \to \lambda$, then

$$\Pr[M_0 - (m - n) = k] \to \frac{\lambda^k}{k!} e^{-\lambda} \qquad \text{(Békéssy, 1963).}$$

2. If $n^2m^{-1} \to \infty$ or $m^{-1} \exp(nm^{-1}) \to 0$ (or both), then

$$\Pr\left[\frac{M_0 - E[M_0]}{\sqrt{(\text{var}(M_0))}} < x\right] \to \Phi(x).$$

This result was established under the conditions $0 < c_1 \leq nm^{-1} \leq c_2$ by Weiss (1958). Rényi (1962) proved it under (*a*) $n^2m^{-1} \to \infty$ *and* $nm^{-1} < \infty$, and (*b*) $nm^{-1} \geq c_1 > 0$ *and* $m^{-1} \exp(nm^{-1}) \to 0$. It is a special case of a theorem proved by Sevast'yanov (1964).

3. If $nm^{-1} - \ln m \to \ln \lambda$,

$$\Pr[M_0 = k] \to \frac{\lambda^k}{k!} e^{-\lambda} \qquad \text{(Sevast'yanov and Chistyakov, 1964).}$$

4. For the occupancy distribution corresponding to B–E statistics [see (3.33)], Park (1976) has shown that, if $nm^{-3/4} \to \infty$ but $nm^{-6/5} \to 0$, then

$$\Pr\left[\frac{M_0 - E[M_0]}{\sqrt{\operatorname{var}(M_0)}} < x\right] \to \Phi(x).$$

The proof is obtained by considering the limiting values of factorial cumulants of the distribution of M_0.

Results on the Distribution of M_r (Number of Urns Containing Exactly r Balls Each)

We again assume $p_1 = p_2 = \cdots = p_m = m^{-1}$ and consider limiting cases as $m \to \infty$ and $n \to \infty$.

1. If $nm^{-1} = a_m + \ln m + r \ln \ln m$, with $a_m \to a$, then

$$\Pr[M_r = k] \to \frac{[(1/r!)e^{-a}]^k}{k!} \exp - \left(\frac{1}{r!} e^{-a}\right) \qquad \text{(Erdös and Rényi, 1961).}$$

[A similar result was established by von Mises (1939) and Castoldi (1955).]

2. For $r \geq 2$, with $a_r = (\alpha^r/r!)e^{-\alpha} \to 0$ (where $\alpha = n/m$) and $\lambda_r = ma_r$,

$$\Pr[M_r = k] = \frac{\lambda_r^k e^{-\lambda_r}}{k!} [1 + \tfrac{1}{2}m^{-1}\lambda_r\{1 + \alpha^{-1}(\alpha - r)^2\}\{1 - \lambda_r^{-1}(k - \lambda_r)^2\}$$
$$+ O(m^{-1}|k - \lambda_r|(1 + \alpha^{-1}) + m^{-2}\{\lambda_r^2 + (k - \lambda_r)^4\}(\alpha^2 + \alpha^{-2}))],$$

provided $\tfrac{1}{2}m^{-1}\lambda_r\{1 + \alpha^{-1}(\alpha - r)^2\} \leq 1$ (Kolchin, 1966).

3. Kolchin (1966, 1967) has obtained the results shown in Table 6.1 for the distribution of M_r, with $r \geq 2$. In this table we use (for convenience) the symbols

$$\alpha = nm^{-1} \tag{6.1a}$$

$$a_r = \frac{\alpha^r}{r!} e^{-\alpha} \quad (r > 1), \qquad a_1 = e^{-\alpha}, \tag{6.1b}$$

$$\lambda_r = ma_r. \tag{6.1c}$$

TABLE 6.1

Summary of Asymptotic Distributions

$\alpha \to$	$\lambda_r \to$	Other conditions	Asymptotic distribution of M_r or $(M_r - \lambda_r)\sigma_{rr}^{-1/2}$ $(r \geq 2)$
∞	0	$m = o(\alpha)$	Poisson (λ_r)
∞	Bounded	$\alpha = o(m)$	Poisson (λ_r)
∞	∞		Poisson (λ_r), Normal (0, 1)
$\alpha_1 \leq \alpha \leq \alpha_2 < \infty$	∞		— Normal (0, 1)
0	∞		Poisson (λ_r), Normal (0, 1)
0	Bounded		Poisson (λ_r)

NOTE: Except when $\alpha \to 0$, these results are also valid for M_0 and M_1.

We also define

$$\sigma_{rr} = a_r\{1 - a_r - a_r\alpha^{-1}(\alpha - r)^2\} \tag{6.1d}$$

and

$$\sigma_{rt} = -a_r a_t\{1 + \alpha^{-1}(\alpha - r)(\alpha - t)\}, \qquad r \neq t. \tag{6.1e}$$

If $\alpha \to 0$, then $M_0 - m + n$ has an asymptotic Poisson distribution with parameter $me^{-nm^{-1}} - m + n$. If this parameter tends to infinity, then M_0 has an asymptotic normal distribution. M_1 is an exceptional case when $\alpha \to 0$. For $\lambda_{10} = \frac{1}{2}[n - m\exp(-nm^{-1})] = o(n^{1/2})$ the distribution is asymptotically concentrated at the points $n - 2j$ for $j = 0, 1, \ldots, [\frac{1}{2}n]$, with

$$\Pr[M_1 = n - 2j] = \frac{\lambda_{10}^j}{j!} e^{-\lambda_{10}}\{1 + o(1)\}.$$

For $n = o(\lambda_{10}^2)$ the distribution is asymptotically concentrated at points $j = 0, 1, \ldots, n$, with

$$\Pr[M_1 = n - 2j] = \Pr[M_1 = n - 2j - 1] = \frac{1}{2}\frac{\lambda_{10}^j}{j!} e^{-\lambda_{10}}\{1 + o(1)\}.$$

If λ_{10} is of the order $O(n^{1/2})$, then if also

(*a*) $n^3m^{-3} \to \infty$, the distribution is asymptotically concentrated on $n - j$ $(j = 0, 1, \ldots, n)$.

(*b*) $n^3m^{-2} \to 0$, it is concentrated on $n - 2j$ $(j = 0, 1, \ldots, [\frac{1}{2}n]$.)

Kolchin (1966) also showed that

$$\sup_{0<\alpha<\infty} \min_i \sum_{k=0}^{\infty} |\Pr[M_r = k] - \pi_{ir}(k)| = \left(\frac{c_1 c_2^2}{9n}\right)^{1/3} (\ln n + (3r+1)\ln\ln n)^{2/3} \times \left\{1 + O\left(\frac{1}{\ln n}\right)\right\} \qquad (\text{for } r = 0, 2, 3, \ldots, i = 1, 2.)$$

while (for $r = 1$)

$$\sup_{\sqrt{n^{-1}\ln n} \leq \alpha < \infty} \min_i \sum_{k=0}^{\infty} |\Pr[M_1 = k] - \pi_{i1}(k)| = \left(\frac{c_1 c_2^2}{9n}\right)^{1/3} (\ln n + 4\ln\ln n)^{2/3} \left\{1 + O\left(\frac{1}{\ln n}\right)\right\}, \qquad i = 1, 2,$$

where

$$c_1 = \sqrt{2/(\pi e)} = 0.48394, \qquad c_2 = (3\sqrt{2\pi})^{-1}(1 + 4e^{-3/2}) = 0.25167,$$

$$(c_1 c_2^2/9)^{1/3} = 0.15045, \qquad \pi_{1r}(k) = (\lambda_r^k/k!)e^{-\lambda_r},$$

$$\pi_{2r}(k) = (\sqrt{2\pi}\,\sigma_{rr})^{-1}\phi((k - \lambda_r)\sigma_{rr}^{-1/2}).$$

We now no longer assume that $p_1 = p_2 = \cdots = p_m$.

The results for the asymptotic distribution of M_0 remain valid in nearly the same form. There are, in addition, some new results.

If $mp_j \leq c < \infty$ for $j = 1, \ldots, m$ and $\frac{1}{2}n^2 \sum_{j=1}^m p_j^2 \to \lambda$ as $n, m \to \infty$, then

$$\Pr[M_0 - m + n = k] \to \frac{\lambda^k}{k!} e^{-k} \qquad \text{(Chistyakov, 1964).}$$

If, however, the condition $\frac{1}{2}n^2 \sum_{j=1}^m p_j^2 \to \lambda$ is replaced by $0 < \alpha_0 \leq nm^{-1} \leq \alpha_1 < \infty$, then

$$\Pr\left[\frac{M_0 - E[M_0]}{\sqrt{(\mathrm{var}(M_0))}} < x\right] \to \Phi(x) \qquad \text{(Chistyakov, 1967).}$$

Holst (1971) has obtained this result with the condition $0 < \alpha_0 \leq nm^{-1}$ replaced by the conditions $nm^{-1} \to 0$ and $n^2 m^{-1} \to \infty$.

If the p's (i.e., the urns) are ordered so that $p_1 \leq p_2 \leq p_3 \leq \cdots \leq p_m$, and

$$E[M_0] = \sum_{j=1}^{m} (1 - p_j)^n \to \mu \qquad \text{and} \qquad (1 - p_j)^n \to \gamma_j,$$

then the pgf of M_0, $E[t^{M_0}]$, tends to

$$\exp\{-\lambda(t-1)\} \prod_{j=1}^{\infty} (1 - \gamma_j + \gamma_j t), \tag{6.2}$$

where $\lambda = \mu - \sum_{j=1}^{\infty} \gamma_j \geq 0$.

Chistyakov (1967) proved this result under the further assumption $nm^{-2} \to 0$. We discuss this problem in Section 6.2.2.

Results on Asymptotic Joint Distribution of $M_0, M_1, M_2, \ldots$

Assuming $p_1 = p_2 = \cdots = p_m = m^{-1}$, we have, if $0 < \alpha_0 \leq nm^{-1} \leq \alpha_1 < \infty$,

$$\Pr\left[\bigcap_{j=1}^{s} (M_{r_j} = m_{r_j})\right] = \frac{\exp\left\{-\frac{1}{2}\sum_{i=1}^{s}\sum_{j=1}^{s}\delta_{ij}u_i u_j\right\}}{(2\pi m)^{s/2}|\boldsymbol{\Delta}|^{1/2}}\{1 + O(m^{-1/2})\}$$

uniformly in $|u_i| \leq c$ and $\alpha_0 \leq nm^{-1} \leq \alpha_1$, where $\Delta = |(\sigma_{r_i,r_j})|$, δ_{ij} is the minor of σ_{r_i,r_j} in (σ_{r_i,r_j}), and $u_i = (m_{r_i} - \lambda_{r_i})/(|\boldsymbol{\Delta}|m)^{1/2}$ (Sevast'yanov and Chistyakov, 1964). (σ_{ij} and λ_r are defined in (6.1).)

If we do not assume equality among the p_j's, we can still obtain a "global central limit" result, namely, if $0 < \alpha_0 \leq nm^{-1} \leq \alpha_1 < \infty$ and $mp_{r_j} \leq c$ $(j = 1, 2, \ldots, s)$, then

$$\left|\Pr\left[\bigcap_{i=1}^{s}\left(\frac{M_{r_i} - E[M_{r_i}]}{\sqrt{m}} \leq x_i\right)\right] - \Phi(x_1, \ldots, x_s; (\sigma_{r_i,r_j}))\right| \to 0$$

uniformly with respect to $x_1, x_2, \ldots, x_s$ and $p_1, p_2, \ldots, p_m$, where $\Phi(x_1, \ldots, x_s; \mathbf{V})$ is the joint cdf of multinormal variables with a zero expected value vector and a variance-covariance matrix $\mathbf{V}$ (Chistyakov and Viktorova, 1965).

6.2.2. Poisson-Type Limits

We now consider limiting distributions of the M–B (Maxwell–Boltzmann) classical occupancy models. In other words, n indistinguishable balls are randomly distributed among m urns labeled $1, 2, \ldots, m$. Using the simplest, most straightforward methods yet derived, we will arrive at results similar to the central limit theorem for independent discrete random variables. These results are interesting from a mathematical viewpoint, because here we are dealing with dependent discrete random variables. However, they should not be considered as mere exercises. With the current availability of computers, we are able to check the accuracy of the limiting distributions. Moreover, theorems are now available, which present bounds on the speeds of convergence in these limit theorems (see e.g., Kolchin, 1966).

The methods used in the proofs include the somewhat artificial but ingenious concept of epgf's [see (1.166b)] and complex contour integration, especially saddle-point techniques. In general, the methods are not inherent in the problem. They are merely tools with which the desired results are obtained. We begin with the simplest and most direct solution of the following general problem.

A certain number of (indistinguishable) balls are distributed into a certain number of (labeled) urns according to a *specific* random mechanism. We study the probability distribution of the number of empty urns after the balls have been distributed. In all that follows we use the notation: n = number of balls, m = number of urns, and M_r = number of urns containing exactly r balls after n balls have been distributed among m urns. (As previously stated, we first consider only M_0, and the dependence on n and m are not explicitly indicated when no confusion arises.) We also define $\alpha = n/m$.

The first theorem (Sevast'yanov, 1972) is interesting, because it applies to the case where there are unequal probabilities of falling into the different urns (though, for a given urn, the probability is the same for each ball). In the proof M_0 is represented as a *sum* of dependent indicator random variables. This clearly shows the analogy between the present theorem and classic central limit theorems.

Theorem 6.1. Let p_i be the probability of a ball falling into the ith urn ($i = 1, 2, \ldots, m$), $p_i > 0$; $\sum_{i=1}^{m} p_i = 1$. If, as $n, m \to \infty$

(i) $\max_{1 \le i \le m} (1 - p_i)^n \to 0$, and
(ii) $\sum_{i=1}^{m} (1 - p_i)^n \to \lambda > 0$, then
(iii) $\lim_{n,m\to\infty} \Pr[M_0 = k] = (\lambda^k/k!)e^{-\lambda}$ $(k = 0, 1, 2, \ldots)$, that is, the limiting distribution of M_0 is Poisson with expected value λ.

Observe that, if $\lambda > 0$, conditions (i) and (ii) imply $n\min_{1 \le i \le m} p_i \to \infty$ as $n, m \to \infty$. Moreover, since $\min_{1\le i\le m} p_i \le m^{-1}$, we have $\alpha \to \infty$ or $m = o(n)$.

Proof. Let $M_0 = Y_1 + Y_2 + \cdots + Y_m$, where

$$Y_i = \begin{cases} 1 & \text{if the } i\text{th urn is empty,} \\ 0 & \text{otherwise.} \end{cases}$$

Let $b_i = \Pr[Y_i = 1]$, $b_{i_1 i_2 \cdots i_r} = \Pr[Y_{i_1} = Y_{i_2} = \cdots = Y_{i_r} = 1]$, where $i_k \neq i_g$ for $k \neq g$. Then $b_i = (1 - p_i)^n$, because $1 - p_i$ is the probability that a ball does *not* fall into the ith urn, hence $(1 - p_i)^n = \Pr$[none of the n balls fall into the ith urn] $= \Pr[i$th urn is empty$] = b_i$.

Analogously,

$$b_{i_1 i_2 \cdots i_r} = (1 - p_{i_1} - p_{i_2} - \cdots - p_{i_r})^n. \tag{6.3}$$

(Note that $p_{i_1} + \cdots + p_{i_r} = \Pr$[ball falls into one of the i_1th, $\ldots$, i_rth urns].) To prove this theorem, we use the method of moments. Recall that, if X is a Poisson random variable with parameter λ, the rth factorial moment of X

is λ^r, [cf. (2.13)]. Therefore, to show conclusion (iii) it is necessary to show only that

$$\lim_{n,m\to\infty} E[M_0^{(r)}] = \lambda^r.$$

Now it can be shown that

$$E[M_0^{(r)}] = \sum_{i_1}\sum_{i_2}\cdots\sum_{i_r} b_{i_1\cdots i_r}, \tag{6.4}$$

where the summation is taken over all collections $i_1, \ldots, i_r$ such that $h \neq k \Rightarrow i_h \neq i_k$. This can be seen by considering the special case $m = 2$ for $r = 1$ and $r = 2$.

For $r = 1$,

$$E(M_0) = E(Y_1 + Y_2) = E(Y_1) + E(Y_2) = b_1 + b_2.$$

For $r = 2$,

$$\begin{aligned} E[M_0^{(2)}] &= E(M_0(M_0 - 1)) = E[(Y_1 + Y_2)(Y_1 + Y_2 - 1)] \\ &= E[(Y_1 + Y_2)^2 - (Y_1 + Y_2)] \\ &= E[Y_1^2] + 2E[Y_1 Y_2] + E[Y_2^2] - E[Y_1] - E[Y_2]. \end{aligned}$$

Since Y_i takes only the values 0 and 1, $E[Y_i^2] = E[Y_i]$. Also, $Y_1 Y_2 \neq 0$ only when $Y_1 = Y_2 = 1$, in which case it is equal to 1. The probability of this is $b_{12} = b_{21}$.

Thus we have $E[M_0^{(2)}] = b_1 + 2b_{12} + b_2 - b_1 - b_2 = b_{12} + b_{21}$. Using induction, the reader can prove (6.4) as an exercise.

Now we define the sets

$$I_1(n) = \{i \in \{1, 2, \ldots, m\} : p_i > n^{-3/4}\} \qquad \text{(the exceptional set)}$$

and

$$I_r(n) = \{(i_1, \ldots, i_r)\text{: at least one } i_k \in I_1(n)\}.$$

Using this definition, we can break the right-hand side of (6.4) into two disjoint sums.

$$\sum_{i_1}\sum_{i_2}\cdots\sum_{i_r} b_{i_1\cdots i_r} = \sum_{I_r^c(n)} b_{i_1\cdots i_r} + \sum_{I_r(n)} b_{i_1\cdots i_r}. \tag{6.5}$$

Observe that by the definition of $I_1(n)$ and the properties of e^{-x} we have

$$\sum_{I_1(n)} (1 - p_i)^n \leq \sum_{I_1(n)} e^{-np_i} \leq \sum_{I_1(n)} e^{-n^{1/4}} \leq me^{-n^{1/4}} = o(ne^{-n^{1/4}}) \to 0 \tag{6.6}$$

as $n \to \infty$. [The last equality is due to the fact that $m = o(n)$.] Therefore, for all $r \geq 2$, we have

$$0 \leq \sum_{I_r(n)} (1 - p_{i_1})^n (1 - p_{i_2})^n \cdots (1 - p_{i_r})^n$$
$$\leq \left[r \left\{ \sum_{k=1}^{m} (1 - p_k)^n \right\}^{r-1} \right] \sum_{I_1(n)} (1 - p_i)^n \to 0.$$

The second inequality can be seen by the fact that every term on the left-hand side can be found on the right-hand side. Then the convergence takes place because the bracketed factor on the right-hand side tends to $r\lambda^{r-1}$ by condition (ii), and the other factor tends to 0 by (6.6). Substituting, we have

$$\lim_{n, m \to \infty} \sum_{I_r(n)} b_{i_1} b_{i_2} \cdots b_{i_r} = 0. \tag{6.7}$$

Analogously,

$$0 \leq \sum_{I_r(n)} (1 - p_{i_1} - \cdots - p_{i_r})^n \leq \sum_{I_r(n)} e^{-(p_{i_1} + \cdots + p_{i_r})n}$$
$$\leq r \left(\sum_{i=1}^{m} e^{-p_i n} \right)^{r-1} \sum_{I_1(n)} e^{-p_i n} \to 0,$$

whence

$$\lim_{n, m \to \infty} \sum_{I_r(n)} b_{i_1 \cdots i_r} = 0. \tag{6.8}$$

Therefore the second sum on the right-hand side of (6.5) vanishes, and we need consider only the first sum. Evaluating its limit is facilitated by the following preliminary arguments.

For $0 \leq x \leq \frac{1}{2}$, $-x - x^2 \leq \log(1 - x) \leq -x$; also, $p_i \leq \frac{1}{2}$, for $n \geq 3$, if $i \notin I_1(n)$ ($i \notin I_1(n) \Rightarrow p_i \leq n^{-3/4} \leq 3^{-3/4} < \frac{1}{2}$). Hence, for $i \notin I_1(n)$,

$$e^{-p_i n - p_i^2 n} = e^{n(-p_i - p_i^2)} \leq e^{n \log(1 - p_i)} = (1 - p_i)^n. \tag{6.9}$$

This implies $e^{-p_i n} \leq e^{p_i^2 n}(1 - p_i)^n$

Since

$$i \notin I_1(n) \Rightarrow p_i \leq n^{-3/4} \Rightarrow p_i^2 \leq n^{-3/2} \Rightarrow np_i^2 \leq n^{-1/2},$$

we have

$$e^{-p_i n} \leq e^{1/\sqrt{n}}(1 - p_i)^n.$$

Thus

$$\sum_{i=1}^{m} e^{-p_i n} = \sum_{I_1^c(n)} e^{-p_i n} + \sum_{I_1(n)} e^{-p_i n} \leq e^{1/\sqrt{n}} \sum_{i=1}^{m} (1 - p_i)^n + o(1) = O(1), \tag{6.10}$$

where the second sum goes to $o(1)$ or 0 by the definition of $I_1(n)$ and the first sum goes to λ by condition (ii).

Since $e^{-x} \geq (1 - x/n)^n$, it follows that

$$b_{i_1 \cdots i_r} \leq \exp\left(-n \sum_{j=1}^{r} p_{i_j}\right).$$

Also,

$$b_{i_1} b_{i_2} \cdots b_{i_r} \geq \exp\left\{-n \sum_{j=1}^{r} (p_{i_j} + p_{i_j}^2)\right\},$$

by (6.9). Also,

$$\begin{aligned} b_{i_1 \cdots i_r} &= [1 - (p_{i_1} + \cdots + p_{i_r})]^n \\ &\geq \exp[-n\{(p_{i_1} + \cdots + p_{i_r}) + (p_{i_1} + \cdots + p_{i_r})^2\}], \end{aligned}$$

and

$$b_{i_1} b_{i_2} \cdots b_{i_r} \leq e^{-np_{i_1}} e^{-np_{i_2}} \cdots e^{-np_{i_r}} = e^{-n(p_{i_1} + \cdots + p_{i_r})},$$

since $(1 - x)^n \leq e^{-nx}$. Together these inequalities yield

$$\exp\left[-n\left(\sum_{i=1}^{r} p_{i_k}\right)^2\right] \leq \frac{b_{i_1 \cdots i_r}}{b_{i_1} b_{i_2} \cdots b_{i_r}} \leq \exp\left[n \sum_{k=1}^{r} p_{i_k}^2\right]. \tag{6.11}$$

Thus, for any $(i_1, \ldots, i_r) \notin I_r(n)$, $n \geq 3$, we have

$$\exp(-r^2 n^{-1/2}) \leq \frac{b_{i_1 \cdots i_r}}{b_{i_1} b_{i_2} \cdots b_{i_r}} \leq \exp(rn^{-1/2})$$

Since $\exp(-r^2 n^{-1/2})$ and $\exp(rn^{-1/2})$ both converge uniformly to 1 as $n \to \infty$, we have, for any $i_1, \ldots, i_r \notin I_r(n)$,

$$\lim_{n, m \to \infty} \frac{b_{i_1 \cdots i_r}}{b_{i_1} b_{i_2} \cdots b_{i_r}} = 1 \qquad \text{uniformly in } I_r^c(n). \tag{6.12}$$

It follows that, for any $\varepsilon > 0$, one can find n_0 such that for all $n > n_0$ and all $(i_1, \ldots, i_r) \notin I_r(n)$,

$$b_{i_1} \cdots b_{i_r}(1 - \varepsilon) \leq b_{i_1 \cdots i_r} \leq b_{i_1} \cdots b_{i_r}(1 + \varepsilon). \tag{6.13}$$

Summing all three parts of (6.13) over $I_r^c(n)$. we obtain

$$(1 - \varepsilon) \sum_{I_r^c(n)} b_{i_1} \cdots b_{i_r} \leq \sum_{I_r^c(n)} b_{i_1 \cdots i_r} \leq (1 + \varepsilon) \sum_{I_r^c(n)} b_{i_1} \cdots b_{i_r}. \tag{6.14}$$

Therefore, for large n, it is only necessary to show that

$$\sum_{I_r^c(n)} b_{i_1} \cdots b_{i_r} \to \lambda^r.$$

To do this, we use the following expansion:

$$\sum_{I_r^c(n)} b_{i_1} \cdots b_{i_r} = \left(\sum_{i=1}^{m} b_i\right)^r - \sum_{I_r(n)} b_{i_1} \cdots b_{i_r} - \sum_{(i_1, \ldots, i_r)}^{*} b_{i_1} \cdots b_{i_r}, \tag{6.15}$$

where the last sum is taken over all sets $(i_1, \ldots, i_r)$ in which at least one number appears more than once.

Equation (6.15) is an algebraic identity. The right-hand side takes $(\sum_{i=1}^{m} b_i)^r$ and then removes all terms with (1) r different subscripts such that $(i_1, i_2, \ldots, i_r) \in I_r(n)$ and (2) any subscripts repeated. On the right-hand side of (6.15), the first term tends to λ^r [by condition (ii)], the second term tends to 0 [by (6.7)], and the third term tends to 0 [by condition (i)]. Hence $\lim_{n,m\to\infty} E[M_0^{(r)}] = \lambda^r$, and the theorem is proved.

A similar theorem was originally proved, under somewhat different conditions, by Chistyakov (1967). Chistyakov assumed that $nm^{-2} \to 0$ and $\frac{1}{2}n^2 \sum_{j=1}^{m} p_j^2 \to \lambda > 0$. It should be noted that, if $\sum_{i=1}^{m} (1 - p_i)^n$ is expanded using the binomial theorem and the terms are reordered, it appears in the form

$$m - n + \frac{n^{(2)}}{2} \sum_{j=1}^{m} p_j^2 - \frac{n^{(3)}}{3!} \sum_{j=1}^{m} p_j^3 + \cdots. \tag{6.16}$$

The ratio of the third term to $(n^2/2) \sum_{j=1}^{m} p_j^2$ (which tends to λ under Chistyakov's conditions) tends to 1 as $n \to \infty$. Condition (ii) in Theorem 6.1 requires that $\sum_{i=1}^{m} (1 - p_i)^n$ tend to a finite limit as $n, m \to \infty$. In view of the difference in the assumptions, Chistyakov's result is for the limiting distribution of the *centered* quantity $M_0 - (m - n)$, that is, $\Pr[M_0 - (m - n) = k] \to (\lambda^k/k!)e^{-\lambda}$. [Observe the leading two terms in (6.16).]

Chistyakov (1967) also proved the following theorem.

Theorem 6.2. If $n \to \infty$, $m \to \infty$, with $nm^{-2} \to 0$, $E[M_0] \to \mu < \infty$, and $(1 - p_j)^n \to \gamma_j < \infty$ $(j = 1, 2, \ldots)$, then

$$E[t^{M_0}] \to e^{\lambda(t-1)} \prod_{j=1}^{\infty} (1 - \gamma_j + \gamma_j t), \tag{6.17}$$

where $\lambda = \mu - \sum_{j=1}^{\infty} \gamma_j$.

This implies that the limiting distribution of M_0 is that of the sum of a random variable X_0 having a Poisson distribution and random variables $X_1, X_2, \ldots$, with $\Pr[X_i = 1] = \gamma_j = 1 - \Pr[X_i = 0]$, all the variables being mutually independent.

The theorem was proved using the saddle-point method.

Partially for historical reasons, but also on account of its methodological interest, we mention a result obtained by von Mises (1939). This states that

$$\lim_{\substack{n\to\infty \\ m\to\infty}} \left\{ \Pr[M_r = k] - \frac{\lambda(r; m; \rho)}{k!} e^{-\lambda(r; m; \rho)} \right\} = 0 \tag{6.18}$$

[where $\lambda(r; m; \rho) = me^{-\rho}\rho^r/r!$, with $\rho = n/m$] provided $\lambda(r; m; \rho)$ is bounded as $n \to \infty$, $m \to \infty$.

Castoldi (1955) has extended this to the joint distribution of $M_{r_1}, M_{r_2}, \ldots, M_{r_s}$. He shows that

$$\lim_{\substack{n\to\infty \\ m\to\infty}} \left\{ \Pr\left[\bigcap_{j=1}^{s} (M_{r_j} = k_j) \right] - \prod_{j=1}^{s} \left[\frac{\lambda(r_j; m_j; \rho_j)}{k_j!} e^{-\lambda(r_j; m_j; \rho_j)} \right] \right\} = 0, \tag{6.19}$$

where $m_j = m - \sum_{i=1}^{j} k_i$, $n_j = n - \sum_{i=1}^{j} k_i r_i$, $\rho_j = n_j/m_j$, provided each $\lambda\cdot;\cdot;\cdot)$ is bounded as $n \to \infty$, $m \to \infty$. These conditions imply some asymptotic constraints on the k_j's.

Joint distributions of several M_r's are discussed in Section 6.2.6.

We now turn our attention to the more general question of the limiting distribution of M_r for $r > 0$. Recall that M_r is the number of urns containing exactly r balls after n indistinguishable balls have been randomly distributed into m urns. We begin this section with an elementary result from the end of a paper by Erdös and Rényi (1961), the main body of which attacked the inverse of our current problem, namely, the distribution of N_k, the number of balls needed to provide at least k balls in each urn.

Recall from (3.15) that the actual distribution function of M_r is

$$\Pr[M_r = k] = \sum_{j=k}^{m} (-1)^{j+k} \binom{j}{k} S_j, \tag{6.20}$$

where

$$S_j = \binom{m}{j} \frac{n!}{(r!)^j (n - rj)!} \frac{(m-j)^{n-rj}}{m^n}. \tag{6.21}$$

Von Mises (1939) shows, using an adaptation of the *continuity theorem of moments* that, if $n, m \to \infty$ in such a way that $\alpha = n/m \to c < \infty$, and such that $E[M_r] \to \lambda < \infty$, then $\Pr[M_r = k] \to \lambda^k e^{-\lambda}/k!$.

The condition $E(M_r) \to \lambda$ necessitates $\alpha = O(\ln m)$. Erdös and Rényi (1961) proved the following theorem.

Theorem 6.3. If

$$n = n(m) = m \ln m + rm \ln \ln m + nx + o(m), \tag{6.22}$$

which implies $n = O(m \ln m)$, then as $n \to \infty$

$$\Pr[M_r = k] \to \frac{(e^{-x}/r!)^k \exp(-e^{-x}/r!)}{k!}. \tag{6.23}$$

[Note that x is defined via (6.22) and the limiting distribution of M_r is Poisson with expected value $e^{-x}/r!$.]

[As we have seen in Section 3.1.2, the explicit expression for $E[M_r]$ is

$$m\binom{n}{r}\left(\frac{1}{m}\right)^r\left(1-\frac{1}{m}\right)^{n-r}.\Bigg]$$

Outline of Proof. First it is shown that

$$\lim_{n, m\to\infty} S_j = \frac{(e^{-x}/r!)^j}{j!}, \tag{6.24}$$

where the n in (6.24) is

$$n(m) = m \ln m + rm \ln \ln m + xm + o(m). \tag{6.22$'$}$$

Now, by (6.21),

$$S_j = \frac{1}{j!(r!)^j}\, m^{(j)} \frac{n!}{(n-rj)!}\, m^{-rj}\left(1-\frac{j}{m}\right)^{n-rj}.$$

Therefore, to prove (6.24), it is only necessary to show that

$$m^{(j)} \frac{n!}{(n-rj)!}\frac{1}{m^{rj}}\left(1-\frac{j}{m}\right)^{n-rj} \to e^{-jx}. \tag{6.25}$$

Note that

$$\left(1-\frac{j}{m}\right)^{n-rj}\Big/ e^{-(n-rj)j/m} \to 1 \text{ and}$$

$$e^{-(n-rj)j/m} = e^{-nj/m}e^{rj^2/m} = e^{-j(\ln m + r\ln\ln m + x + o(m)/m)}e^{rj^2/m}$$

$$= \frac{1}{m^j}\frac{1}{(\ln m)^{rj}}\, e^{-jx}e^{-j(o(m)/m)+rj^2/m} \to \frac{1}{m^j}\frac{1}{(\ln m)^{rj}}\, e^{-jx}.$$

Hence the whole expression on the left-hand side of (6.25) approaches

$$m^{(j)}n^{(rj)}\frac{1}{m^{rj}}\frac{1}{m^j}\frac{1}{(\ln m)^{rj}}\, e^{-jx} \to \frac{m^j}{m^j}\frac{n^{rj}}{m^{rj}}\frac{1}{(\ln m)^{rj}}\, e^{-jx} = \frac{(n/m)^{rj}}{(\ln m)^{rj}}\, e^{-jx}$$

$$= \left(\frac{\ln m + r\ln\ln m + x + o(m)/m}{\ln m}\right)^{rj} e^{-jx} \to e^{-jx},$$

and (6.25) is proved. Hence (6.24) is also proved.

Now, by (6.20) and (6.24),

$$\Pr[M_r = k] \to \sum_{j=k}^{m} (-1)^{j+k} \binom{j}{k} \frac{(e^{-x}/r!)^j}{j!} = \sum_{j=0}^{m-k} (-1)^j \binom{j+k}{k} \frac{(e^{-x}/r!)^{j+k}}{(j+k)!}$$

$$= \frac{(e^{-x}/r!)^k}{k!} \left\{ \sum_{j=0}^{m-k} (-1)^j \frac{1}{j!} \left(\frac{e^{-x}}{r!} \right)^j \right\}$$

$$= \frac{(e^{-x}/r!)^k}{k!} \left(e^{-(e^{-x}/r!)} - \sum_{j=m-k+1}^{\infty} (-1)^j \frac{1}{j!} \left(\frac{e^{-k}}{r!} \right)^j \right)$$

$$\to \frac{(e^{-x}/r!)^k}{k!} e^{-(e^{-x}/r!)}.$$

Thus (6.23) is proved.

Remark. In this, and in Theorem 6.1, it was assumed that $\alpha \to \infty$. If $\alpha \to 0$, with $\alpha^r(e^{-\alpha}/r!)m \to c > 0$, this result holds only for $r \geq 2$. However, there are separate Poisson results for $r = 0$ and $r = 1$. In particular, if $\alpha \to 0$, and $\lim_{n,m\to\infty} m\alpha^2 = \gamma$, we have

$$\Pr[M_0 - m + n = k] \to \frac{(\gamma/2)^k e^{-\gamma/2}}{k!} \tag{6.26}$$

and

$$\Pr\left[\frac{n - M_1}{2} < x\right] \to \sum_{k=0}^{[x]} \frac{(\gamma/2)^k e^{-\gamma/2}}{k!} \quad \text{(Békéssy, 1963).} \tag{6.27}$$

6.2.3. Normal Limit Distributions

In the preceding section we saw that under many circumstances Poisson limit distributions can be obtained. However, there is a considerably wider range of circumstances under which asymptotically normal distributions are encountered. A pioneer in this respect was Sherman (1950). He proved that, when n balls are randomly distributed among m urns, then, in the equiprobable case, as $n, m \to \infty$, with

$$\alpha = nm^{-1} \to c > 0,$$

the limit standardized distribution corresponding to the number of empty urns is normal. (See also Okamoto, 1952.) Weiss (1956) devised a new proof of the same result. Sherman, Okamoto, and Weiss all used the method of moments in their proofs. They showed that the moment ratios of M_0 converge to the corresponding moment ratios of the normal distribution.

Harris and Park (1971) proved a slightly stronger result, also using the method of moments. They allow α to tend to zero, provided $nm^{-5/6} \to \infty$, and allow α to tend to infinity, provided $\alpha - \frac{1}{3}\ln m \to -\infty$. An even stronger result was obtained by Rényi (1962). This allows α to tend to zero, provided only $nm^{-1/2} \to \infty$, and allows α to tend to infinity, provided only $\alpha - \ln m \to -\infty$. Although Rényi's work preceded that of Harris and Park, the latter authors used simpler methods.

Their method consists of estimating the factorial cumulants [see (1.163)] of the standardized random variable

$$M_0' = \frac{M_0 - E[M_0]}{(\text{var}(M_0))^{1/2}}.$$

These in turn determine the cumulants [see (1.164a)].

Using this method, Harris and Park (1971) showed that (under their conditions) the cumulants of M_0' tend to the values $\kappa_r = 0\ (r > 2)$ appropriate to the unit normal distribution. Since the normal distribution is characterized by its moments (cf. Section 2.2.2) this suffices to show that the limit distribution of M_0' is normal. [David and Barton (1962) used cumulants to prove the theorem under the condition $\alpha \to c > 0$.]

We now give some details of the proof as given in Harris and Park (1971).

Theorem 6.4.

$$M_0' = \frac{M_0 - E[M_0]}{(\text{var}(M_0))^{1/2}}$$

has a normal limit distribution when $m \to \infty$ and

(i) $nm^{-1} \to c > 0$, or
(ii) $nm^{-1} \to 0$ and $nm^{-5/6} \to \infty$, or
(iii) $nm^{-1} \to \infty$ and $3nm^{-1} - \ln m \to -\infty$.

Proof. From (3.12) the factorial moment generating function of M_0 is

$$\begin{aligned}\phi^*_{M_0}(s) &= \sum_{j=0}^{m} \left(\frac{m-j}{m}\right)^n \frac{m^{(j)}}{j!} s^j \\ &= \sum_{j=0}^{m} \binom{m}{j}\left(1 - \frac{j}{m}\right)^n s^j. \end{aligned} \tag{6.28}$$

Since $\kappa_r(M_0') = \kappa_r(M_0)\{\kappa_2(M_0)\}^{-(1/2)r}$, we need to prove that

$$\kappa_r(M_0)\{\kappa_2(M_0)\}^{-(1/2)r} \to 0$$

for all integers r greater than 2. We need to use two lemmas, which are now stated.

Lemma 6.1. If $P(s)$ is a polynomial of degree p in s, with p real roots in $[-1, 0]$, then $(sD)^h P(s)$ has p real roots in $[-1, 0]$ for h any positive integer.

Lemma 6.2. $\kappa_r(M_0) = O(m)$, for $r = 1, 2, \ldots$. This is in itself quite remarkable, because it says that the order of magnitude of $\kappa_r(M_0)$ does not depend on n, the number of balls.

Proof of Lemma 6.1. Clearly, $sDP(s) = sP'(s)$ is of degree p and, by Rolle's theorem, the roots of $P'(s)$ fall in the same interval $[-1, 0]$ as those of $P(s)$. The factor s introduces a root at zero. The result follows by induction on h.

Proof of Lemma 6.2. Define

$$P(s) = (1 + s)^m = \sum_{j=0}^{m} \binom{m}{j} s^j$$

and (noting that $(sD)^n \equiv (sD)(sD) \cdots (sD)$)

$$P_n(s) = (sD)^n P(s) = \sum_{j=1}^{m} \binom{m}{j} j^n s^j. \tag{6.29}$$

$P(s)$ has m zeros at $s = -1$, and $P_n(s)$ has a simple root at zero for $n \geq 1$. The remaining $(m - 1)$ roots of $P_n(s)$ must therefore lie in $[-1, 0)$. Thus we can write

$$P_n(s) = m^n s \prod_{j=1}^{m-1} (s - \gamma_j), \tag{6.30}$$

with $-1 \leq \gamma_j < 0$. From (6.28) and (6.30),

$$\begin{aligned}\phi^*_{M_0}(s) = \sum_{j=0}^{m} \binom{m}{j} \left(\frac{m - j}{m}\right)^n s^j &= \sum_{g=0}^{m} \binom{m}{m - g} \left(\frac{g}{m}\right)^n s^{m-g} \\ &= \sum_{g=0}^{m} \binom{m}{g} \left(\frac{g}{m}\right)^n s^{m-g} \\ &= m^{-n} s^m P_n(s^{-1}).\end{aligned} \tag{6.31}$$

From (6.30) and (6.31),

$$\phi^*_{M_0}(s) = s^{m-1} \prod_{j=1}^{m-1} (s^{-1} - \gamma_j) = \prod_{j=1}^{m-1} (1 - \gamma_j s). \tag{6.32}$$

For $|s| < 1$ and so $|\gamma_j s| < 1$ $(j = 1, 2, \ldots, m - 1)$,

$$\ln \phi^*_{M_0}(s) = \sum_{j=1}^{m-1} \ln(1 - \gamma_j s) = -\Gamma_1 s - \frac{\Gamma_2}{2} s^2 - \frac{\Gamma_3}{3} s^3 - \cdots, \tag{6.33}$$

where $\Gamma_u = \sum_{j=1}^{m-1} \gamma_j^u$.

Since also (by definition)

$$\ln \phi^*_{M_0}(s) = \sum_{j=1}^{\infty} \frac{\kappa_{(j)}(M_0)}{j!} s^j, \tag{6.34}$$

we see that

$$\frac{\kappa_{(r)}(M_0)}{r!} = -\frac{\Gamma_r}{r}, \tag{6.35}$$

and (because $|\gamma_j| \le 1$)

$$\frac{|\kappa_{(r)}(M_0)|}{r!} \le \frac{1}{r}\sum_{j=1}^{m-1} |\gamma_j^r| \le \frac{m-1}{r}. \tag{6.36}$$

From (1.164a) and (6.36),

$$|\kappa_r(M_0)| \le \sum_{j=1}^{r} |\Delta^j \mathbf{0}^r|(m-1)j^{-1} = (m-1)\sum_{j=1}^{r} j^{-1}|\Delta^j \mathbf{0}^r|,$$

and since the last summation does not depend on n, Lemma 6.2 is established.

In view of this result, Theorem 6.4 will be proved if we can prove that $m\{\kappa_2(M_0)\}^{-(1/2)r} \to 0$ for $r = 3, 4, \ldots$. It suffices to show that $m\{\kappa_2(M_0)\}^{-3/2} \to 0$ [since if this is so, we must have $\kappa_2(M_0) \to \infty$].

From (3.13),

$$\begin{aligned}\kappa_2(M_0) &= m(m-1)(1-2m^{-1})^n + m(1-m^{-1})^n - m^2(1-m^{-1})^{2n}\\ &= m^2\{(1-2m^{-1})^n - (1-m^{-1})^{2n}\}\\ &\quad + m\{(1-m^{-1})^n - (1-2m^{-1})^n\}.\end{aligned}$$

Under any of the conditions (i) through (iii) of the theorem, $\alpha^k = n^k m^{-k} = o(m)$ for any positive integer k. Hence, under any of these conditions,

$$\begin{aligned}\kappa_2(M_0) &= -m\alpha e^{-2\alpha} + O(\alpha^2) + me^{-\alpha}(1-e^{-\alpha}) + O(\alpha)\\ &= me^{-\alpha}(1-e^{-\alpha}-\alpha e^{-\alpha}) + O[\max(\alpha, \alpha^2)].\end{aligned}$$

We now take each condition separately:

(i) If $\alpha c > 0$, then $m\{\kappa_2(M_0)\}^{-3/2} = O(m^{-1/2}) \to 0$.

(ii) If $\alpha \to 0$, then $\{\kappa_2(M_0)\}^3 m^{-2} = me^{-3\alpha}(1-e^{-\alpha}-\alpha e^{-\alpha})^3$ + other terms. This quantity tends to infinity [and so $m\{\kappa_2(M_0)\}^{-3/2} \to 0$] if

$$me^{-3\alpha}(1-e^{-\alpha}-\alpha e^{-\alpha})^3 \to \infty, \tag{6.37}$$

that is, if

$$m(\alpha^2)^3 \to \infty$$

or

$$\frac{n^6}{m^5} \to \infty;$$

that is,

$$\frac{n}{m^{5/6}} \to \infty.$$

(iii) If $\alpha \to \infty$, then (6.37) is satisfied if

$$me^{-3\alpha} \to \infty,$$

that is,

$$\ln m - 3\alpha \to \infty.$$

Harris and Park (1971) also point out that, from (6.32) and the fact that the factorial moment generating function of a random variable Y_j with

$$\Pr[Y_j = 1] = -\gamma_j = 1 - \Pr[Y_j = 0] \tag{6.38}$$

is

$$1 - \gamma_j s,$$

it follows that we have the representation

$$M_0 = Y_1 + Y_2 + \cdots + Y_{m-1}, \tag{6.39}$$

with the Y's *mutually independent*. (cf. (6.3).)

Park (1976) has applied a similar method to establish conditions for asymptotic normality when B–E statistics apply. An outline of his analysis follows.

For B–E statistics [see (3.33b)],

$$E[M_0^{(k)}] = \frac{m^{(k)}\binom{n+m-k-1}{n}}{\binom{n+m-1}{n}}. \tag{6.40}$$

The factorial moment generating function of M_0 is

$$\phi^*_{M_0}(s) = \sum_{j=0}^{m-1} \frac{\binom{m}{j}\binom{n+m-j-1}{n}}{\binom{n+m-1}{n}} s^j. \tag{6.41}$$

Park shows that the rth cumulant of M_0 is $O(m)$ (as in the M–B case) and

$$\kappa_2(M_0) = m\alpha^2(1+\alpha)^{-3} + O(\alpha(1+\alpha)^{-1}).$$

As before, we require $\{\kappa_3(M_0)\}^2\{\kappa_2(M_0)\}^{-3} \to 0$ or, equivalently,

$$m^{-2}\{\kappa_2(M_0)\}^3 \to \infty.$$

This will be so if

1. $\alpha \to c > 0$, or
2. $\alpha \to 0$ and $m\alpha^6 \to \infty$ (i.e., $nm^{-5/6} \to \infty$), or
3. $\alpha \to \infty$ and $m\alpha^{-3} \to \infty$ (i.e., $mn^{-3/4} \to \infty$).

We now proceed to give some details of a method of proof used by Rényi (1962). As already indicated, Rényi uses a different method of proof involving the introduction of complex variable techniques.

We introduce the characteristic function

$$\phi_{M_0}(s) = E[\exp(iM_0 s)] \tag{6.42}$$

and the characteristic function generating function

$$G_m(s, z) = \sum_{j=0}^{\infty} \theta_{j,m}(s)\frac{(mz)^j}{j!}, \tag{6.43}$$

where $\theta_{j,m}(s)$ is the characteristic function of M_0 when j balls are used (with m urns).

The plan of the proof is as follows. We have [cf. (3.81)]

$$G_m(s, z) = (e^{is} + e^z - 1)^m.$$

By Cauchy's formula,

$$\theta_{n,m}(s) = \frac{n!}{m^n 2\pi i}\oint_{|z|=c}\frac{(e^{is}+e^z-1)^m}{z^{n+1}}\,dz, \tag{6.44}$$

for any $c > 0$.

The integral is approximated by transforming to polar coordinates and splitting the range of integration $[-\pi, \pi]$ of the resultant line integral into parts, one of which is $[-n^{1/7-1/2}\pi, n^{1/7-1/2}\pi]$. [Some details of a similar method used by Holst (1971) are given in connection with Theorem 6.12.]

In this way we obtain an asymptotic expansion for $\theta_{n,m}(s\{\text{var}(M_0)\}^{-1/2})$, hence for

$$\{\exp[-isme^{-\alpha}\{\text{var}(M_0)\}^{-1/2}]\}\theta_{n,m}(s\{\text{var}(M_0)\}^{-1/2}), \tag{6.45}$$

which is asymptotically equivalent to $\phi_{M_0}(s)$, because $E[M_0] = m(1-m^{-1})^n \sim me^{-\alpha}$. The asymptotic expression for $\phi_{M_0}(s)$ is found to be $\exp(-\frac{1}{2}s^2)$, the characteristic function of the unit normal distribution.

Rényi (1962) gives two other proofs which also establish the result under conditions (1) $nm^{-1} \to \infty$, $nm^{-1} - \ln m \to -\infty$, and (2) $nm^{-1} \to 0$, $n^2m^{-1} \to \infty$, respectively. These two sets of conditions can be combined as

$$nm^{-1} \to 0 \text{ or } \infty \qquad \text{and} \qquad m \operatorname{var}(M_0) = me^{-\alpha}(1 - e^{-\alpha} - \alpha e^{-\alpha}) \to \infty. \tag{6.46}$$

Convergence to normal limit distribution for M_r under less restrictive conditions on the asymptotic behavior of nm^{-1} has been established by Békéssy (1963).

Theorem 6.5 (Békéssy, 1963).

(i) If $n, m \to \infty$ and

$$m\left(\frac{n+1}{m}\right)^r \exp\left(-\frac{n+1}{m}\right) \to \infty, \tag{6.47}$$

for $r \geq 2$, then

$$\lim_{n\to\infty} \Pr\left[\frac{M_r - \dfrac{m}{r!}\left(\dfrac{n+1}{m}\right)^r \exp\left(-\dfrac{n+1}{m}\right)}{D_r} \leq x\right] = \Phi(x), \tag{6.48}$$

where

$$D_r^2 = \left\{\frac{m}{r!}\left(\frac{n+1}{m}\right)^r \exp\left(-\frac{n+1}{m}\right)\right\}$$
$$\times\left[1 - \left\{\frac{1}{r!}\left(\frac{n+1}{m}\right)^r \exp\left(-\frac{n+1}{m}\right)\right\}\left\{1 + \frac{m}{n+1}\left(\frac{n+1}{m} - r\right)^2\right\}\right].$$

Note that D_r is a function of n and m, as well as r.

(ii) For $r = 0, 1$, the same result holds if condition (6.47) is replaced by $m^{-1}(n+1)^2 \to \infty$. [*Note* that Condition (6.47) ensures that $D_r \to \infty$ as $n \to \infty$, for $r \geq 2$.]

Outline of Proof. From (3.82), the epgf of M_r is

$$\left\{e^z + (x-1)\frac{z^r}{r!}\right\}^m.$$

We introduce the notation $M_r(n, m)$ to denote the M_r variable for specified values of n and m, where such specification is essential for clarity of the argument.

The characteristic function of $M_r(n, m)$ is, by Cauchy's theorem.

$$\phi(n, m; r, t) = \frac{1}{2\pi i}\frac{n!}{m^n} \oint \left\{e^z + (e^{it} - 1)\frac{z^r}{r!}\right\}^m \frac{dz}{z^{n+1}}, \tag{6.49}$$

where the path of integration may be along any circle with center 0.

Using a result of Curtiss (1942), Békéssy regards t as a pure imaginary quantity. If this is so, there is at least one real positive saddle point, say b.

We define

$$\phi(n, m; r, -it') = \psi(n, m; r, t').$$

Using Curtiss' theorem, all we need to show is that

$$\lim_{n\to\infty} \psi\left(n, m; r, \frac{u}{D_r}\right)\exp\left(-\frac{uE_r}{D_r}\right) = e^{(1/2)u^2}$$

$$(= e^{-(1/2)(-iu)^2}),$$

where

$$E_r = \frac{m}{r!}\left(\frac{n+1}{m}\right)^r \exp\left(-\frac{n+1}{m}\right)$$

and D_r is as defined in (6.48). This is equivalent to

$$\lim_{n\to\infty} \frac{n!}{m^n}\frac{1}{2\pi i} \oint \left\{e^z + (e^{u/D_r} - 1)\frac{z^r}{r!}\right\}^m \frac{dz}{z^{n+1}} = \exp\left(\frac{uE_r}{D_r} + \frac{u^2}{2}\right). \tag{6.50}$$

At the saddle point $z = b$ we must have

$$\frac{\partial}{\partial z}\left[\left\{e^z + (e^{u/D_r} - 1)\frac{z^r}{r!}\right\}^m z^{-(n+1)}\right] = 0. \tag{6.51}$$

After some algebraic reduction, we obtain, from (6.51),

$$b = \alpha + (\alpha - r)(e^{u/D_r} - 1)\frac{b^r e^{-b}}{r!}, \tag{6.52}$$

where $\alpha = (n + 1)m^{-1}$.

On the left-hand side of (6.50), make the change in variables $z = by$, leading to

$$\frac{n!}{m^n}\frac{1}{\sqrt{2\pi n}}\frac{\sqrt{n}}{i\sqrt{2\pi}} \oint \left\{e^{by} + (e^{u/D_r} - 1)\frac{(by)^r}{r!}\right\}^m \frac{b\,dy}{(by)^{n+1}}$$

$$= \left[\frac{n!}{m^n}\frac{1}{\sqrt{2\pi n}}\left\{e^b + (e^{u/D_r} - 1)\frac{b^r}{r!}\right\}^m b^{-n}\right]$$

$$\times \frac{\sqrt{n}}{i\sqrt{2\pi}} \oint \left\{\frac{e^{b(y-1)} + (e^{u/D_r} - 1)(by)^r e^{-b}/r!}{1 + (e^{u/D_r} - 1)b^r e^{-b}/r!}\right\}^m \frac{dy}{y^{n+1}}. \tag{6.53}$$

The quantity in square brackets is $(n!/m^n)(1/\sqrt{2\pi n}) \times$ (the value of the integrand at saddle point b).

Békéssy shows that the remaining factor in (6.53) tends to 1 as $n \to \infty$ and [using (6.52)] that (6.53) tends to $e^{-(1/2)u^2}$, by expanding in powers of b and using Stirling's theorem to approximate $n!$.

6.2.4. Limit Theorems for Modifications of Occupancy Distributions

The classical occupancy distribution may be modified by supposing that the number of balls N is a random variable, instead of being fixed, all other conditions remaining unchanged. The randomized classical occupancy distribution described in Section 3.3.2 corresponds to such a case, with N having a binomial distribution with parameters n, p. (The distribution of N may depend on m.)

Generally, from (3.19) we have, for given m,

$$E[M_r^{(k)} | N = n, m] = \frac{m^{(k)}}{m^{rk}(r!)^k} n^{(rk)}(1 - km^{-1})^{n-rk}.$$

Hence

$$\begin{aligned} E[M_r^{(k)} | m] &= \frac{m^{(k)}}{m^{rk}(r!)^k} E_N[N^{(rk)}(1 - km^{-1})^{N-rk}] \\ &= \frac{m^{(k)}}{m^{rk}(r!)^k} D^{rk} G_m(s)|_{s=1-km^{-1}}, \end{aligned} \tag{6.54}$$

where $G_m(s) = \sum_{n=0}^{\infty} \Pr[N = n|m]s^n$ is the pgf of N when there are m urns.

Theorem 6.6. (Ivchenko et al., 1967). If the distribution function of Nm^{-1} is weakly convergent to a distribution function $F(x)$ as $m \to \infty$ [i.e., $\lim_{m\to\infty} \Pr[Nm^{-1} \le x] = F(x)$ at all points of continuity of $F(x)$], then the distribution function of $M_r m^{-1}$ is weakly convergent, as $m \to \infty$, to the distribution of the random variable $(X^r/r!)e^{-X}$, where

$$\Pr[X \le x] = F(x).$$

Proof. We use the method of moments, evaluating

$$\lim_{m\to\infty} \left[\frac{E[M_r^{(k)}|m]}{m^k}\right]$$

and using the fact that, if this is finite, then it also equals $\lim_{m\to\infty} E[(M_r/m)^k | m]$, for all k.

The function

$$G_m(e^{-s/m}) = \gamma_m(s) \tag{6.55}$$

is a moment generating function of N/m, for given m. The condition stated in the theorem implies that, for $s > 0$,

$$\lim_{m\to\infty} \gamma_m(s) = \gamma(s) = \int_0^\infty e^{-sx}\, dF(x). \tag{6.56}$$

[$\gamma(s)$ is a moment generating function of X.]

It follows that, for any bounded continuous function $h(x)$,

$$\lim_{m\to\infty} \int_0^\infty h(x)\, dF_m(x) = \int_0^\infty h(x)\, dF(x).$$

In particular, taking $h(x) = x^k e^{-sx}$, we have

$$\lim_{m\to\infty} \int_0^\infty x^k e^{-sx}\, dF_m(x) = \int_0^\infty x^k e^{-sx}\, dF(x);$$

that is,

$$\lim_{m\to\infty} D^k\gamma_m(s) = D^k\gamma(s) \qquad (\text{for } k = 0, 1, 2, \ldots).$$

Note that

$$D^k\gamma_m(s) = D^k G_m(e^{-s/m}) = (-1)^k m^{-k} \sum_{n=0}^{\infty} \Pr[N = n \mid m] n^k e^{-sn/m}. \tag{6.57}$$

From (6.54),

$$\lim_{m\to\infty} \left\{\frac{E[M_r^{(k)} \mid m]}{m^k}\right\} = \frac{1}{(r!)^k} \lim_{m\to\infty} \frac{m^{(k)}}{m^k} m^{-kr} D^{kr} G_m(s)\big|_{s=1-km^{-1}}.$$

$$= \frac{1}{(r!)^k} \lim_{m\to\infty} m^{-kr} D^{kr} G_m(s)\big|_{s=1-km^{-1}},$$

since $\lim_{m\to\infty} m^{-k} m^{(k)} = 1$. (Differentiation is with respect to s.)

From (6.57),

$$D^{kr}\gamma_m(k) = (-1)^{kr} m^{-kr} D^{kr} G_m(e^t)\big|_{t=-k/m}.$$

(Differentiation is with respect to t.)

Ivchenko et al. claim that

$$\lim_{m\to\infty} \frac{D_s^{kr} G_m(s)\big|_{s=1-km^{-1}}}{D_t^{kr} G_m(e^t)\big|_{t=-km^{-1}}} = 1,$$

whence

$$\lim_{m\to\infty} \left\{\frac{E[M_r^{(k)} \mid m]}{m^k}\right\} = \frac{(-1)^{kr}}{(r!)^k} \lim_{m\to\infty} D^{kr}\gamma_m(k) = \frac{(-1)^{kr}}{(r!)^k} D^{kr}\gamma(k).$$

This is also the limit of the kth moment of $M_r m^{-1}$ as $m \to \infty$.

Constructing a moment generating function from these limiting moments, we obtain

$$\sum_{j=0}^{\infty} \frac{(-1)^j}{j!} \left\{ s^j \lim_{m\to\infty} E\left[\left(\frac{M_r}{m}\right)^j \middle| m\right]\right\} = \sum_{j=0}^{\infty} \frac{(-1)^j s^j}{j!} \frac{(-1)^{jr}}{(r!)^j} D^{jr}\gamma(j)$$

$$= \sum_{j=0}^{\infty} \frac{(-1)^j s^j}{j!(r!)^j} \int_0^{\infty} x^{jr} e^{-jx}\, dF(x)$$

$$= \int_0^{\infty} \sum_{j=0}^{\infty} \frac{(-1)^j}{j!} \left[\frac{sx^r e^{-x}}{r!}\right]^j dF(x)$$

$$= \int_0^{\infty} \exp\left(-s\frac{x^r e^{-x}}{r!}\right) dF(x)$$

which is a moment generating function of $X^r e^{-X}/r!$.

Ivchenko and Medvedev (1965) point out that a multivariate generalization of this result can be found using the epgf of $M_0, M_1, \ldots, M_n$ [see (3.82)]. The result so obtained states that the joint limit distribution of $\{m^{-1}M_{r_j}\}$ $(j = 1, \ldots, p)$ is that of $M_{r_j} m^{-1} = X^{r_j} e^{-X}/r_j!$, with X distributed as in Theorem 6.6. [Note that the limit distribution is degenerate, since the value of each $M_r m^{-1}$ is determined by $M_0 m^{-1}$.]

From this result we obtain the following interesting corollary.

If $\Pr[Nm^{-1} \leq x] \to (r!)^{-1} \int_0^x y^r e^{-y} dy$ as $m \to \infty$, then the limit distribution of $(M_0 + M_1 + \cdots + M_r)m^{-1}$ (i.e., the proportion of urns containing r or fewer balls) is uniform over the interval [0, 1].

The proof of this, given the result stated, follows immediately by noting that

$$F(X) = (r!)^{-1} \int_0^X y^r e^{-y}\, dy = \sum_{j=0}^{r} \left(\frac{X^j e^{-X}}{j!}\right)$$

and that $F(X)$ has a standard uniform distribution.

There are other conditions under which Poisson limit distributions are obtained. Proofs can be found in Ivchenko et al. (1967).

Theorem 6.7. For $r \geq 2$, if the distribution of $(N^r/r!)m^{-(r-1)}$ converges weakly to a distribution function $F(x)$ as $m \to \infty$, then

$$\lim_{m\to\infty} \Pr[M_r = j] = (j!)^{-1} \int_0^{\infty} x^j e^{-x}\, dF(x). \tag{6.58}$$

[For $r = 0, 1$, if $\lim_{m\to\infty} \Pr[\frac{1}{2}N^2 m^{-1} \leq x] \to F(x)$, then $(M_0 - m + N)$ and $\frac{1}{2}(N - M_1)$ each have the limit distribution (6.58) as $m \to \infty$.]

Theorem 6.8. If $\Pr[(N - m\ln m - rm\ln\ln m)m^{-1} < x] \to F(x)$ as $m \to \infty$, then

$$\lim_{m\to\infty} \Pr[M_r = j] = (j!)^{-1} \int_0^\infty \left(\frac{1}{r!}e^{-x}\right)^j \exp\left(-\frac{1}{r!}e^{-x}\right) dF(x).$$

We now find conditions under which M_r has a normal limit distribution.

Theorem 6.9. If the distribution of Nm^{-1} is asymptotically normal with expected value a and variance $m^{-1}\sigma^2$, so that

$$\lim_{m\to\infty} \Pr\left[\frac{\sqrt{m}(Nm^{-1} - a)}{\sigma} < x\right] = \Phi(x), \tag{6.59}$$

then

$$\lim_{m\to\infty} \Pr\left[\frac{\sqrt{m}\{M_r m^{-1} - p_r(a)\}}{[\sigma_r^2(a) + \sigma^2\{p_r'(a)\}^2]^{1/2}} \le x\right] = \Phi(x), \tag{6.60}$$

where

$$p_r(a) = \frac{a^r e^{-a}}{r!}, \qquad p_r'(a) = \frac{dp_r(a)}{da}, \tag{6.61}$$

and

$$\sigma_r^2(a) = p_r(a) - \{p_r(a)\}^2\{1 + a^{-1}(a - r)^2\} \tag{6.62}$$

[cf. (6.1b) and (6.1d)].

Proof. We write $Nm^{-1} = A$ and note that

$$A' = \frac{(A - a)\sqrt{m}}{\sigma}$$

has a unit normal limit distribution as $m \to \infty$. We also write

$$M_r' = \frac{M_r - mp_r(A)}{\sigma_r(A)\sqrt{m}}.$$

If we (for the moment) regard $Nm^{-1} = A$ as fixed for each m and let $m \to \infty$, then, from Theorem 6.4 (i), the limit distribution of M_r' will be unit normal. The probability that the condition $0 < \alpha_0 \le A \le \alpha_1 < \infty$ (for suitable *fixed* α_0 and α_1) is satisfied tends to 1 as $m \to \infty$.

Therefore A and M_r' are asymptotically independent, hence so are A' and M_r'.

We now put

$$A = a + A'\sigma m^{-1/2} \tag{6.63a}$$

$$M_r = mp_r(A) + M_r'\sigma_r(A)m^{1/2} \tag{6.63b}$$

and expand $p_r(A)$ and $\sigma_r(A)$ in Taylor series about $A = a$. This yields

$$\begin{aligned} M_r &= m\{p_r(a) + p_r'(a)(A - a) + O((A - a)^2)\} \\ &\quad + M_r' m^{1/2}\{\sigma_r(a) + O(|A - a|)\} \\ &= mp_r(a) + m^{1/2}\{p_r(a)\sigma A' + M_r'\sigma_r(a)\} + O(A'^2) + O(|A'M_r'|). \end{aligned}$$

Equivalently,

$$\frac{M_r - mp_r(a)}{[m\{\sigma^2[p_r'(a)]^2 + \sigma_r^2(a)\}]^{1/2}} = \frac{p_r'(a)\sigma A' + \sigma_r(a)M_r'}{\{\sigma^2[p_r'(a)]^2 + \sigma_r^2(a)\}^{1/2}} + O(A'^2 m^{-1/2}) + O(|A'M_r'|m^{-1/2}). \tag{6.64}$$

Since A' and M_r' are asymptotically distributed as independent unit normal variables, it follows that the first term on the right-hand side of (6.64) also has an asymptotically unit normal distribution. Furthermore, the remainder terms

$$O(A'^2 m^{-1/2}) \qquad \text{and} \qquad O(|A'M_r'|m^{-1/2})$$

converge in probability to zero as $m \to \infty$.

A further variant of the classical occupancy distribution is obtained by supposing that balls are assigned to the m urns not individually but in "bunches" of size Q ($\leq m$), *no more than one ball in each bunch being assigned to the same urn*. The number of possible selections of Q urns is $\binom{m}{Q}$, and each selection is assumed to have the same probability $\binom{m}{Q}^{-1}$ of being chosen to be assigned one ball each. Q is supposed to be a random variable with

$$\Pr[Q = q] = \omega_q, \qquad \sum_{q=1}^{m} \omega_q = 1. \tag{6.65}$$

The classical case, considered so far in this section, is $\omega_1 = 1$. Sevast'yanov (1967) and Ivchenko and Medvedev (1965) have shown that, if $\omega_q = 1$ (i.e., bunches are of fixed size q), then if $m \to \infty$ and $n = (m/q)(\ln m + x + o(1))$ for some fixed x, the distribution of M_0 (the number of empty urns) has a limiting Poisson distribution with parameter e^{-x}.

Ivchenko (1973b) has obtained results on the limiting distribution of M_0 in the general case (6.65). Before stating and proving his theorem we first define the rth moment about zero of Q/m as μ_r'. Since $1 \geq Q/m > 0$, we must have $\mu_1' \geq \mu_2' \geq \cdots \geq \mu_r'$ and

$$\mu_1' \leq \mu_2'^{1/2} \leq \cdots \leq \mu_r'^{1/r} \leq \cdots, \tag{6.66a}$$

and so

$$\mu_1'^{r+1} \leq \mu_{r+1}' \leq \mu_r'. \tag{6.66b}$$

Theorem 6.10. If $\mu_1' \to 0$ as $m \to \infty$, and $\delta_r = \mu_r' - \mu_1'^r = o(\mu_1'/\ln m)$ uniformly in r, then the limiting distribution of M_0 is Poisson with parameter e^{-x}, where x is a real number defined by

$$n = \frac{1}{\mu_1'}\left\{-(\ln m)\frac{\mu_1'}{\ln(1-\mu_1')} + x + o(1)\right\}. \tag{6.67}$$

Proof. We introduce indicator random variables X_i with

$$X_i = \begin{cases} 1 & \text{if the } i\text{th urn is empty after assigning } n \text{ bunches,} \\ 0 & \text{otherwise } (i = 1, \ldots, m). \end{cases}$$

Then

$$M_0 = \sum_{i=1}^{m} X_i.$$

We show that the kth moment of M_0 tends to the kth moment of a Poisson distribution.

From (1.138), we have

$$E[M_0^k] = \sum_{j=1}^{k} \tau_j \Delta^j \mathbf{0}^k, \tag{6.68}$$

where

$$\tau_j = \sum_{1 \leq i_1 \leq \cdots \leq i_j \leq m} \cdots \sum E\left[\prod_{g=1}^{j} X_{i_g}\right]. \tag{6.69}$$

From Exercise 2.4, the kth moment about zero of a Poisson random variable with parameter e^{-x} is

$$\sum_{j=1}^{k} \frac{e^{-jx}}{j!} \Delta^j \mathbf{0}^k. \tag{6.70}$$

Comparing (6.68) and (6.70), we see that the stated result will be proved if we can establish

$$\tau_j \to \frac{e^{-jx}}{j!} \qquad \text{(for all } j = 1, 2, \ldots).$$

Now

$$\tau_j = \binom{m}{j} \Pr\left[\bigcap_{i=1}^{j} (X_i = 1)\right]$$

$$= \binom{m}{j}\left\{\sum_{q=1}^{m-j} \omega_q \frac{\binom{m-j}{q}}{\binom{m}{q}}\right\}^n = \binom{m}{j}\left\{\sum_{q=1}^{m} \omega_q \frac{(m-q)^{(j)}}{m^{(j)}}\right\}^n. \tag{6.71}$$

Note that $(m-q)^{(j)} = 0$ if $q > m - j$. From Section 1.1.3, we find that

$$(m-q)^{(j)} = \sum_{g=0}^{j} a_g q^g,$$

with

$$a_g = \frac{(-1)^g m^{(j)}}{g!} \sum \sum_{i_1 \neq i_2 \neq \cdots \neq i_g}^{j-1} \cdots \sum \left\{\prod_{h=1}^{g} (m - i_h)\right\}^{-1}$$

$$= (-1)^g m^{(j)} \binom{j}{g} m^{-g}\{1 + O(m^{-1})\}. \tag{6.72}$$

(Note that, for $g = 0$, the $O(m^{-1})$ term is zero.) The lower limit for each i is zero and there are $\binom{j}{g}$ terms in the multiple summation.

From (6.71) and (6.72), we have

$$\tau_j = \binom{m}{j}\left[\sum_{q=1}^{m} \omega_q \sum_{g=0}^{j} q^g (-1)^g \binom{j}{g} m^{-g}\{1 + O(m^{-1})\}\right]^n$$

$$= \binom{m}{j}\left[\sum_{g=0}^{j} (-1)^g \binom{j}{g} \sum_{q=1}^{m} \omega_q \left(\frac{q}{m}\right)^g \{1 + O(m^{-1})\}\right]^n$$

$$= \binom{m}{j}\left[\sum_{g=0}^{j} (-1)^g \binom{j}{g} \mu'_g \{1 + O(m^{-1})\}\right]^n$$

$$= \binom{m}{j}\left[\sum_{g=0}^{j} (-1)^g \binom{j}{g} \left\{\mu_1'^g + o\left(\frac{\mu'_1}{\ln m}\right)\right\}\{1 + O(m^{-1})\}\right]^n$$

$$= \binom{m}{j}\left\{\sum_{g=0}^{j} (-1)^g \binom{j}{g} \mu_1'^g + o\left(\frac{\mu'_1}{\ln m}\right)\right\}^n$$

[noting that (1) the $O(m^{-1})$ term is zero for $g = 0$, and (2) $\delta_g = \mu'_g - \mu_1'^g = o(\mu'_1/\ln m)$]. Hence

$$\tau_j = \binom{m}{j}(1 - \mu'_1)^{jn}[1 + o(1)], \tag{6.73}$$

because from (6.67) $n\mu'_1(\ln m)^{-1} \to 1$ as $n, m \to \infty$.

Again using (6.67),

$$\begin{aligned}(1 - \mu'_1)^{jn} &= \exp\{jn \ln(1 - \mu'_1)\} \\ &= \exp\{-j \ln m - jx + o(1)\},\end{aligned}$$

hence

$$\tau_j = \left\{\binom{m}{j}m^{-j}\right\}e^{-jx}\{1 + o(1)\},$$

so that

$$\lim_{m\to\infty} \tau_j = (j!)^{-1}e^{-jx}.$$

This completes the proof of Theorem 6.10.

Turning now to cases with general values for $p_1, p_2, \ldots, p_m$, we note that Chistyakov (1964) proved the following result.

Theorem 6.11. If $n \to \infty$, $m \to \infty$, with $0 < \alpha_0 \le nm^{-1} \le \alpha_1 < \infty$ and $0 \le np_j \le c < \infty$, then the standardized M_0 distribution is asymptotically standard normal.

Chistyakov used some asymptotically equivalent (under the conditions of Theorem 6.11) expressions for the mean and variance of M_0. We have [from (3.10) and (3.11)]

$$\begin{aligned}E[M_0] = \sum_{j=1}^{m}(1 - p_j)^n &= \sum_{j=1}^{m}\exp[\alpha m \log(1 - p_j)] \\ &= \sum_{j=1}^{m}\{e^{-\alpha m p_j} + O(n^{-1})\} \qquad (\text{since } np_j \le c) \\ &= \sum_{j=1}^{m} e^{-\alpha m p_j} + O(mn^{-1})\end{aligned} \tag{6.74a}$$

$$\begin{aligned}\operatorname{var}(M_0) &= \sum_{j=1}^{m}(1 - p_j)^n\{1 - (1 - p_j)^n\} + 2\sum_{j<j'}^{m}\sum(1 - p_j - p_{j'})^n \\ &= \sum_{j=1}^{m} e^{-\alpha m p_j}(1 - e^{-\alpha m p_j}) - \alpha m\left(\sum_{j=1}^{m} p_j e^{-\alpha m p_j}\right)^2 + O(mn^{-1}).\end{aligned} \tag{6.74b}$$

It is noteworthy that Chistyakov's theorem does not impose any direct conditions on the asymptotic behavior of either $m^{-1}E[M_0]$ or $m^{-1}\operatorname{var}(M_0)$.

We do not present the proof in detail, since stronger results have been obtained by Holst (1971, 1976a) whose proofs we do present. Holst proved the following theorem.

Theorem 6.12. If $n \to \infty$, $m \to \infty$, with $np_j \le c < \infty$ for all j, and either

(i) $n^4 m^{-3} \to \infty$ and $nm^{-1} \to 0$, or
(ii) $n^2 m^{-1} \to \infty$ and $n^5 m^{-4} \to 0$,

then M_0 is asymptotically normally distributed with expected value $\mu_n = \sum_{j=1}^m \exp(-np_j)$ and variance $\sigma_n^2 = \frac{1}{2}n^2 \sum_{j=1}^m p_j^2$, that is,

$$\lim \Pr\left[\frac{M_0 - \sum_{j=1}^m \exp(-np_j)}{n\left\{\frac{1}{2}\sum_{j=1}^m p_j^2\right\}^{1/2}} < x\right] = \Phi(x).$$

Note that, in place of $E[M_0] = \sum_{j=1}^m (1-p_j)^n$, the asymptotically equivalent formula [see (6.74a)] $\sum_{j=1}^m \exp(-np_j)$ is used.

Proof.

(i) The characteristic function of M_0 is representable as

$$E[e^{itM_0}] = \frac{n!}{2\pi i m^n}\oint_{|z|=n/m} z^{-n-1}e^{mz}\prod_{j=1}^m \{1 + (e^{it}-1)e^{-mp_j z}\}\,dz \qquad (6.75)$$

(cf. Chistyakov, 1964). Hence

$$E\left[\exp\left\{\frac{it(M_0-\mu_n)}{\sigma_n}\right\}\right]$$
$$= e^{-it\mu_n/\sigma_n}\frac{n!}{2\pi i m^n}\oint_{|z|=n/m} z^{-n-1}e^{mz}\prod_{j=1}^m \{1 + (e^{it/\sigma_n}-1)e^{-mp_j z}\}\,dz,$$

where μ_n is as stated in the theorem and $\sigma_n^2 = \sum_{j=1}^m e^{-np_j}(1-e^{-np_j}) - \{\sum_{j=1}^m p_j e^{-np_j}\}^2$.

Using Stirling's theorem (for $n!$), and introducing polar coordinates ($z = (n/m)\exp(i\theta)$) we find $\lim_{n\to\infty} E[\exp\{it(M_0-\mu_n)/\sigma_n\}]$ is equal to

$$\lim_{n\to\infty} e^{-it\mu_n/\sigma_n}\left(\frac{n}{2\pi}\right)^{1/2}\int_{-\pi}^{\pi}\exp\{n(e^{i\theta}-1-i\theta)\}$$
$$\times \prod_{j=1}^m \{1 + (e^{it/\sigma_n}-1)\exp(-np_j e^{i\theta})\}\,d\theta. \qquad (6.76)$$

We now show that (6.76) tends to $\exp(-\frac{1}{2}t^2)$ as $n, m \to \infty$, using the following two lemmas. These show that, if the range of integration is split into two parts, $d \leq \theta \leq \pi$, $0 < |\theta| < d$, the second part tends to $\exp(-\frac{1}{2}t^2)$ and the first to zero, *whatever the value of d*.

Lemma 6.3. If $n^3m^{-2} \to \infty$ (which certainly happens if $n^4m^{-3} \to \infty$ and $nm^{-1} \to 0$), then, for $0 < d \leq \pi$,

$$\lim_{n,m\to\infty} \left(\frac{n}{2\pi}\right)^{1/2} \int_{d\leq|\theta|\leq\pi} \exp\{n(e^{i\theta} - 1 - i\theta)\}$$

$$\times \prod_{j=1}^{m} \{1 + (e^{it/\sigma_n} - 1)\exp(-np_je^{i\theta})\}\, d\theta = 0. \qquad (6.77)$$

Proof. Using Taylor series expansions,

$$\ln \prod_{j=1}^{m} \{1 + (e^{it/\sigma_n} - 1)\exp(-np_je^{i\theta})\} = \sum_{j=1}^{m} \{(e^{it/\sigma_n} - 1)\exp(-np_je^{i\theta}) + O(\sigma_n^{-2})\}$$

$$= mit\sigma_n^{-1} + O(n\sigma_n^{-1}) + O(m\sigma_n^{-2}).$$

Remembering that $|\exp(mit\sigma_n^{-1})| = |\exp(-ni\theta)| = |\exp(ni \sin\theta)| = 1$, we see that the modulus of the integrand in (6.77) is less than

$$\exp[-n(1 - \cos\theta + K_1\sigma_n^{-1} + K_2n^{-1}m\sigma_n^{-2})]$$

(K_1 and K_2 are constants).

Since

$$\sigma_n^2 = \tfrac{1}{2}n^2 \sum_{j=1}^{m} p_j^2 \geq \tfrac{1}{2}n^2m^{-1}$$

and so

$$\sigma_n^{-2} \leq 2n^{-2}m,$$

it follows that the integrand in (6.77) is less than

$$\exp[-n\{(1 - \cos\theta) + K_1'n^{-1}m^{1/2} + K_2'n^{-3}m^2\}] \qquad (6.78)$$

(K_1' and K_2' are constants).

For $d \leq |\theta| \leq \pi$,

$$1 - \cos\theta = 2\sin^2 \tfrac{1}{2}\theta \leq 2\sin^2 \tfrac{1}{2}d.$$

Since $n^{-3}m^2 \to 0$ and *a fortiori* $n^{-1}m^{1/2}[=(n^{-4}m^2)^{1/4}] \to 0$, it follows that for sufficiently large m, n (6.78) is less than

$$\exp[-n\sin^2 \tfrac{1}{2}d],$$

and so the left-hand side of (6.77) is less than

$$K_3 n^{1/2} \exp(-n \sin^2 \tfrac{1}{2}d)$$

(K_3 is a positive constant).

Lemma 6.3 is thus established, since $\lim_{n\to\infty} n^{1/2} \exp(-n \sin^2 \frac{1}{2}d) = 0$.

Lemma 6.4. If $n^4 m^{-3} \to \infty$ and $nm^{-1} \to 0$ as $n, m \to \infty$, then

$$\lim_{n,m\to\infty} \exp\left(\frac{-it\mu_n}{\sigma_n}\right)\left(\frac{n}{2\pi}\right)^{1/2} \int_{|\theta|<d} [\exp\{n(e^{i\theta} - 1 - i\theta)\}] \times \prod_{j=1}^{m} \{1 + (e^{it/\sigma_n} - 1)\exp(-np_j e^{i\theta})\}\, d\theta = e^{-(1/2)t^2}, \tag{6.79}$$

whatever the value of d.

Proof. We split up the range of integration into two parts:

$$0 \le |\theta| < g(n) = (n^{-4}m^3)^{1/7} n^{1/6-1/2}$$

and

$$g(n) \le |\theta| < d.$$

[Since $g(n) \to 0$ as $n \to \infty$, $g(n) < d$ for sufficiently large n.]

We first show that the integral over $g(n) \le |\theta| < d$ tends to zero as $n \to \infty$ (for all d), and then that the integral over $0 \le |\theta| < g(n)$ tends to $e^{-(1/2)t^2}$ (for all d).

Following an argument similar to that in the proof for Lemma 6.3 and noting that

$$\cos\theta - 1 \le -K_d \sigma^2$$

for some $K_d > 0$ and all $|\theta| \le d$, we find

$$\left|\left(\frac{n}{2\pi}\right)^{1/2} \int_{g(n)\le\theta<d} [\exp\{n(e^{i\theta} - 1 - i\theta)\}] \prod_{j=1}^{m} \{1 + (e^{it/\sigma_n} - 1)\exp(-np_j e^{i\theta})\}\, d\theta\right|$$

$$\le \left(\frac{n}{2\pi}\right)^{1/2} \int_{g(n)\le\theta<d} \exp(-n\theta^2 K_d + m^{1/2}|\theta| K_1' + K_2')\, d\theta$$

(K_1', K_2' are constants).

Making the transformation $y = n^{1/2}\theta$, this last expression is equal to

$$K_3 \int_{n^{1/2}g(n)\le y< dn^{1/2}} \exp(-y^2(K_d + o(1))\, dy \tag{6.80}$$

(K_3 is a constant).

Since

$$n^{1/2}g(n) = n^{1/6}(n^{-4}m^3)^{1/7} \\ = n^{1/42}(n^{-1}m)^{3/7}, \tag{6.81}$$

it follows that $\lim_{n\to\infty} n^{1/2}g(n) = \infty$ and (6.80) tends to zero as $n \to \infty$.

We now consider the integral in (6.77) over $0 \leq \theta < g(n)$. A further Taylor series expansion gives us

$$\ln \prod_{j=1}^{m} \{1 + (e^{it/\sigma_n} - 1)\exp(-np_j e^{i\theta})\}$$

$$= \left(\frac{it}{\sigma_n}\right) \sum_{j=1}^{m} \exp(-np_j e^{i\theta}) + \frac{1}{2!}\left(\frac{it}{\sigma_n}\right)^2 \sum_{j=1}^{m} \{\exp(-np_j e^{i\theta}) - \exp(-2np_j e^{i\theta})\}$$

$$+ \frac{1}{3!}\left(\frac{it}{\sigma_n}\right)^3 \{-ne^{i\theta} + O(n^2m^{-1})\} + O\left(\frac{n}{\sigma_n^4}\right) + O\left(\frac{m}{\sigma_n^5}\right)$$

$$= \left(\frac{it}{\sigma_n}\right) \sum_{j=1}^{m} e^{-np_j} + \left(\frac{t\theta}{\sigma_n}\right) \sum_{j=1}^{m} np_j e^{-np_j} + \theta^2 O\left(\frac{n}{\sigma_n}\right)$$

$$- \frac{1}{2!}\left(\frac{t}{\sigma_n}\right)^2 \sum_{j=1}^{m} e^{-np_j}(1 - e^{-np_j}) + |\theta| O\left(\frac{n}{\sigma_n^2}\right) + o(1).$$

(Note that $n/\sigma_n^4 = O(n^{-3}m^2) \to 0$;

$$\frac{m}{\sigma_n^5} = O[(n^{-10}m^7)^{1/2}] = O[(n^{-4}m^3)^{3/2}(nm^{-1})] \to 0,$$

and

$$\exp(-np_j e^{i\theta}) = e^{-np_j}\{1 - i\theta np_j + O(np_j\theta^2)\} \\ = e^{-np_j}\{1 - i\theta np_j + \theta^2 O(1)\} \\ = e^{-np_j}(1 - i\theta np_j) + \theta^2 O(1),$$

since $np_j < c < \infty$.)

The function of the left-hand side of (6.77), with d replaced by $g(n)$, can then be written

$$\lim_{n\to\infty} e^{o(1)} e^{-(1/2)t^2} \left(\frac{n}{2\pi}\right)^{1/2} \int_{0 \leq |\theta| < g(n)} \exp(h(\theta))\, d\theta,$$

with

$$h(\theta) = -\frac{1}{2} n\left(\theta - \frac{t}{\sigma_n} \sum_{j=1}^{m} p_j e^{-np_j}\right)^2 + \theta^2 O\left(\frac{n}{\sigma_n}\right) + |\theta| O\left(\frac{n}{\sigma_n^2}\right) + \theta^3 O(n).$$

Making the transformation $y = n^{1/2}\theta$, we obtain

$$\lim_{n\to\infty} e^{o(1)}e^{-(1/2)t^2}\left(\frac{1}{2\pi}\right)^{1/2}\int_{0\le|\theta|<n^{1/2}g(n)} \exp(-\tfrac{1}{2}y^2)\,dy = e^{-(1/2)t^2}$$

(remembering that $n^{1/2}g(n) \to \infty$).

(ii) Coming now to the case $n^5m^{-4} \to 0, n^2m^{-1} \to \infty$, we use another form of the characteristic function of M_0, also due to Chistyakov (1967). This is

$$E\left[\exp\left\{\frac{it(M_0 - \mu_n)}{\sigma_n}\right\}\right]$$

$$= e^{-it\mu_n/\sigma_n}\frac{n!}{2\pi i m^n}\oint_{|z|=n/m} z^{-n-1}\prod_{j=1}^{m}[1 + \{\exp(mp_j z e^{it/\sigma_n}) - 1\}e^{-it/\sigma_n}]\,dz.$$

This can be shown to be equal to

$$\left[\exp\left\{\frac{-it}{\sigma_n}\sum_{j=1}^{m}(e^{-np_j} - 1 + np_j)\right\}\right] \times e^{o(1)}\left(\frac{n}{2\pi}\right)^{1/2}\int_{-\pi}^{\pi}[\exp\{-n(1 + i\theta)\}]$$

$$\times \prod_{j=1}^{m}\{1 + \{\exp(np_j e^{i(\theta + t/\sigma_n)} - 1)\}e^{-it/\sigma_n}\}\,d\theta. \qquad (6.82)$$

We show that this tends to $e^{-(1/2)t^2}$ as $n, m \to \infty$, with $n^5m^{-4} \to 0, n^2m^{-1} \to \infty$, by methods similar to those used for case (i).

The range of integration is first broken into

$$0 \le |\theta| < d \qquad \text{and} \qquad d \le |\theta| \le \pi.$$

It is shown that the integral over $d \le |\theta| \le \pi$ tends to zero as $n, m \to \infty$. Then it is shown that the integral over $n^{1/7-1/2}\pi \le |\theta| \le d$ also tends to zero, and finally that, if the limits of integration in (6.82) are replaced by $\pm n^{1/7-1/2}\pi$, the limiting value of (6.82) as $n, m \to \infty$ is indeed $e^{-(1/2)(t^2)}$.

Recently, Holst (1976a) developed an alternative method which may lead to a simpler proof for this theorem.

An interesting example of a situation in which observed phenomena suggested a theoretical result that was subsequently verified is described in Hill (1970). Let n denote the number in a set of biological species and m denote the number of (nonempty) genera into which these are classified. We denote by Y_i the number of species classified into the ith genus ($Y_1 + Y_2 + \cdots + Y_m = n$) and let M_r denote the number of genera containing exactly r species. It was noted by Willis (1922) that, in many observed cases, the relative frequencies $m^{-1}M_r$ are approximately proportional to $r^{-(1+\beta)}$, where r is large. (β is a positive constant.)

If we regard the species as n balls and the genera as m urns, then, with B–E statistics [Section 3.2.2],

$$\Pr[\mathbf{Y} = \mathbf{y} \mid n, m] = \binom{n-1}{m-1}^{-1} \tag{6.83}$$

$(\mathbf{Y} = (Y_1, \ldots, Y_m);\ \mathbf{y} = (y_1, \ldots, y_m))$ with $y_i \geq 1$ and $y_1 + y_2 + \cdots + y_m = n$.

We now replace m by a random variable M with a distribution depending on n in such a way that as $n \to \infty$

$$\Pr[n^{-1}M \leq t] \to F(t), \tag{6.84}$$

and $F(x)$ is a proper cdf $[\lim_{x\to\infty} F(x) = 1]$ with $F(0) = 0$.

Hill (1970) shows that

1. $\lim_{n\to\infty} E[M^{-1}M_r] = \int_0^1 t(1-t)^{r-1}dF(t) \qquad (r = 1, 2, \ldots).$ (6.85)
2. The limit distribution of $M^{-1}M_r$ is that of $T(1-T)^{r-1}$, where T has the cdf $F(t)$.

If we make yet another assumption—that

$$F(t) = t^{\beta} \qquad (0 < t < 1;\ \beta > 0), \tag{6.86}$$

then

$$\lim_{n\to\infty} E[M^{-1}M_r] = \int_0^1 t(1-t)^{r-1}\beta t^{\beta-1}\,dt = \beta\int_0^1 t^{\beta}(1-t)^{r-1}\,dt$$

$$= \frac{\beta\Gamma(\beta+1)\Gamma(r)}{\Gamma(\beta+r+1)}. \qquad \text{[from (1.52)]}$$

For r large,

$$\frac{\Gamma(r)}{\Gamma(\beta+r+1)} \sim r^{-(1+\beta)}, \tag{6.87}$$

so that

$$\lim_{n\to\infty} E[M^{-1}M_r] \propto r^{-(1+\beta)}$$

approximately when r is large, which agrees with the observed phenomenon.

EXERCISE 6.1. Find $\lim_{n\to\infty} E[M^{-1}M_r]$, when $\beta = 1$ in (6.86). (*Answer*: $\{r(r+1)\}^{-1}$.)

The agreement with observation is interesting, although it does not demonstrate the correctness of the model, which is based on several assumptions. It is generally the case that many models can be constructed that give fair agreement with observed data. Such agreement is necessary for validity of the model and may justify further consideration of it, but it is not sufficient, on its own, to establish practical validity. [See Ijiri and Simon (1975) for further discussion of this point.]

Ijiri and Simon (1975) have derived a similar result to (6.87) by envisioning a sequential process in which at each stage either a new urn or a new ball is added, with probabilities α, $1 - \alpha$, respectively. The new urn is supposed to be empty. In this case we clearly have $n^{-1}M \to \alpha(1 - \alpha)^{-1}$. The method of proof used by these authors is the derivation of difference equations for the proportion of urns containing r balls at successive stages and solution for a limiting stable state.

Hill (1970) obtains expressions for the moments of $M^{-1}M_r(n, M)$, averages them over the distribution of M, and then finds the limits of these averages. In the case of M–B statistics, Hill (1970) obtained two results:

1. $$\lim_{n\to\infty} E[M^{-1}M_r] = \{(r - 1)!\}^{-1} \int_0^1 \left(\frac{t}{1 - t}\right)^{r-1} \left\{\exp\left(-\frac{t}{1 - t}\right)\right\} dF(t). \tag{6.88}$$

2. the limit distribution of $M^{-1}M_r$ is that of

$$\frac{1}{(r - 1)!}\left(\frac{T}{1 - T}\right)^{r-1} \exp\left(-\frac{T}{1 - T}\right), \tag{6.89}$$

where T has cdf $F(t)$.

Holst (1976b) has obtained some interesting results on the limiting distributions (as $m, n \to \infty$ with $\beta = mn^{-1} \to \rho > 0$) of certain functions of the numbers $N_1, N_2, \ldots, N_m$ of *balls* in the m urns when B–E statistics apply.

In particular he finds that (for fixed x)

$$\lim_{n\to\infty} \Pr[\max(N_1, \ldots, N_m) \ln \beta - \ln m \le x] = \exp(-e^{-x}) \tag{6.90}$$

which is the cumulative distribution function of the Type 1 extreme value distribution [Johnson and Kotz, 1970, Chapter 21].

Holst also shows that $\max(N_1, N_2, \ldots, N_m)$ and the number of nonempty urns (i.e., number of nonzero N_j's) are asymptotically independent.

These results are obtained by applying general theorems on the limit characteristic function of a general function $h(N_1, N_2, \ldots, N_m)$ of $N_1, N_2, \ldots, N_m$.

6.2.5. Limit Laws for Sequential Occupancy (Waiting-Time) Distributions

We first recall some problems in sequential occupancy from Section 3.5.

We suppose we have m urns, labeled $1, 2, \ldots, m$, and define random variables $X_1, X_2, \ldots$ by

$$X_j = k \text{ if the } j\text{th ball is placed in the urn labeled } k.$$

In the equiprobable case it is assumed that

$$\Pr[X_j = k] = m^{-1} \qquad (k = 1, \ldots, m).$$

We continue dropping balls into the urns until every urn contains at least r balls. [The case $r = 1$ corresponds to the single dixie cup problem (see Section 3.5) and $r = 2$ to the double dixie cup problem (see Newman and Shepp, 1960).]

The distribution of the number of balls needed to achieve this—say $N_r(m)$—is a *sequential occupancy* or *waiting-time* distribution.

Newman and Shepp (1960) showed that

$$E[N_r(m)] = m \ln m + (r - 1)m \ln \ln m + mK_m + o(m), \tag{6.91}$$

where K_m is a constant.

Erdös and Rényi (1961) obtain a limit distribution for $N_r(m)$ in the form

$$\lim_{m\to\infty} \Pr[m^{-1}N_r(m) < \ln m + (r - 1) \ln \ln m + x] = \exp\left\{-\frac{e^{-x}}{(r-1)!}\right\}. \tag{6.92}$$

EXERCISE 6.2. Check that the expected value of formula (6.91) is consistent with (6.92), putting $K_m = \gamma - \ln(r - 1)!$, where γ $(=0.577216)$ is Euler's constant.

We now give an *outline* (only) of the proof of (6.92). We let $Y_k(m, n)$ denote the number of balls in the urn labeled k after n balls have been distributed among the m urns, and define

$$Y(m, n) = \min(Y_1(m, n), Y_2(m, n), \ldots, Y_m(m, n)).$$

We also suppose that n is a function $n(m)$ of m, satisfying

$$n(m) = m \ln m + (r - 1)m \ln \ln m + xm + o(m)$$

[cf. (6.22)].

From the definitions of $N_r(m)$ and $n(m)$,

$$\Pr[N_r(m) > n(m)] = \Pr[Y(m, n(m)) < r].$$

All we need to do therefore is to show that

$$\lim_{n\to\infty} \Pr[Y(m, n(m)) < r] = 1 - \exp\left\{-\frac{e^{-x}}{(r-1)!}\right\}. \tag{6.93}$$

For $j \leq r - 2$,

$$\Pr[Y(m, n(m)) = j] \leq m\binom{n(m)}{j}\left(\frac{1}{m}\right)^j\left(1 - \frac{1}{m}\right)^{n(m)-j} = O((\ln m)^{-(r-1-j)}),$$

and so

$$\lim_{n\to\infty} \Pr[Y(m, n(m)) < r - 1] = 0. \tag{6.94}$$

Now denote by $A_r(m)$ the event that at least one of the urns contains *exactly* $(r - 1)$ balls when $n(m)$ balls have be assigned [i.e., at least one of $Y_1(m, n(m)), \ldots, Y_m(m, n(m))$ equals $(r - 1)$].

If $A_r(m)$ is valid, then $Y(m, n(m)) < r$. Hence

$$\Pr[Y(m, n(m)) < r] \geq \Pr[A_r(m)].$$

But also,

$$\begin{aligned}\Pr[Y(m, n(m)) < r] &- \Pr[Y(m, n(m)) < r - 1]\\ &= \Pr[Y(m, n(m)) = r - 1] \leq \Pr[A_r(m)].\end{aligned}$$

Hence

$$0 \leq \Pr[Y(m, n(m)) < r] - \Pr[A_r(m)] \leq \Pr[Y(m, n(m)) < r - 1]. \tag{6.95}$$

From (6.94) and (6.95), we see that

$$\lim_{n\to\infty} \Pr[Y(m, n(m)) < r] = \lim_{n\to\infty} \Pr[A_r(m)]. \tag{6.96}$$

From (1.88), we have

$$\Pr[A_r(m)] = \sum_{j=1}^{m}\binom{m}{j}(-1)^{j-1}W_{j,r}(m), \tag{6.97}$$

where

$$\begin{aligned}W_{j,r}(m) &= \Pr[j\ \textit{prescribed}\ \text{urns each contain exactly } (r-1) \text{ balls}]\\ &= \frac{n(m)!}{\{(r-1)j\}!\{n(m)-(r-1)j\}!}\frac{1}{m^{(r-1)j}}\left(1 - \frac{j}{m}\right)^{n(m)-(r-1)j}.\end{aligned}$$

By direct calculation we find

$$\lim_{m\to\infty}\binom{m}{j}W_{j,r}(m) = \left\{\frac{e^{-x}}{(r-1)!}\right\}^j \frac{1}{j!},$$

and substituting in (6.97) gives

$$\lim_{m\to\infty} \Pr[A_r(m)] = 1 - \exp\left\{-\frac{e^{-x}}{(r-1)!}\right\}. \tag{6.98}$$

From (6.93), (6.96), and (6.98), (6.92) is established.

The limit distribution is an *extreme value* (*type-1*) distribution (see (6.90)). For any r we have

$$\lim_{m\to\infty} \Pr\left[\frac{N_r(m) - E[N_r(m)]}{m} < x\right] = \exp(-e^{-x-\gamma}). \tag{6.99}$$

[Of course, this does not mean that the limit distribution of $N_r(m)$ does not depend on r, because $E[N_r(m)]$ depends on r, as can be seen from (6.91).]

EXERCISE 6.3. Show that

$$\lim_{n\to\infty} \Pr\left[\frac{N_1(m) - m \ln m}{m} < x\right] = \exp(-e^{-x}),$$

and compare with the results in Section 3.5.

Brayton (1963) improved the formula (6.91) for $E[m^{-1}N_r(m)]$ to

$$E[m^{-1}N_r(m)] = \gamma - \ln\{(r-1)!\} + \ln m + (r-1)\ln\ln m + O\left(\frac{\ln\ln m}{\ln m}\right), \tag{6.100}$$

and also obtained the formula

$$\operatorname{var}[m^{-1}N_r(m)] = \tfrac{1}{6}\pi^2 + O\left(\frac{\ln\ln\ln m}{\ln\ln m}\right). \tag{6.101}$$

A more general problem is the distribution of waiting time until k out of m urns each contain at least r balls. We denote the needed number of balls by the random variable $N_r(m, k)$. If the ratio km^{-1} tends to a constant λ, which is *less than* 1, then the standardized variable corresponding to $N_r(m, k)$ has a limit unit normal distribution as $m, k \to \infty$ (r remaining constant). [Note that (6.99) is not inconsistent with this result, since in this case $\lambda = 1$.]

Békéssy (1964) has studied this problem and obtained the following:

1. If $m \to \infty$, with $(m - k)$ constant and r constant,

$$\lim \Pr[m^{-1}N_r(m, k) - \ln m - (r-1)\ln\ln m + \ln(r-1)! < x]$$

$$= \exp(-e^{-x}) \sum_{j=0}^{m-k} \frac{e^{-jx}}{j!}. \tag{6.102}$$

2. If $m \to \infty$ with $m^{-1}k \to \lambda$, then

$$\lim \Pr\left[\frac{N_r(m, k) - E}{D} < x\right] = \Phi(x), \tag{6.103}$$

where

$$\begin{aligned} E &= mu_r(1 - \lambda), \\ D^2 &= m\lambda(1 - \lambda)[\{u_r'(1 - \lambda)\}^2 - u_r(1 - \lambda)], \\ u_r(y) &= K_r^{-1}(y), \\ K_r(u) &= e^{-u} \sum_{j=0}^{r-1} \frac{u^j}{j!} = \{(r - 1)!\}^{-1} \int_u^\infty t^{r-1}e^{-t}\, dt. \end{aligned}$$

Samuel-Cahn (1974) has obtained the following results on the limiting distribution of $N_k(m)$ (the number of balls needed so that k urns out of m are occupied) for the randomized occupancy case (Section 3.3). She allows for the possibility that k and p (the probability of staying in the urn) may each depend on m, the number of urns. To indicate this, we use (when necessary) the symbols k_m and p_m.

The results are:

(i) If k_m is fixed $(=k)$ and $p_m \to p > 0$, then the limit distribution of $N_k(m)$ is negative binomial with parameters k, p.

(ii) If k_m is fixed $(=k)$ and $p_m \to 0$, then the limit distribution of $2p_m N_k(m)$ is χ^2 with $2k$ degrees of freedom.

(iii) If $k_m \to \infty$ with $\lim_{m\to\infty} k_m(1 - p_m) = \mu$, $\lim_{m\to\infty} k_m^2 m^{-1} = \lambda$, and $(\mu + \lambda)$ is finite, then the limit distribution of $[N_{k_m}(m) - k_m)$ is Poisson with expected value $\mu + \frac{1}{2}\lambda$.

(iv) Under the same conditions as (iii) but with $\mu + \lambda = \infty$, and $m - k_m \to \infty$, the limit distribution of the standardized $N_{k_m}(m)$ variable is unit normal.

(v) If $k_m = m - b \to \infty$, with b a positive constant, then the limit distribution of

$$2m \exp\{-m^{-1}p_m N_{k_m}(m)\}$$

is χ^2 with $2(b + 1)$ degrees of freedom.

EXERCISE 6.4. Using the relationship (3.92) between $N_r(m)$ and M_r, obtain results analogous to (i) through (v) for limit distributions of M_r in the randomized occupancy case. [*Hint*: See Samuel-Cahn (1974).]

6.2.6. Limit Multivariate Occupancy Distributions

We have encountered multivariate occupancy distributions in the form of joint distributions of $M_{r_1}, M_{r_2}, \ldots$, the numbers of urns containing r_1 balls,

r_2 balls, ..., in Section 3.4. It is the limit distributions of these that are the primary topic of this section, though toward the end of the section we draw attention to other multivariate distributions associated with the model we are using.

Sevast'yanov and Chistyakov (1964) showed that in the equiprobable situation $M_{r_1}, M_{r_2}, \ldots, M_{r_s}$ have a limit multinormal distribution as $n \to \infty$. These authors use a multivariate epgf [as do Rényi (1962) and Békéssy (1963)] and obtain an expression for $\Pr[\bigcap_{j=1}^{s} (M_{r_j} = m_{r_j})]$ as the product of three factors. The first two factors are expressions in n, m, $\{r_j\}$, $\{M_{r_j}\}$, which are approximated by using Stirling's formula [cf. (1.50)]. The third factor is $\Pr[\bigcap_{j=1}^{s} (M_{r_j}(n', m') = 0)]$, where (n', m') is introduced to indicate that the probability is calculated for different numbers of balls n' and urns m' than those actually used in the experiment. The value of this factor is expressed as an integral, using Cauchy's formula and the epgf (already mentioned) evaluated at $\{t_j = 0\}$ $(j = 1, \ldots, s)$. The integral is then approximated by the saddle-point method.

We first recall the definitions (6.1a) through (6.1e) and introduce the following additional notation:

$$u_j = \frac{M_{r_j} - ma_{r_j}}{(|\mathbf{\Delta}|m)^{1/2}}. \tag{6.104}$$

where $\Delta = (\sigma_{r_i r_j})$.

Theorem 6.13. In the equiprobable case, where $0 < c_1 \leq \alpha = nm^{-1} \leq c_2 < \infty$, then the asymptotic expansion

$$\Pr\left[\bigcap_{j=1}^{s}(M_{r_j} = m_{r_j})\right] = \frac{1}{(2\pi m)^{s/2}|\mathbf{\Delta}|^{1/2}}\left[\exp\left(-\frac{1}{2}\sum_{i=1}^{s}\sum_{j=1}^{s}\delta_{ij}u_i u_j\right)\right]$$
$$\times [1 + O(m^{-1/2})] \tag{6.105}$$

holds uniformly with respect to u_i for $|u_i| < K < \infty$ for some positive constant K, where δ_{ij} is the minor of $\sigma_{r_i r_j}$ in Δ.

The proof is lengthy, yet interesting. We give sufficient detail to understand it, but there are some lacunae.

Proof. From (3.82) we have the joint epgf

$$G_{m; r_1, \ldots, r_s}(z; x_1, \ldots, x_s)$$
$$= \sum_{n=0}^{\infty}\sum_{r_1=0}^{\infty}\cdots\sum_{r_s=0}^{\infty}\frac{(mz)^n}{n!}\Pr\left[\bigcap_{j=1}^{m} M_{r_j}(n, m) = m_{r_j})\right]\prod_{j=1}^{m} x_j^{m_{r_j}}$$
$$= \left\{e^z + \sum_{j=1}^{m}\frac{z^{r_j}}{r_j!}(x_{r_j} - 1)\right\}^m. \tag{6.106}$$

From this it can be shown that, if $n = O(m)$, then

$$\lim_{m \to \infty} E[m^{-1} M_{r_j}] = a_{r_j}, \tag{6.107a}$$

$$\lim_{m \to \infty} m \operatorname{var}(M_{r_j}) = \sigma_{r_j r_j}, \tag{6.107b}$$

$$\lim_{m \to \infty} m \operatorname{cov}(M_{r_i}, M_{r_j}) = \sigma_{r_i r_j}, \qquad i \neq j. \tag{6.107c}$$

In the univariate case ($s = 1$), (6.106) gives

$$\begin{aligned} G_{1;r}(z; x) &= \left\{ e^z + \frac{z^r}{r!}(x-1) \right\}^m \\ &= \left(\frac{z^r}{r!}\right)^m \left\{ \frac{e^z r!}{z^r} - 1 + x \right)^m \\ &= \left(\frac{z^r}{r!}\right)^m \sum_{i=0}^{m} \binom{m}{i} \left(\frac{e^z r!}{z^r} - 1 \right)^{m-i} x^i. \end{aligned}$$

The coefficient of x^k is

$$\begin{aligned} \left(\frac{z^r}{r!}\right)^m \binom{m}{k} \left(\frac{e^z r!}{z^r} - 1\right)^{m-k} &= \binom{m}{k} \left(\frac{z^r}{r!}\right)^k \left(e^z - \frac{z^r}{r!}\right)^{m-k} \\ &= \binom{m}{k} \frac{z^{rk}}{(r!)^k} \sum_{n=0}^{\infty} \frac{\{(m-k)z\}^n}{n!} \Pr[M_r(n, m-k) = 0] \end{aligned}$$

(since $G_{1;r}(z; 0) = \Pr[M_r = 0]$).

It follows that

$$\begin{aligned} &\sum_{n=0}^{\infty} \frac{(mz)^n}{n!} \Pr[M_r(n, m) = k] \\ &\quad = \binom{m}{k} \frac{z^{rk}}{(r!)^k} \sum_{n=0}^{\infty} \frac{\{(m-k)z\}^n}{n!} \Pr[M_r(n, m-k) = 0] \\ &\quad = \binom{m}{k} \frac{1}{(r!)^k} \sum_{n=0}^{\infty} \frac{(m-k)^n}{n!} z^{n+rk} \Pr[M_r(n, m-k) = 0]. \end{aligned}$$

Equating coefficients of z^n, we have

$$\frac{m^n}{n!} \Pr[M_r(n, m) = k] = \binom{m}{k} \frac{1}{(r!)^k} \frac{(m-k)^{n-rk}}{(n-rk)!} \Pr[M_r(n-rk, m-k) = 0],$$

whence

$$\Pr[M_r(n, m) = k] = \binom{m}{k} \frac{n!}{(r!)^k (n - rk)!} \left(\frac{1}{m}\right)^{rk} \left(1 - \frac{k}{m}\right)^{n-rk} \Pr[M_r(n - rk, m - k) = 0]. \tag{6.108}$$

The following generalization of (6.108), which we will use, is established in an identical manner:

$$\Pr\left[\bigcap_{j=1}^{s} (M_{r_j}(n, m) = m_{r_j})\right] = \frac{m!}{(m - m')! \prod_{j=1}^{s} m_{r_j}!} \cdot \frac{n!}{(n - t)! \prod_{j=1}^{s} (r_j!)^{m r_j}} \left(\frac{1}{m}\right)^t \left(1 - \frac{m'}{m}\right)^{n-t} \times \Pr\left[\bigcap_{j=1}^{s} (M_{r_j}(n - t, m - m') = 0)\right], \tag{6.109}$$

where $m' = \sum_{j=1}^{s} m_{r_j}$, $t = \sum_{j=1}^{s} r_j m_{r_j}$. Taking logarithms,

$$\ln \Pr\left[\bigcap_{j=1}^{s} (M_{r_j}(n, m) = m_{r_j})\right] = \ln \mathscr{P}_1 + \ln \mathscr{P}_2 + \ln \mathscr{P}_3, \tag{6.110}$$

where

$$\mathscr{P}_1 = \frac{m!}{(m - m')! \prod_{j=1}^{s} m_{r_j}!}, \tag{6.111a}$$

$$\mathscr{P}_2 = \frac{n!}{\prod_{j=1}^{s} (r_j!)^{m_{r_j}}} \left(\frac{1}{m}\right)^n, \tag{6.111b}$$

$$\mathscr{P}_3 = \frac{(m - m')^{n-t}}{(n - t)!} \Pr\left[\bigcap_{j=1}^{s} (M_{r_j}(n - t, m - m') = 0)\right]. \tag{6.111c}$$

From our definitions,

$$\begin{aligned} n - t &= m\left(\alpha - \sum_{j=1}^{s} a_{r_j} r_j\right) + m \sum_{j=1}^{s} a_{r_j} r_j - \sum_{j=1}^{s} m_{r_j} r_j \\ &= m\left(\alpha - \sum_{j=1}^{s} a_{r_j} r_j\right) - \sum_{j=1}^{s} (m_{r_j} - m a_{r_j}) r_j \\ &= m\left(\alpha - \sum_{j=1}^{s} a_{r_j} r_j\right) - \left(\sum_{j=1}^{s} u_j r_j\right)(|\boldsymbol{\Delta}| m)^{1/2} \end{aligned} \tag{6.112}$$

and

$$m - m' = m\left(1 - \sum_{j=1}^{s} a_{r_j}\right) - u'(|\mathbf{\Delta}|m)^{1/2}, \tag{6.113}$$

where

$$u' = \sum_{j=1}^{s} u_j.$$

Using (6.112) and (6.113) in the expression (6.111a), we obtain

$$\ln \mathscr{P}_1 = \ln m! - \ln\left[\left\{m\left(b - u'\frac{|\mathbf{\Delta}|^{1/2}}{m^{1/2}}\right)\right\}!\right] - \sum_{j=1}^{s} \ln\left[\left\{m\left(a_{r_j} + u_j\frac{|\mathbf{\Delta}|^{1/2}}{m^{1/2}}\right)\right\}!\right],$$

where $b = 1 - \sum_{j=1}^{s} a_{r_j}$.

By applying Stirling's formula, we obtain

$$\begin{aligned}\ln \mathscr{P}_1 = &-\tfrac{1}{2}s\ln(2\pi m) - \tfrac{1}{2}\ln\left(b\prod_{j=1} a_{r_j}\right)\\ &- m\left(\sum_{j=1}^{s} a_{r_j}\ln a_{r_j} + b\ln b\right) + |\mathbf{\Delta}|^{1/2}m^{1/2}\left(u'\ln b - \sum_{j=1}^{s} u_j \ln a_{r_j}\right)\\ &- \tfrac{1}{2}|\mathbf{\Delta}|\left(\sum_{j=1}^{s} u_j^2 a_{r_j}^{-1} + u'^2 b^{-1}\right) + O(m^{-1/2}).\end{aligned} \tag{6.114}$$

By a similar analysis, we obtain

$$\begin{aligned}\ln \mathscr{P}_2 = \tfrac{1}{2}\ln(2\pi\alpha m) &+ m\left\{\left(\alpha - \sum_{j=1}^{s} a_{r_j} r_j\right)\ln\alpha - \alpha b + \sum_{j=1}^{s} a_{r_j}\ln a_{r_j}\right\}\\ &- |\mathbf{\Delta}|^{1/2}m^{1/2}\left\{\left(\sum_{j=1}^{s} u_j r_j\right)\ln\alpha - \alpha u' - \sum_{j=1}^{s} u_j \ln a_{r_j}\right\} + O(m^{-1/2}).\end{aligned} \tag{6.115}$$

$\mathscr{P}_3$ is the coefficient of z^{n-t} in

$$\left\{e^z - \sum_{j=1}^{s} (r_j!)^{-1} z^{r_j}\right\}^{m-m'}.$$

By Cauchy's formula,

$$\mathscr{P}_3 = \frac{1}{2\pi i}\oint_{|z|=c}\left\{e^z - \sum_{j=1}^{s} (r_j!)^{-1} z^{r_j}\right\}^{m-m'} z^{-(n-t+1)}\,dz$$

(where c is a positive constant).

Putting

$$(m - m')^{-1}(n - t) = \bar{\beta}, \tag{6.116}$$

we have

$$\mathscr{P}_3 = \frac{1}{2\pi i} \oint_{|z|=c} z^{-1} \left\{ \frac{A(z)}{z^{\bar{\beta}}} \right\}^{m-m'} dz \tag{6.117}$$

where

$$A(z) = e^z - \sum_{j=1}^{s} (r_j!)^{-1} z^{r_j}. \tag{6.118}$$

$A(z)$ is an entire (integral) function. Its Maclaurin series is

$$A(z) = \sum_{j=0}^{\infty} a_j z^j, \tag{6.119}$$

with all a_j's real and with an infinite number of a_j's positive; further, the greatest common divisor of the pairwise differences $(j - j')$, for which neither a_j nor a'_j is 0, is equal to 1. We introduce the function

$$f(z) = \ln A(z) - \rho \ln z \tag{6.120}$$

in the right half-plane in such a way that it is real on the real axis.

If $a_j = 0$ for $j < n_1$, but $a_{n_1} > 0$, then the equation

$$f'(z) = 0$$

has one real positive root Z_ρ, if $\rho > n_1$. If $\rho' > \rho$, then $Z_{\rho'} > Z_\rho$. If further $f''(Z_\rho) \geq d > 0$ for all ρ in the interval $[\rho_0, \rho_1]$, where $n_1 < \rho_0 \leq \rho_1 < \infty$, then for any $\lambda > 0$, Z_ρ is a saddle point of the integrand in

$$I_\lambda = (2\pi i)^{-1} \int_{|z|=Z_\rho} z^{-1} \left\{ \frac{A(z)}{z^\rho} \right\}^\lambda dz$$

and

$$I_\lambda = \frac{1}{Z_\rho \{2\pi\lambda f''(Z_\rho)\}^{1/2}} e^{\lambda f(Z_\rho)} \{1 + O(\lambda^{-1/2})\} \tag{6.121}$$

uniformly with respect to ρ over $[\rho_0, \rho_1]$. (Sevast'yanov and Christyakov (1964, Theorem 3).)

We choose $\rho = \bar{\beta}$ and take

$$f(z) = \ln A(z) - \bar{\beta} \ln z. \tag{6.122}$$

Denote the positive root of $f'(z) = 0$ by $Z_{\bar{\beta}} = \bar{\alpha}$. This is a saddle point of the integrand in (6.117), and so it is also a solution of the equation

$$\frac{\partial}{\partial z}\left[z^{-1}\left\{\frac{A(z)}{z^{\bar{\beta}}}\right\}^{m-m'}\right] = 0. \tag{6.123}$$

Substituting $z = \bar{\alpha}$ in (6.123), we obtain

$$\bar{\beta} = \left(\bar{\alpha} - \sum_{j=1}^{s} \bar{a}_{r_j} r_j\right)\bar{b}^{-1}, \tag{6.124}$$

where $\bar{a}_r = \bar{\alpha}^r e^{-\bar{\alpha}}/r!$ and $\bar{b} = 1 - \sum_{j=1}^{s} \bar{a}_{r_j}$.

In particular, when $\bar{\alpha} = \alpha$, we define

$$\beta = \left(\alpha - \sum_{j=1}^{s} a_{r_j} r_j\right)b^{-1}. \tag{6.125}$$

Using (6.121) and also (6.117) and (6.124), Sevast'yanov and Chistyakov show that

$$\ln \mathscr{P}_3 = \ln[\bar{\alpha}\{2\pi(m - m')f''(\bar{\alpha})\}^{1/2}] + (m - m')f(\bar{\alpha}) + O((m - m')^{-1/2}), \tag{6.126}$$

where $f(z)$ is defined by (6.120).

Expanding in powers of $m^{-1/2}$, they obtained

$$\begin{aligned}\ln \mathscr{P}_3 = {} & m(\alpha b + b \ln b - \beta b \ln \alpha) - (|\Delta| m)^{1/2}\left\{u'(\alpha + \ln b) - \left(\sum_{j=1}^{s} u_j r_j\right)\ln \alpha\right\} \\ & - \tfrac{1}{2}\alpha^{-1} b\left\{\prod_{j=1}^{s} a_{r_j}\right\}\left(\beta u' - \sum_{j=1}^{s} u_j r_j\right)^2 - \tfrac{1}{2}\ln(2\pi m \alpha \Delta) \\ & + \tfrac{1}{2}\ln b + \frac{1}{2}\sum_{j=1}^{s} \ln a_{r_j} + O(m^{-1/2}).\end{aligned} \tag{6.127}$$

Finally, combining the asymptotic formulas for $\ln \mathscr{P}_1$, $\ln \mathscr{P}_2$, and $\ln \mathscr{P}_3$, we obtain the result stated in the theorem.

Another kind of multivariate distribution arises if we consider the sequence of variables

$$M_r(n_1, m), M_r(n_1 + n_2, m), \ldots, M_r(n_1 + n_2 + \cdots + n_s, m),$$

representing the number of urns (among m in all) containing exactly r balls after $n_1, (n_1 + n_2), \ldots, (n_1 + n_2 + \cdots + n_s)$ balls have been assigned in the *same* sequence of trials. Asymptotic results are obtained by letting each n_i $(i = 1, 2, \ldots, s)$ and the number of urns tend to infinity simultaneously.

We discuss only the case $r = 0$ and establish a result due to Bolotnikov (1968a). We first find the joint epgf of the s variables $M_0(n_1, m)$, $M_0(n_1 + n_2, m), \ldots, M_0(n_1 + n_2 + \cdots + n_s, m)$. This is

$$\begin{aligned} G_m(x_1, \ldots, x_s; z_1, \ldots, z_s) &= \sum_{n_1=0}^{\infty} \cdots \sum_{n_t=0}^{\infty} \sum_{m_1=0}^{\infty} \cdots \sum_{m_t=0}^{\infty} \left\{ \prod_{j=1}^{s} \frac{(mz_j)^{n_j}}{n_j!} x_j^{m_j} \right\} \\ &\quad \times \Pr\left[\bigcap_{j=1}^{s} \left\{ M_0\left(\sum_{g=1}^{j} n_g, m \right) = m_j \right\} \right] \\ &= \left\{ \sum_{j=1}^{s} \left(\prod_{g=1}^{j-1} x_g \right) (e^{z_j} - 1) \exp\left(\sum_{h=g+1}^{s} z_h \right) + \prod_{j=1}^{g} x_j \right\}^m, \end{aligned} \tag{6.128}$$

with the convention $\prod_{g=1}^{0} x_g = 1$.

[Formula (6.128) can be established by an argument similar to that used in deriving (3.82).]

Theorem 6.14 (Bolotnikov, 1968a). If $n_1, n_2, \ldots, n_s, m \to \infty$ in such a way that (for each $j = 1, \ldots, s$)

(i) $n_j m^{-1} \to \infty$.
(ii) $n_j m^{-1} - \ln m \to -\infty$.
(iii) There is a number α_m such that $\alpha_m \to \infty$ but $\theta_j = (\sum_{g=1}^{j} n_g)m^{-1} - \alpha_m$ is bounded.

Then the random variables

$$M_0'(\theta_j) = \frac{M_0(n_1 + \cdots + n_j, m) - E[M_0(n_1 + \cdots + n_j, m)]}{(me^{-\alpha})^{1/2}}$$

have a limit multinormal joint distribution with covariance function

$$\lim_{m \to \infty} \operatorname{cov}(M_0'(\theta_j), M_0'(\theta_h)) = \exp\{-\max(\theta_j, \theta_h)\}.$$

(*Note*: The asymptotic *correlation* between $M_0'(\theta_j)$ and $M_0'(\theta_h)$ is $\exp\{-\tfrac{1}{2}|\theta_j - \theta_h|\}$.)

Outline of Proof. The joint characteristic function of $M_0'(\theta_1), M_0'(\theta_2), \ldots, M_0'(\theta_s)$ is

$$\phi_{n_1, n_2, \ldots, n_s; m}(t_1, t_2, \ldots, t_s)$$

$$= E\left[\exp\left\{i \sum_{j=1}^{s} t_j M_0'(\theta_j)\right\}\right]$$

$$= \frac{\prod_{j=1}^{s} n_j!}{(2\pi i)^s m^{\sum_{j=1}^{s} n_j}} \oint \cdots_{C_s} \oint \left(\prod_{j=1}^{s} z_j^{n_j+1}\right)^{-1} \{G_m(e^{it_1\mu}, \ldots, e^{it_s\mu}; z_1, \ldots, z_s)\}^m$$

$$\times \exp\left(-i \sum_{j=1}^{s} t_j \mu^{-1} e^{-\theta_j}\right) dz_1 dz_2 \cdots dz_s, \tag{6.129}$$

where C_s are appropriate circles and $\mu = (me^{-\alpha})^{-1/2}$. (Note that $\lim_{m\to\infty} E[m^{-1}M_0(\sum_{g=1}^{j} n_j, m)] = e^{-\alpha-\theta_j}$.)

Note the similarity to (6.75) used in the proof of Theorem 6.12. The proof follows lines similar to those of Theorem 6.12. The region of integration is split up; the first factor, $(2\pi i)^{-s} m^{-\sum_1^s n_j} \prod_{j=1}^{s} n_j!$, is approximated using Stirling's formula, and changes in variables are introduced to approximate the integrals. The restrictions on the behavior of $n_j m^{-1}$ are somewhat lighter than those on nm^{-1} in Theorem 6.12, but the present theorem is restricted to equal urn probabilities.

With different kinds of restrictions on the behavior of n and m we obtain Theorem 6.15, which is due to Bolotnikov (1968a).

Theorem 6.15. If $n_1, n_2, \ldots, n_s, m \to \infty$ in such a way that $(\sum_{j=1}^{k} n_j)m^{-1} = \alpha\theta_k$ (θ_k a constant, $\alpha \to 0$) and $m\alpha^2 \to \infty$ for $k = 1, \ldots, s$ (with the *same* α for each k), then the limit joint distribution of

$$M_0'(\theta_j) = (\alpha\sqrt{\tfrac{1}{2}m})^{-1}\{M_0(n_1 + \cdots + n_j, m) - E[M_0(n_1 + \cdots + n_j, m)]\}$$

is multinormal with covariances

$$\lim_{m\to\infty} \operatorname{cov}(M_0'(\theta_j), M_0'(\theta_h)) = \min(\theta_j^2, \theta_h^2).$$

[*Note*: The asymptotic *correlation* between $M_0'(\theta_j)$ and $M_0'(\theta_h)$ is $\min(\theta_j\theta_h^{-1}, \theta_j^{-1}\theta_h)$.]

Outline of Proof. The method used is similar to that of Rényi (1962) in proving part (iii) of Theorem 6.4 (page 336). The random variables $N_k(m)$, equal to the number of balls needed to have *exactly* k cells occupied by *at least* one ball, are introduced, and the relation [cf. (3.92)]

$$\Pr\left[\bigcap_{j=1}^{s}\left\{M_0\left(\sum_{h=1}^{j} n_h, m\right) < m - m_j\right\}\right] = \Pr\left[\bigcap_{j=1}^{s}\left\{N_{m_j}(m) < \sum_{h=1}^{j} n_h\right\}\right] \quad (6.130)$$

is used.

Asymptotic multinormality of $(N_{m_1}(m), N_{m_2}(m), \ldots, N_{m_h}(m))$ is established using the representation of these variables as sums of independent random variables. Putting $Y_g = N_g(m) - N_{g-1}(m)$ and noting that $\Pr[N_1(m) = 1] = 1$, we have

$$N_k(m) = \sum_{g=1}^{k} Y_g = 1 + \sum_{g=2}^{k} Y_g,$$

with $Y_2, \ldots, Y_k$ mutually independent and with

$$E[Y_g] = m(m - g + 1)^{-1}, \qquad \operatorname{var}(Y_g) = m(g - 1)(m - g + 1)^{-2},$$

whence

$$E[N_k(m)] = \sum_{h=0}^{k-1} (m - h)^{-1} = -m \ln(1 - km^{-1}) + O(1) \quad (6.131a)$$

$$\operatorname{var}(N_k(m)) = m \sum_{h=1}^{k-1} h(m - h)^{-2} = h(1 - km^{-1})^{-1} + m \ln(1 - km^{-1}) + O(1). \quad (6.131b)$$

Also (see Rényi, 1962), if $km^{-1/2} \to \infty$ as $n, m \to \infty$,

$$\{\operatorname{var}(N_k(m))\}^{-1} \sum_{g=1}^{k} E[\,|Y_g - m(m - g + 1)^{-1}|^3] \to 0. \quad (6.131c)$$

This satisfies Lyapounov's condition and so, if $km^{-1/2} \to \infty$, $N_k(m)$ is asymptotically normal. Similarly, $N_k(m) - N_g(m)$ is asymptotically normal if $|k - g|m^{-1/2} \to \infty$.

Now introduce numbers x_k ($k = 1, 2, \ldots, s$) such that

$$\sum_{j=1}^{k} n_j = -m \ln(1 - m^{-1}m_k) + x_k\{(1 - m^{-1}m_k)^{-1} + m \ln(1 - m^{-1}m_k)\}^{1/2} + O(1). \quad (6.132)$$

Solving (6.132) for m_k, Bolotnikov (1968a) claims

$$m_k = m(1 - e^{-\alpha\theta_k}) - x_k\alpha\theta_k(\tfrac{1}{2}m)^{1/2}[1 + o(1)] + O(1) \doteqdot m\alpha\theta_k, \quad (6.133)$$

and from (6.131b)

$$\operatorname{var}(N_k(m)) \doteqdot \tfrac{1}{2}m\alpha^2\theta_k^2. \tag{6.134}$$

Substituting in (6.130),

$$\Pr\left[\bigcap_{j=1}^{s}\left\{\frac{M_0(n_1 + \cdots + n_j, m) - me^{-\alpha\theta_j}}{\alpha\sqrt{\frac{1}{2}m}} < x_j\theta_j\right\}\right]$$
$$= \Pr\left[\bigcap_{j=1}^{s}\left\{\frac{N_{m_j}(m) - E[N_{m_j}(m)]}{\alpha\sqrt{\frac{1}{2}m}} < x_j\theta_j\right\}\right]. \tag{6.135}$$

The $km^{-1/2} \to \infty$ conditions are satisfied, because from (6.133) $m_1 m^{-1/2} \doteqdot \alpha\theta_1 m^{1/2} \to \infty$, $(m_h - m_{h-1})m^{-1/2} \doteqdot \alpha(\theta_h - \theta_{h-1})m^{1/2} \to \infty$ for $h > 1$, and so the variables

$$N_{m_1}(m), \quad N_{m_h}(m) - N_{m_{h-1}}(m) \qquad (h > 1)$$

are asymptotically independent and normally distributed.

The result now follows from (6.135).

6.2.7. Miscellaneous

1. It is of interest to consider under what conditions a Poisson or a normal limit distribution is to be expected. The latter is not expected unless (at least) the variance of the quantities [M_r, $N_r(m)$, etc.] discussed in earlier sections tends to infinity as $n, m \to \infty$. Conversely, if a Poisson limit is to be obtained, this variance should be bounded.

These are not sufficient conditions; finiteness or nonfiniteness of the limit of variance does not in itself guarantee Poisson or normal limits.

One way of looking at this problem is to express M_r, for example, in terms of variables $X_1, X_2, \ldots, X_m$, where X_j = number of balls in the jth urn. Clearly, $\sum_{j=1}^m X_j = n$, and M_r = number of X_j's equal to r.

We can consider what properties of the joint distributions of $X_1, X_2, \ldots, X_m$ determine a Poisson or a normal limit distribution for M_r. This mode of approach was used by Kolchin (1968). He considered a model which can be formally defined in the following way.

Let $X_{m1}, X_{m2}, \ldots, X_{mm}$ ($m = 1, 2, 3, \ldots$) be a sequence of sets of integer-valued random variables. Within each set the variables are independent, conditional on $\sum_{j=1}^m X_{mj} = n(m)$. Let $\mathscr{A}_{k,r}^{(m)}$ denote the event that exactly k of the random variables in the set with index m are equal to r. The distributions of the random variables $M_r(n, m)$ are defined by

$$\Pr[M_r(n, m) = k] = \Pr\left[\mathscr{A}_{k,r}^{(m)} \,\middle|\, \sum_{j=1}^m X_{mj} = n\right]. \tag{6.136}$$

With this formulation, it is possible to include in the analysis cases where n is a random variable.

In the (equiprobable case) classical occupancy problem, the conditional joint distribution of $X_{m1}, \ldots, X_{mm}$ is multinomial with parameters $n; m^{-1}, m^{-1}, \ldots, m^{-1}$. For B–E statistics, the conditional joint distribution ascribes equal probabilities $\binom{m+n-1}{n-1}^{-1}$ to every possible different event $\bigcap_{j=1}^{m} (X_{mj} = x_j)$, with $\sum_{j=1}^{m} x_j = n$.

We do not give here details of Kolchin's results, but his method of approach is noteworthy.

2. A powerful theorem due to Moran (1973) is helpful in establishing several results on limit distributions.

Theorem 6.16. (Moran, 1973). $N_1, N_2, \ldots, N_m$ have a joint symmetric multinomial distribution with parameters $n; m^{-1}, m^{-1}, \ldots, m^{-1}$, so that

$$\Pr\left[\bigcap_{j=1}^{m} (N_j = n_j)\right] = m^{-n} n! \left(\prod_{j=1}^{m} n_j!\right)^{-1}. \tag{6.137}$$

$X_1, X_2, \ldots, X_m$ are random variables which, *conditional on* $\bigcap_{j=1}^{m} (N_j = n_j)$, are mutually independent and have cdf's $F(x \mid n_j)$ determined by $n_1, n_2, \ldots, n_m$, respectively.

Defining

$$\mu'_h(X_j \mid n_j) = E[X_j^h \mid N_j = n_j] \qquad (h = 1, 2; j = 1, 2, \ldots, m)$$

and

$$\nu_3(X_j \mid n_j) = E[\,|X_j|^3 \mid N_j = n_j] \qquad (j = 1, 2, \ldots, m),$$

and assuming that, for suitable positive constants A, B,

$$|\mu'_h(X_j \mid n_j)| < A \exp(Bn_j) \tag{6.138a}$$

$$\nu_3(X_j \mid n_j) < A \exp(Bn_j), \tag{6.138b}$$

then $S_m = \sum_{j=1}^{m} X_j$ has an asymptotically normal limit distribution as $m \to \infty$ if $nm^{-1} = a$, with $0 < a$ and $n^{-1} \operatorname{var}(S_m) \to b$ $(0 < b < \infty)$.

Sketch of Proof. The method used is similar to that employed by Rényi (1962) in establishing the asymptotic normality of $M_r(n, m)$, with modifications due to Holst (1971, 1972b).

The characteristic function of S_m is

$$\begin{aligned}\phi_m(t) &= E_N[E[\exp(itS_m)|N_1, N_2, \ldots, N_m]] \\ &= \sum_{n_1} \cdots \sum_{n_m} m^{-n} n! \left(\prod_{j=1}^{m} n_j! \right)^{-1} \prod_{j=1}^{m} E[\exp(itX_j)|n_j] \\ &= \frac{n!}{2\pi i m^n} \oint_{|z|=a/m} \frac{e^{mz}\{\psi(t, z)\}^m}{z^{n+1}} dz, \end{aligned} \tag{6.139}$$

where

$$\psi(t, z) = \sum_{g=0}^{\infty} \left(\frac{z^r e^{-z}}{r!} \right) \int_{-\infty}^{\infty} e^{itx} \, dF(x|g).$$

Changing to polar coordinates by the transformation $z = (n/m)e^{i\omega} = ae^{i\omega}$, we obtain

$$\phi_m(t) = \frac{n!}{2\pi} \int_{-\pi}^{\pi} n^{-n}\{\exp(ne^{i\omega} - ni\omega)\}\{\psi(t, ae^{i\omega})\}^m \, d\omega. \tag{6.140}$$

Using Stirling's formula (1.50),

$$\begin{aligned}\lim_{n,m\to\infty} \phi_m(t) = \lim_{n,m\to\infty} \left(\frac{n}{2\pi} \right)^{1/2} \\ \times \int_{-\pi}^{\pi} \left[1 + \sum_{g=0}^{\infty} \left\{ \frac{(ae^{i\omega})^g}{g!} \exp(-ae^{i\omega})\{\psi_1(t|g) - 1\} \right\} \right]^m \\ \times \exp\{n(e^{i\omega} - 1 - i\omega)\} \, d\omega, \end{aligned} \tag{6.141}$$

where $\psi_1(t|g) = \int_{-\infty}^{\infty} e^{itx} \, dF(x|g)$.

Replacing t by $t\sigma_m^{-1}$ [where $\sigma_m^2 = \text{var}(S_m)$] in (6.141) gives the limit of the characteristic function of $S_m \sigma_m^{-1}$.

In following Holst (1971) (see discussion of proof of Theorem 6.12), the range of integration ($0 \le |\omega| \le \pi$) is split into three parts:

(i) $d \le |\omega| < \pi$.
(ii) $d_1 = n^{1/7-1/2}\pi \le |\omega| \le d$.
(iii) $0 \le |\omega| \le d_1$.

(d is a fixed number such that $d_1 < d < \pi$.)

The remainder of the proof follows lines similar to Holst's (see also Holst (1972b), where a similar result is obtained for the special case when $F(x|n_j)$ is degenerate [$=0$ for $x < g(n_j)$, 1 for $x \ge g(n_j)$]).

EXERCISE 6.5. Apply this theorem to establish asymptotic normality of the distribution of $M_r(n, m)$ when $n/m \to a$ $(0 < a < \infty)$. (*Hint*: Take X_j to be an indicator variable with

$$X_j = \begin{cases} 1 & \text{if there are } r \text{ balls in the } j\text{th urn,} \\ 0 & \text{otherwise,} \end{cases}$$

and let n_j = number of balls in the jth urn. Care is especially necessary in establishing the condition

$$\lim_{n\to\infty} n^{-1} \operatorname{var}(M_r|(n, m)) < \infty).$$

3. Limit distributions of linear functions of variables having joint multinomial distribution (such as $N_1, N_2, \ldots, N_m$ in the notation we have just been using) have been studied by Ivchenko (1971), Harris and Park (1971b), and Ivchenko and Levin (1976), among others. In the last-mentioned paper, a rather more general situation is considered.

Balls of s different colors $\mathscr{C}_1, \mathscr{C}_2, \ldots, \mathscr{C}_s$ are distributed randomly among m urns. It is supposed that there are n_j balls of color $\mathscr{C}_j$, with $\sum_{j=1}^{s} n_j = n$. The joint distribution of random variables

$$L_h = \sum_{g=1}^{m} f_{g,h;m}(X_{g1}, \ldots, X_{gs}) \qquad (h = 1, \ldots, k),$$

where X_{gj} = number of balls of color $\mathscr{C}_j$ in the gth urn, and $f_{g,h;m}(\cdot)$ is a given function, is studied.

It is shown that sufficient conditions for the limit joint distribution of $L_1, L_2, \ldots, L_k$ to be multinormal are

(i) $m, n_1, \ldots, n_s \to \infty$.
(ii) $n_j m^{-1} \to \alpha_j, 0 < \alpha_j < \infty \qquad (j = 1, \ldots, s)$.
(iii) $mp_{gj} \le c < \infty$, for some c and all m, g, and j.
(iv) $|f_{g,h;m}(x_{g1}, \ldots, x_{gs})| < A \exp\{B \sum_{i=1}^{s} x_{gi}\}$, for some positive A and B and all m, g, and h.
(v) for all h and i [writing $(\mathbf{X}_g)$ for $(X_{g1}, \ldots, X_{gs})$],

$$c_{hi} = \lim_{m\to\infty} m^{-1} \left[\sum_{g=1}^{m} \operatorname{cov}\{f_{g,h;m}(\mathbf{X}_g), f_{g,i,m}(\mathbf{X}_g)\} - \sum_{u=1}^{s} n_u^{-1} \prod_{t=h,i} \left\{ \sum_{g=1}^{m} \operatorname{cov}(X_{gu}, f_{g,t;m}(\mathbf{X}_g)) \right\} \right]$$

exists and is finite, with $c_{hh} > 0$.

The proof uses a saddle-point method of approximation to the joint characteristic function expressed as a contour integral.

4. Models in which the number of urns m is assumed to be actually infinite (as opposed to limiting cases as $m \to \infty$) have been considered by Karlin (1967), but there has been little further work in this direction until recently (Holst (1976d), Kaplan (1976)). It is supposed that the probability of assignment to the kth urn is p_k, with $\sum_{k=1}^{\infty} p_k = 1$.

Karlin considers a situation in which the number of urns for which the assignment probability p_k is at least w^{-1} is $w^{\gamma}L(w)$, where $0 < \gamma \leq 1$ and $\lim_{t\to\infty}\{L(tx)/L(t)\} \to 1$, for all $x > 0$, that is, $L(w)$ is "slowly varying at infinity." He shows that, denoting the number of *occupied* urns by $\overline{M}_0$, and with

$$B_n = \begin{cases} (2^{\gamma} - 1)\Gamma(1-\gamma)\{n^{\gamma}L(n)\} & (\text{for } 0 < \gamma < 1) \\ n\displaystyle\int_0^{\infty} t^{-1}e^{-1/t}L(nt)\,dt & (\text{for } \gamma = 1), \end{cases}$$

$$\lim_{n\to\infty} \Pr[\{\overline{M}_0 - E[\overline{M}_0]\}B_n^{-1/2} < x] = \Phi(x). \tag{6.142}$$

Denoting the number of urns containing an *odd* number of balls by $\overline{M}_0^*$, Karlin shows that

$$\lim_{n\to\infty} \Pr[2\{\overline{M}_0^* - E[\overline{M}_0^*]\}\{(2^{\gamma} - 1)^{-1}B_{4n}\}^{-1/2} < x] = \Phi(x). \tag{6.143}$$

EXERCISE 6.6.
(i) Show that

$$E[\overline{M}_0] = \sum_{k=1}^{\infty}\{1 - (1 - p_k)^n\}.$$

(ii) Derive a formula for $E[\overline{M}_0^*]$.
(*Hint*: Use the relation $E[\sum_{j=1}^{\infty} Y_j] = \sum_{j=1}^{\infty} E[Y_j]$ if both sides exist.)

5. Urn model formulations of the committee problems discussed in Section 3.6 can be constructed in the following way.

We suppose there are m urns (corresponding to individuals). There are w_1 balls of color $\mathscr{C}_1$, w_2 of color $\mathscr{C}_2 \ldots w_k$ of color $\mathscr{C}_k$ ($w_j \leq m$ for all j). (These correspond to k committees of sizes $w_1, w_2, \ldots, w_k$.) We define $p_j = w_j/m$.

The set of w_j balls of color $\mathscr{C}_k$ is assigned to w_j urns chosen at random, one ball to each urn ($j = 1, 2, \ldots, k$).

Under the conditions $m \to \infty$, $n_j \to \infty$,

$$m\prod_{j=1}^{k} p_j \to \lambda > 0 \qquad (j = 1, 2, \ldots, k).$$

Holst (1976c) has shown that the limiting distribution of the number of urns containing balls of all k colors is Poisson with parameter λ.

6.3. PÓLYA-EGGENBERGER AND RELATED LIMIT DISTRIBUTIONS

6.3.1. Elementary Results

We recall, from Section 4.1, that the Pólya–Eggenberger distribution has the probability function

$$\Pr[X = j] = \binom{n}{j} \frac{\left(\dfrac{b}{s}\right)^{[j]} \left(\dfrac{w}{s}\right)^{[n-j]}}{\left(\dfrac{b+w}{s}\right)^{[n]}} \qquad (j = 0, 1, \ldots, n). \tag{6.144}$$

(This distribution was obtained by supposing an urn to contain b black and w white balls initially. A ball is drawn at random from the urn and returned with s additional balls of the same color. The operation is repeated n times, and X represents the total number of times a black ball is drawn.)

The following limit distributions can be established by direct and elementary arguments:

1. If $b, w \to \infty$ with $b(b + w)^{-1} \to \bar{p}$ $(0 < \bar{p} < 1)$, n remaining fixed, then

$$\Pr[X = j] \to \binom{n}{j} \bar{p}^j (1 - \bar{p})^{n-j} \qquad (j = 0, 1, \ldots, n). \tag{6.145}$$

The limit distribution of X is binomial (Section 2.2.1) with parameters n, $\bar{p}$.

2. If $b, w \to \infty$ with $b(b + w)^{-1} \to 0$, and *also* $n \to \infty$ with $nb(b + w)^{-1} \to \bar{\lambda}$ $(0 < \bar{\lambda} < \infty)$ and $(ns)^{-1}(b + w) \to \bar{\gamma}$ $(0 < \bar{\gamma} < \infty)$, then

$$\Pr[X = j] \to \binom{\bar{\lambda}\bar{\gamma} + j - 1}{j} \left(\frac{\bar{\gamma}}{1 + \bar{\gamma}}\right)^{\bar{\lambda}\bar{\gamma}} \left(\frac{1}{1 + \bar{\gamma}}\right)^j \qquad (j = 0, 1, \ldots). \tag{6.146}$$

The limit distribution of X is the second form negative binomial (Section 2.4) with parameters $\bar{\lambda}\cdot\bar{\gamma}$, $(1 + \bar{\gamma})^{-1}$.

3. If $n \to \infty$ with $n^{-2}s^{-1}(b + w) \to \bar{\theta}$ $(0 < \bar{\theta} < \infty)$ and $nb(b + w)^{-1} \to \bar{\lambda}$ $(0 < \bar{\lambda} < \infty)$, then

$$\Pr[X = j] \to \frac{\bar{\lambda}^j}{j!} e^{-\bar{\lambda}} \qquad (j = 0, 1, 2, \ldots). \tag{6.147}$$

The limit distribution is Poisson with parameter $\bar{\lambda}$.

3.′ The same Poisson limit distribution is obtained if $n \to \infty$ with $n^{-2}s^{-1}(b + w) \to 0$, $n^{-3}s^{-2}(b + w)^2 \to \infty$, and $nb(b + w)^{-1} \to \bar{\lambda}$ $(0 < \bar{\lambda} < \infty)$.

3.″ The same Poisson limit distribution is also obtained if $n \to \infty$ with $n^{-2}s^{-1}(b + w) \to \infty$, $n^{-3}s^{-1}(b + w) \to 0$, and $nb(b + w)^{-1} \to \bar{\lambda}$ $(0 < \bar{\lambda} < \infty)$ (Śródka, 1964).

Pólya (1931) describes the Pólya–Eggenberger urn scheme as a model of *contagion* in contrast to the Markov scheme (see Section 5.9) in which the probability of drawing a black ball depends only on the color of the last ball drawn, and which Pólya called a model of *heredity*.

Pólya used the following scheme of classification for "contagion" models:

	Verbal Description	
Class No.	Dependence	Type of Event
1a	Independent	Common
1b	Independent	Rare
2a	Weak	Common
2b	Weak	Rare
3a	Strong	Common
3b	Strong	Rare
4a	Very Strong	Common
4b	Very Strong	Rare
t	Transition	—

	Conditions on Parameters		
Class No.	$\omega = b(b + w)^{-1}$	$\delta = s(b + w)^{-1}$	Other
1a	$0 < \omega < 1$	$\delta = 0$	
1b	$n\omega \to c > 0$	$\delta = 0$	
2a	$0 < \omega < 1$	$n\delta \to d > 0$	$b + ns > 0$
2b	$n\omega \to c > 0$	$n\delta \to d > 0$	
3a	$0 < \omega < 1$	$\delta > 0$	
3b	$n\omega \to c > 0$	$\delta > 0$	
4a	$0 < \omega < 1$	$\delta = \infty$	
4b	$\omega = 0$	$\delta = \infty$	
t	$n^\theta \omega \to c > 0$	$n^\theta \delta \to d > 0$	$0 < \theta < \frac{1}{2}$

Note: Where $\to$ appears, "as $n \to \infty$" is implied.

Classes 4a and 4b are extraordinary in that after the first drawing there is no further random variation.

The paper by Pólya (1931) is a classical one, and Bricas (1949) gives a careful discussion of some limiting cases.

EXERCISE 6.7. Obtain appropriate limiting distributions for classes 1b, 2b, and 3b.

6.3.2. Rate of Convergence

Śródka (1964) gives several results illustrating the rate of convergence to the limit distributions in each of the above cases. We now give a summary, together with a demonstration in one case. (The methods are quite similar in all cases.) In the following summary p, λ, θ, and γ (without bars) denote the actual values (for finite parameter values) of the quantities which tend to $\bar{p}$, $\bar{\lambda}$, $\bar{\theta}$, and $\bar{\gamma}$, respectively.

1. If $\bar{p} \neq n^{-1}[j \pm \{j(n-j)(n-1)^{-1}\}^{1/2}]$, then

$$\lim_{b+w\to\infty}\left[(b+w)\left\{\Pr[X=j]-\binom{n}{j}p^j(1-p)^{n-j}\right\}\right]=\binom{n}{j}\bar{p}^{\,j}(1-\bar{p})^{n-j}s \times \frac{n(n+1)\bar{p}^2-2j(n-1)\bar{p}+j(j-1)}{2\bar{p}(1-\bar{p})}. \tag{6.148}$$

If $\bar{p} = n^{-1}[j \pm \{j(n-j)(n-1)^{-1}\}^{1/2}]$, convergence is faster, and

$$\lim_{b+w\to\infty}\left[(b+w)^2\left\{\Pr[X=j]-\binom{n}{j}p^j(1-p)^{n-j}\right\}\right]=\binom{n}{j}\bar{p}^j(1-\bar{p})^{n-j}s^2 \times \frac{-n(n-1)(2n-1)\bar{p}^3(2-\bar{p})+6(n-1)j(n-j)\bar{p}^2-j(j-1)(2j-1)(1-2\bar{p})}{12\bar{p}^2(1-\bar{p})^2} \tag{6.149}$$

2. If $\bar{\gamma} \neq (\bar{\lambda}-2j)\{j-(\bar{\lambda}-j)^2\}^{-1}$, $\quad j \neq 1$ or $\bar{\lambda} \neq 2$, then

$$\lim_{n\to\infty}\left[n\left\{\Pr[X=j]-\binom{\lambda\gamma+j-1}{j}\left(\frac{1}{1+\gamma}\right)^j\left(\frac{\gamma}{1+\gamma}\right)^{\lambda\gamma}\right\}\right] = \binom{\bar{\lambda}\bar{\gamma}+j-1}{j}\left(\frac{\bar{\gamma}}{1+\bar{\gamma}}\right)^{\bar{\lambda}\bar{\gamma}}\left(\frac{1}{1+\bar{\gamma}}\right)^j-\frac{\{j-(j-\bar{\lambda})^2\}\bar{\gamma}+2j-\bar{\lambda}}{2(\bar{\gamma}+1)}. \tag{6.150}$$

If both $j=1$ *and* $\bar{\lambda}=2$, then

$$\lim_{n\to\infty}\left[n^2\left\{\Pr[X=1]-2\left(\frac{\gamma}{1+\gamma}\right)^{2\gamma+1}\right\}\right]=-2\left(\frac{\bar{\gamma}}{1+\bar{\gamma}}\right)^{2\bar{\gamma}+1}\frac{(2\bar{\gamma}+1)^2}{6\bar{\gamma}(1+\bar{\gamma})}. \tag{6.151}$$

The convergence is also of order n^{-2} if $\bar{\gamma}=(\bar{\lambda}-2j)/[j-(\bar{\lambda}-j)^2]$, $\bar{\lambda}\neq j\pm\sqrt{j}$, and either $j\neq 1$ or $\bar{\lambda}\neq 2$, but the limit [given by Śródka

(1964)] of $n^2 \times$ (difference between $\Pr[X = j]$ and the limit value) is much more complicated.

3. If $\bar{\theta} \neq \bar{\lambda}^{-1}$ and $\bar{\lambda} \neq j \pm \sqrt{j}$, then

$$\lim_{n\to\infty}\left[n\left\{\Pr[X = j] - \frac{\lambda^j e^{-\lambda}}{j!}\right\}\right] = \frac{\bar{\lambda}^j e^{-\bar{\lambda}}}{j!}\,\tfrac{1}{2}\{(j-\bar{\lambda})^2 - j\}\{(\bar{\theta}\bar{\lambda})^{-1} - 1\}. \tag{6.152}$$

If $\bar{\theta} = \bar{\lambda}^{-1}$ or $\bar{\lambda} = j \pm \sqrt{j}$, then

$$\lim_{n\to\infty}\left[\left(\frac{b+w}{ns}\right)^2\left\{\Pr[X = j] - \frac{\lambda^j e^{-\lambda}}{j!}\right\}\right]$$
$$= \frac{\bar{\lambda}^j e^{-\bar{\lambda}}}{j!}\cdot\frac{6\bar{\lambda}^2(j-\bar{\lambda})(j-\bar{\lambda}+1) - (1+\bar{\lambda}^2\bar{\theta}^2)\{j(j-2)(2j-1) + 2\bar{\lambda}^2(2\bar{\lambda}-3j)\}}{12\bar{\lambda}^2}. \tag{6.153}$$

3′. If $\bar{\lambda} \neq j \pm \sqrt{j}$, then

$$\lim_{n\to\infty}\left[\frac{b+w}{ns}\left\{\Pr[X = j] - \frac{\lambda^j e^{-\lambda}}{j!}\right\}\right] = \frac{\bar{\lambda}^j e^{-\bar{\lambda}}}{j!}\cdot\frac{(j-\bar{\lambda})^2 - j}{2\bar{\lambda}}. \tag{6.154}$$

If $\bar{\lambda} = j \pm \sqrt{j}$, then

$$\lim_{n\to\infty}\left[\left(\frac{b+w}{ns}\right)^2\left\{\Pr[X = j] - \frac{\lambda^j e^{-\lambda}}{j!}\right\}\right] = -\frac{\bar{\lambda}^j e^{-\bar{\lambda}}}{j!}\cdot\frac{3\sqrt{j} \pm 1}{12(\sqrt{j} \pm 1)}. \tag{6.155}$$

3″. If $\bar{\lambda} \neq j \pm \sqrt{j}$, then

$$\lim_{n\to\infty}\left[\frac{b+w}{ns}\left\{\Pr[X = j] - \frac{\lambda^j e^{-\lambda}}{j!}\right\}\right] = \frac{\bar{\lambda}^j e^{-\bar{\lambda}}}{j!}\cdot\frac{j - (j-\bar{\lambda})^2}{2}. \tag{6.156}$$

If $\bar{\lambda} = j \pm \sqrt{j}$, then

$$\lim_{n\to\infty}\left[\left(\frac{b+w}{ns}\right)^2\left\{\Pr[X = j] - \frac{\lambda^j e^{-\lambda}}{j!}\right\}\right] = -\frac{\bar{\lambda}^j e^{-\bar{\lambda}}}{j!}\cdot\frac{j(3\sqrt{j} \pm 1)(\sqrt{j} \pm 1)}{12}. \tag{6.157}$$

Special cases arise when the limit term is zero, and so convergence is of higher order. These results of Śródka are of special interest in that the actual magnitude of the error term is estimated.

We now give an account of the analysis leading to result 1. Equation (6.144) can be rewritten

$$\Pr[X = j] = \binom{n}{j} \frac{\Gamma\left(\frac{b+w}{s}\right)\Gamma\left(\frac{b}{s}+j\right)\Gamma\left(\frac{w}{s}+n-j\right)}{\Gamma\left(\frac{b}{s}\right)\Gamma\left(\frac{w}{s}\right)\Gamma\left(\frac{b+w}{s}+n\right)} \qquad (6.144')$$

(see Section 4.1). We take $s > 0$, though the result also holds for $s < 0$. For each of the gamma function in (6.144′) we use Stirling's formula [cf. (1.50)] in the form

$$\Gamma(b) = \sqrt{2\pi}\, b^{b-1/2} \exp\{-b + \tfrac{1}{12}b^{-1} + O(b^{-3})\}.$$

After some manipulations we obtain

$$\begin{aligned}\Pr[X = j] = \binom{n}{j} p^j (1-p)^{n-j} \exp\Bigg[&\left(\frac{b}{s} + j - \frac{1}{2}\right)\ln\left(1 + \frac{js}{b}\right) \\ &+ \left(\frac{w}{s} + n - j - \frac{1}{2}\right)\ln\left(1 + \frac{(n-j)s}{w}\right) \\ &- \left(\frac{b+w}{s} + n - \frac{1}{2}\right)\ln\left(1 + \frac{ns}{b+w}\right) \\ &+ \frac{1}{12}\Bigg\{\left(\frac{b}{s} + j\right)^{-1} + \left(\frac{w}{s} + n - j\right)^{-1} \\ &+ s\left(\frac{1}{b+w} - \frac{1}{b} - \frac{1}{w} - \frac{1}{(b+w)+ns}\right)\Bigg\} + O((b+w)^{-3})\Bigg].\end{aligned} \qquad (6.158)$$

Using the relations

$$\ln(1+x) = x - \tfrac{1}{2}x^2 + O(x^3) \qquad (\text{for } -1 < x < 1)$$

and

$$e^x = 1 + x + \tfrac{1}{2}x^2 + O(x^3),$$

we ultimately obtain

$$\Pr[X = j] = \binom{n}{j} p^j (1-p)^{n-j}\left[1 + \frac{A_1}{b+w} + \frac{A_2}{(b+w)^2} + O((b+w)^{-3})\right]. \qquad (6.159)$$

A_1 and A_2 are functions of p, as well as n and j, but do not otherwise depend on $b + w$. Hence

$$\lim_{n\to\infty}\left[(b+w)\left\{\Pr[X=j]-\binom{n}{j}p^j(1-p)^{n-j}\right\}\right]=\lim_{n\to\infty}\binom{n}{j}p^j(1-p)^{n-j}A_1$$
$$=\binom{n}{j}\bar{p}^j(1-\bar{p})^{n-j}\bar{A}_1 \tag{6.160}$$

where $\bar{A}_1$ is the value of A_1 for $p = \bar{p}$.

If $\bar{A}_1 = 0$ (which corresponds to the special case in the result quoted) we have

$$\lim_{n\to\infty}\left[(b+w)^2\left\{\Pr[X=j]-\binom{n}{j}p^j(1-p)^{n-j}\right\}\right]=\binom{n}{j}\bar{p}^j(1-\bar{p})^{n-j}\bar{A}_2, \tag{6.161}$$

where $\bar{A}_2$ is the value of A_2 for $p = \bar{p}$.

(The actual values of $\bar{A}_1$ and $\bar{A}_2$ are given in the statement of the results.)

6.3.3. Limit Distributions of Proportion of Black Balls for Large Sample Size

In the Pólya–Eggenberger urn model (as described in Section 4.1) it can be established (Pólya, 1931) that (if $s > 0$) the limit $P = \lim_{n\to\infty} n^{-1}X$ exists with probability 1 and is of course also equal to the limit of the proportion of black balls in the urn. The quantity P is in general a random variable (i.e., although it exists with probability 1, it does not have the same value in *all* sequences).

The distribution of P is clearly also the limit distribution of $n^{-1}X$ as $n \to \infty$, with b, w, and s remaining constant.

We now establish this limit distribution. From (4.3), the hth factorial moment of X is

$$E[X^{(h)}] = n^{(h)}\left(\frac{b}{s}\right)^{[h]}\left\{\left(\frac{b+w}{s}\right)^{[h]}\right\}^{-1}. \tag{6.162}$$

Hence

$$E\left[\frac{X^{(h)}}{n^{(h)}}\right] = \left(\frac{b}{s}\right)^{[h]}\left\{\left(\frac{b+w}{s}\right)^{[h]}\right\}^{-1} = \frac{\Gamma\left(\frac{b}{s}+h\right)}{\Gamma\left(\frac{b}{s}\right)}\,\frac{\Gamma\left(\frac{b+w}{s}\right)}{\Gamma\left(\frac{b+w}{s}+h\right)}.$$

Also

$$\lim_{n\to\infty} E\left[\frac{X^{(h)}}{n^{(h)}}\right] = \lim_{n\to\infty}\left[E\left\{\frac{X}{n}\cdot\frac{X-1}{n-1}\cdots\frac{X-h+1}{n-h+1}\right\}\right]$$
$$= \lim_{n\to\infty}\left[E\left\{\left(\frac{X}{n}\right)^h + R(X,n)\right\}\right],$$

with $R(X, n) =$ (polynomial of degree $h - 1$ in X with fixed coefficients)$/n^{(h)}$. From (6.162), $E[X^h]$ is of order h, and so $E[R(X, n)] \to 0$ as $n \to \infty$. It follows that

$$\lim_{n\to\infty} E[(n^{-1}X)^h] = \frac{\Gamma\left(\dfrac{b}{s}+h\right)\Gamma\left(\dfrac{b+w}{s}\right)}{\Gamma\left(\dfrac{b}{s}\right)\Gamma\left(\dfrac{b+w}{s}+h\right)}, \tag{6.163}$$

which is the hth moment of a standard beta distribution with parameters bs^{-1}, ws^{-1} [see (2.23)].

Since this distribution has a finite range of variation, the limiting distribution of $n^{-1}X$ is standard beta with parameters bs^{-1}, ws^{-1}. (See also, for example, Defays, 1974.)

An illuminating way of looking at this result has been provided by Dwass (1970) whose discussion we now paraphrase. We introduce the indicator variables

$$Y_i = \begin{cases} 1 & \text{if the } i\text{th ball drawn is black,} \\ 0 & \text{otherwise.} \end{cases}$$

Then $X = Y_1 + Y_2 + \cdots + Y_n$.

EXERCISE 6.8. Show that $Y_1, Y_2, \ldots, Y_n$ are exchangeable variates (see Section 2.9) though they are not independent. (*Hint*: $\Pr[\bigcap_{i=1}^n (Y_i = y_i)]$ depends only on $\sum_{i=1}^n y_i$.)

Assuming the result stated in Exercise 6.8, it follows, for example, that $E[Y_i] = E[Y_1] = b(b + w)^{-1}$. This result has much broader implications, as well. (Compare with the results in Section 2.9.)

Since $\Pr[\bigcap_{i=1}^n (Y_i = y_i)]$ depends only on $\sum_{i=1}^n y_i$, it is reasonable to see if the joint distribution of $Y_1, Y_2, \ldots, Y_n$ can be derived by supposing these variables, conditional on the value of a random variable P' being equal to p', to be mutually independent with $\Pr[Y_i = 1] = p' = 1 - \Pr[Y_i = 0]$ for all i.

Then,

$$\Pr\left[\bigcap_{i=1}^n (Y_i = y_i)\right] = \int_0^1 p'^s(1-p')^{n-s} f_{P'}(p')\, dp', \tag{6.164}$$

where $f_{P'}(p')$ is the density function of P' and $s = \sum_{i=1}^{n} y_i$. (Note that y_i can only take values 0 or 1.) We find that we obtain the correct value of $\Pr[\bigcap_{i=1}^{n} (Y_i = y_i)]$ from (6.164) by taking

$$f_{P'}(p') = \{B(bs^{-1}, ws^{-1})\}^{-1} p'^{bs^{-1}-1}(1-p')^{ws^{-1}-1} \qquad (0 < p' < 1). \tag{6.165}$$

Since for a given value, p', of P' the mean $\bar{Y}_n = n^{-1}\sum_{j=1}^{n} Y_j$ tends to p' with probability 1, we can see that the limiting distribution of $\lim_{n\to\infty} n^{-1}X$ is the beta distribution (6.165).

Defays (1974) has extended this result to a special *Friedman* urn model (see Section 4.3, case 1). He has shown that, if one adds s_b (s_w) extra black (white) balls when a black (white) ball is drawn, then P tends to 1 or 0 according as $s_b > s_w$ or $s_b < s_w$.

A multivariate generalization is obtained by considering the case when there are c_j balls of color $\mathscr{C}_j (j = 1, \ldots, k)$ initially in the urn. Sampling rules are the same, namely, the chosen ball is returned to the urn together with s balls of the same color. Let $X_{j,n}$ denote the number of balls of color $\mathscr{C}_j$ in the urn after n balls in all have been drawn. The limit distribution of the variables $n^{-1}X_{1n}, n^{-1}X_{2n}, \ldots, n^{-1}X_{kn}$ is Dirichlet [see (2.64)] with parameters $c_1 s^{-1}, c_2 s^{-1}, \ldots, c_k s^{-1}$ (Athreya, 1969) (see Section 4.7.1).

In the special case $c_1 = c_2 = \cdots = c_k = s$ the limit distribution is uniform. [A very thorough discussion of this case appears in Blackwell and Kendall (1964).]

6.4. LIMIT LAWS FOR EXCHANGEABLE VARIABLES*

Interpretation of the concept of exchangeability in terms of urn models has been described in Section 2.9.

In order to obtain limit laws for functions on a finite segment of an infinite sequence of exchangeable random variables, one can use De Finetti's theorem (Theorem 2.4) and limit laws for independent random variables. Some of these results can be made very abstract by making assumptions on the conditioning σ-field $\mathscr{F}$. This is usually not known in applications and therefore, in many practical problems, a direct proof of an asymptotic law is simpler than applying general theorems. Sometimes, however, the assumptions of general theorems are easy to handle. We first note that, if T_n is a nonparametric statistic based on the random variables $X_1, X_2, \ldots, X_n$ and the distribution of T_n is $F_n(t)$ for independent and identically distributed X's, then T_n has the same distribution for segments of infinite sequences of

* This section was written by J. Galambos.

exchangeable random variables. This fact is immediate from Theorem 2.4 because T_n is nonparametric. The actual form of the distribution of X, and thus $\mathscr{F}$ itself, plays no role when one turns to conditional probabilities.

For finite sequences of exchangeable variables, one can argue similarly by an appeal to Theorem 2.2 or to Corollary 2.2. However, for the case of Poisson limits, very simple general theorems are known (Kendall, 1967; Ridler-Rowe, 1967), which we present below, following the general approach of Galambos (1975). In the light of Corollaries 2.1 and 2.2, and Theorem 2.3, we consider a two-parameter family $f_k(a, b)$ of discrete distributions on the nonnegative integers $k = 0, 1, 2, \ldots$. We assume that $f_k(a, b)$ tends to a one-parameter family $g_k(u)$ of distributions as a and b go through certain sequences $\{a_n\}$ and $\{b_n\}$ of real numbers with $n \to \infty$. We further assume that if, as $n \to \infty$, $\lim f_k(a_n, b_n) = g_k(u)$, then, for any $v > 0$, $\lim f_k(va_n, b_n) = g_k(vu)$. Finally, the sequences $\{a_n\}$ and $\{b_n\}$ are assumed to determine each other in the following sense. Given $\{a_n\}$, if for each k, as $n \to +\infty$,

$$\lim f_k(a_n, b_n) = \lim f_k(a_n, b_n'),$$

then $\lim b_n/b_n' = 1$, and conversely, given $\{b_n\}$, a similar property is satisfied by the sequence $\{a_n\}$.

We say that a distribution $g_k(u)$ $(k = 0, 1, 2, \ldots)$ generates an *identifiable mixture* of distributions if a distribution function $T(u)$, with $T(0) = 0$, is uniquely determined by the set of equations

$$\pi_k = \int_0^{+\infty} g_k(u)\, dT(u) \qquad (k = 0, 1, 2, \ldots). \tag{6.166}$$

We can now formulate a result of Galambos (1975).

Theorem 6.17. Let the limiting distribution of $f_k(a, b)$ be $g_k(u)$, where the parameters satisfy the preceding assumptions. If $g_k(u)$ generates an identifiable mixture of distributions, then, for each k,

$$\lim_{n\to\infty} \int_0^{+\infty} f_k(ua_n, b_n)\, dT_n(u) = \pi_k \tag{6.167}$$

exists and $\{\pi_k\}$ is a proper distribution if and only if $T_n(u)$ converges weakly to a proper distribution function $T(u)$. When they exist, $\{\pi_k\}$ and $T(u)$ satisfy (6.166).

Before turning to the proof, we briefly discuss the above result and formulate some important corollaries to it. First of all, let us remark that the Poisson distribution with parameter u generates an identifiable mixture of distributions [see Teicher (1961) and the references therein], and the convergence of the binomial and hypergeometric distributions to the Poisson

satisfy our assumptions on the parameters. Therefore Corollary 2.2 (Section 2.9) and Theorem 6.17 imply the following limit theorem.

Corollary 6.1. Consider a triangular array $A_{j,n}$ $(1 \leq j \leq n, n = 1, 2, \ldots)$ of events where, for fixed n, the $A_{j,n}$ are a segment of $N(n) \geq n$ exchangeable events. Then the random variable $v_n(A)$ defined as the number of $A_{j,n}$'s $(1 \leq j \leq n)$ that occur, has a limiting distribution if and only if the distribution function $T_n(u) = F_n(uN/n)$ converges weakly to a distribution function $T(u)$, where $F_n(m) = F(m)$ is the distribution function occuring in (2.76). The limiting distribution π_k is necessarily of the form

$$\pi_k = \frac{1}{k!} \int_0^{+\infty} u^k e^{-u} \, dT(u). \tag{6.168}$$

Although the proof leading to Corollary 2.2 is constructive for $F_n(m)$, it is very often difficult to carry out the calculations. The following sufficient condition for obtaining a degenerate $T(u)$, and thus a Poisson limit law π_k, is therefore of interest. For finite $N(n)$ it was obtained by Ridler-Rowe (1967), generalizing Kendall's (1967) result for infinite sequences of exchangeable events. We add here that Corollary 6.1 for infinite $N(n)$ is essentially due to Benczur (1968).

Corollary 6.2. We use the notation of Corollary 6.1. Assume that

$$\lim_{n\to\infty} \frac{N(n)}{n} = +\infty, \qquad \lim_{n\to\infty} n\Pr(A_{1,n}) = a > 0,$$

and

$$\lim_{n\to\infty} n^2 \Pr(A_{1,n} \cap A_{2,n}) = a^2.$$

Then $v_n(A)$ has a limiting distribution and the limit is Poisson.

Corollary 6.2 is immediate from Corollary 6.1, since its assumptions imply that the distribution $T_n(u)$ tends to the degenerate law at a. From (6.168), we see that the limit is Poisson. Without going into details of distribution theory, we remark that the limit law π_k is geometric if $T(u)$ is exponential.

Of course Theorem 6.17 has many other possibilities for applications. For these, see for example, Galambos (1975) and also Johnson and Kotz (1969, pp. 31–48, 76–79, 104–114).

We conclude by noting that in Corollary 6.2 exchangeability can be relaxed and arbitrary sequences of events can be considered. This is a consequence of Galambos' (1973) result given in Theorem 2.3. A theoretical implication of this remark is interesting: The class of limiting distributions of $v_n(A)$, when arbitrary events are permitted, is *exactly the same* as for the class of exchangeable events.

We now turn to the proof of Theorem 6.17. We first establish sufficiency. Let

$$\pi_{k,n} = \int_0^\infty f_k(ua_n, b_n)\, dT_n(u).$$

Then the characteristic function corresponding to $\{\pi_{k,n}\}$ is

$$\phi_n(t) = \sum_{k=0}^{\infty} \pi_{k,n} e^{itk} = \int_0^\infty \phi(t; ua_n; b_n)\, dT_n(u), \tag{6.169}$$

where $\phi(t; a; b)$ is the characteristic function of the distribution $\{f_k(a, b)\}$.

If, as assumed in the Theorem, $\{g_k(u)\}$ is the limit distribution of $\{f_k(a, b)\}$, then

$$\lim_{n\to\infty} \phi(t; ua_n, b_n) = \phi(t; u),$$

where $\phi(t; u)$ is the characteristic function of the distribution $\{g_k(u)\}$.

Since $T_n(u)$ converges weakly to $T(u)$, using the Helly-Bray theorem (Loève, 1963, p. 180), we have, for any fixed A,

$$\lim_{n\to\infty} \int_0^A \phi(t; ua_n; b_n)\, dT_n(u) = \int_0^A \phi(t; u)\, dT(u).$$

Also, for A satisfying $2(1 - T(A)) < \varepsilon$ and n sufficiently large,

$$\left| \int_A^\infty \phi(t; ua_n; b_n)\, dT_n(u) \right| \leq 1 - T_n(A) \leq 2(1 - T(A)) < \varepsilon,$$

and so

$$\lim_{n\to\infty} \int_0^\infty \phi(t; ua_n; b_n)\, dT_n(u) = \int_0^\infty \phi(t; u)\, dT(u). \tag{6.170}$$

Since $\phi(t; u)$ is a characteristic function with argument t, the right-hand side of (6.170) is continuous at $t = 0$ and so, by the continuity theorem for characteristic functions, (6.167) holds and $\{\pi_k\}$ is a proper distribution because $T(u)$ is a proper distribution.

This establishes sufficiency of the condition that $T_n(u)$ converges weakly to $T(u)$.

We now assume that (6.167) holds and $\{\pi_k\}$ is a proper distribution. By the compactness of distribution functions (Loève, 1963, p. 179), there is a subsequence $\{T_{n(j)}(u)\}$ of $\{T_n(u)\}$ such that $T_{n(j)}(u)$ converges weakly to a distribution function $T^*(u)$.

Now apply the first (sufficiency) part of the proof to this subsequence. From (6.167) it follows that

$$\pi_k = \int_0^\infty g_k(u)\, dT^*(u).$$

Since $T_n(0) = 0$ for all n, $T^*(n) = 0$; and from the assumption that $\{\pi_k\}$ is a proper distribution, $T^*(\infty) = 1$. Hence $T^*(u)$ is a proper distribution.

Since $g_k(u)$ generates an *identifiable* mixture of distributions, $\{\pi_k\}$ determines $T^*(u)$ uniquely. Therefore $\{T_n(u)\}$ has the *same* limit $T^*(u)$, that is, $T_n(u)$ is weakly convergent and the limit is a proper distribution function.

This completes the proof.

EXERCISE 6.9. In a lot of N items, the number M of defectives varies from time to time in such a way that the percentage M/N tends to zero as N and M tend to $+\infty$. Assume that there is an integer n tending to $+\infty$ such that nM/N is asymptotically exponentially distributed. We select a sample of size n without replacement. Show that the number of defectives to be found in the sample is asymptotically a geometric variate. (*Hint*: Apply Theorem 6.17 with $\{f_k\}$ being a sequence of hypergeometric distributions as defined in Section 2.3.)

References

Athreya, K. B. (1969) On a characteristic property of Pólya's urn, *Stud. Sci. Math. Hung.*, **4**, 31–35.

Baum, L. E. and Billingsley, P. (1965) Asymptotic distributions for the coupon collector's problem, *Ann. Math. Stat.*, **36**, 1835–1839.

Békéssy, A. (1963) On the classical occupancy problems I, *Magy. Tud. Akad. Mat. Kutató Int. Közl*, **8**, 59–71.

Békéssy, A. (1964a) On the classical occupancy problems, II, Sequential occupancy, *Magy. Tud. Akad. Mat. Kutató Int. Közl.*, **9**, 133–141.

Békéssy, A. (1964b) Certain occupancy problems connected with the game of lotto, I, *Mat. Lapok*, **15**, 317–329. (in Hungarian).

Belyaev, P. F. (1965) On the probability of the non-appearance of a given number of s-tuples in a complex Markov chain, *Theory Prob. Appl.*, **10**, 496–499.

Benczur, A. (1968) On a sequence of equivalent events and the compound Poisson process, *Stud. Sci. Math. Hung.*, **3**, 451–458.

Bernstein, S. N. (1946) *Probability Theory*, 4th ed., Moscow, Leningrad: Gostekhizdat.

Blackwell, D. and Kendall, D. G. (1964) The Martin boundary for Pólya's urn scheme and application to stochastic population growth, *J. Appl. Prob.*, **1**, 284–296.

Bolotnikov, Yu. V. (1968a) Convergence to a Gaussian process of the number of empty cells in the classical problem of distributing particles among cells, *Mat. Zametki*, **4**, 97–103 (English translation: *Math. Notes*, **4**, 546–550).

Bolotnikov, Yu. V. (1968b) The convergence of the variables $\mu_r(n)$ to Gaussian and Poisson processes in the classical ball problem, *Theory Prob. Appl.*, **13**, 39–51.

Bolotnikov, Yu. V. (1968c) Limit processes in a model of distribution of particles into cells with unequal probabilities, *Theory Prob. Appl.*, **13**, 504–511.

Bouwkamp, C. J. and Bruijn, N. G. (1969) On some formal power series expansions, *Indag. Math.*, **31**(4), 301–308.

Brayton, R. K. (1963) On the asymptotic behavior of the number of trials necessary to complete a set with random selection, *J. Math. Anal. Appl.*, **7**, 31–61.

Bricas, M. A. (1949) *Le Système de Courbes de Pearson et le Schème d'Urne de Pólya*, Athens: Cristou.

Castoldi, L. (1955) Attorno un problema probabilistico di occupazione, *Atti. Accad. Ligure*, **11**, 119–126.

Chistyakov, V. P. (1964) On the calculation of the power of the test of empty boxes, *Theory Prob. Appl.*, **9**, 648–653.

Chistyakov, V. P. (1967) Discrete limit distributions in the problem of balls falling in cells with arbitrary probabilities, *Mat. Zametki*, 9–16. (English translation: *Math. Notes*, **1**, 6–11.)

Chistyakov, V. P. and Viktorova, I. I., (1965) Asymptotic normality in a problem of balls when probabilities of falling into different boxes are different, *Theory Prob. Appl.*, **10**, 149–154.

Curtiss, J. H. (1942) A note on the theory of moment generating functions, *Ann. Math. Stat.*, **13**, 430–433.

Daniels, H. E. (1954) Saddlepoint approximations in statistics, *Ann. Math. Stat.* **25**, 631–650.

David, F. N. (1950) Two combinatorial tests of whether a sample has come from a given population, *Biometrika*, **37**, 97–110.

David, F. N. and Barton, D. E. (1962) *Combinatorial Chance*, London: Griffin.

Defays, D. (1974) Étude du comportement asymptotique de schémas d'urnes, *Bull. Soc. Roy. Sci. Liège*, **43**(1–2), 26–34.

Dwass, M. (1969) More birthday surprises, *J. Comb. Theory*, **7**, 258–261.

Dwass, M. (1970) *Probability and Statistics*, New York: W. A. Benjamin.

Erdös, P. and Rényi, A. (1959) On the central limit theorem for samples from a finite population, *Publ. Math. Inst. Hung. Acad. Sci.*, **4**, 49–61.

Erdös, P. and Rényi, A. (1961) On a classical problem of probability theory, *Magy. Tud. Akad. Mat. Kutató Int. Közl*, **6**(1–2), 215–220.

Feller, W. (1957) *Probability Theory and its Applications*, 2nd ed., New York: John Wiley and Sons.

Galambos, J. (1973) A general Poisson limit theorem of probability theory, *Duke Math. J.*, **40**, 581–586.

Galambos, J. (1975) Limit laws for mixtures with applications to asymptotic theory of extremes, *Z. Wahrscheinlichkeitstheorie Verw. Geb.*, **32**, 197–207.

Geiringer, H. (1938) On the probability theory of arbitrary linked events, *Ann. Math. Stat.*, **9**, 260–271.

Hafner, R. (1973a) Asymptotische Normalität der Anzahl zufällinger Moleküle, *Sitzungsber. Österr. Akad. Wiss. Math.-Naturwiss. Kl., Abt. 2*, **181**(4–7), 215–251.

Hafner, R. (1973b) Neuer Beweis eines klassischen Besetzungsproblems, *Sitzungsber Österr. Akad. Wiss. Math.-Naturwiss. Kl., Abt. 2*, **181**(8–10), 269–289.

Harris, B. and Park, C. J. (1971a) A note on the asymptotic normality of the distribution of the number of empty cells in occupancy problems, *Ann. Inst. Stat. Math.*, **23**(3), 507–513.

Harris, B. and Park, C. J. (1971b) The distribution of linear combinations of the sample occupancy numbers, *Indag. Math.*, **33**(2), 121–134.

Hill, B. M. (1970) Zipf's law and prior distributions of the composition of a population, *J. Am. Stat. Assoc.*, **65**, 1220–1232.

Hill, B. M. and Woodroofe, M. (1975) Stronger forms of Zipf's law, *J. Am. Stat. Assoc.*, **70**, 212–219.

Holst, L. (1971) Limit theorems for some occupancy and sequential occupancy problems, *Ann. Math. Stat.*, **42**, 1671–1680.

Holst, L. (1972a) Asymptotic results connected with generalizations of occupancy problems, *Abst. Uppsala Dissert. Fac. Sci.*, No. 198.

Holst, L. (1972b) Asymptotic normality and efficiency for certain goodness-of-fit test, *Biometrika*, **59**, 137–145.

Holst, L. (1972c) Asymptotic normality in a generalized occupancy problem, *Z. Wahrscheinlichkeitstheorie Verw. Geb.*, **21**, 109–120.

Holst, L. (1976a) On multinomial sums, Technical Summary Report, Mathematics Research Center, University of Wisconsin, Madison.

Holst, L. (1976b) Some limit theorems for the indistinguishable ball problem with applications in nonparametrics, MRC Technical Summary Report No. 70-040, University of Wisconsin, Madison.

Holst, L. (1976c) Some matrix occupancy problems with dichotomous entries, MRC Technical Summary Report No. 76-078, University of Wisconsin, Madison.

Holst, L. (1976d) Personal communication.

Ijiri, Y. and Simon, H. A. (1975) Some distributions associated with Bose-Einstein statistics, *Proc. Nat. Acad. Sci. U.S.*, **72**(5), 1654–1657.

Ivchenko, G. I. (1971) On limit distributions for order statistics in the multinomial distribution, *Theory Prob., Appl.*, **16**, 94–107 (English translation, 102–115).

Ivchenko, G. I. (1973a) Waiting time and ordered series of frequencies in the polynomial scheme, *Tr. Mosk. Inst. Elektron. Mashinostr.*, **32**, 39–64 (in Russian).

Ivchenko, G. I. (1973b) Some limit theorems in the occupancy model, *Tr. Mosk. Inst. Elektron. Mashinostr.*, **32**, 111–119 (in Russian).

Ivchenko, G. I. and Levin, V. V. (1976) Asymptotic normality of a class of statistics in a polynomial scheme, *Theory Prob. Appl.*, **21**, 190–195 (in Russian).

Ivchenko, G. I. and Medvedev, Yu. I. (1965) Some multidimensional theorems on a classical problem of permutations, *Theory Prob. Appl.*, **10**, 144–149.

Ivchenko, G. I., Medvedev, Yu. I., and Sevast'yanov, B. A. (1967) Distribution of a random number of balls in cells, *Mat. Zametki*, **1**, 549–554 (English translation: *Math. Notes*, **1**, 363–366).

Johnson, N. L. and Kotz, S. (1969) *Distributions in Statistics—Discrete Distributions*, New York: John Wiley and Sons.

Johnson, N. L. and Kotz, S. (1970) *Distributions in Statistics, Continuous Univariate Distributions*, Vol. I, New York: John Wiley and Sons.

Kaplan, N. (1976) A generalization of a result of Erdös and Rényi and a related problem, Institute of Mathematical Statistics, University of Copenhagen.

Karlin, S. (1967) Central limit theorems for certain infinite urn schemes, *J. Math. Mech.*, **17**(4), 373–401.

Kendall, D. G. (1967) On finite and infinite sequences of exchangeable events, *Stud. Sci. Math. Hung.*, **2**, 319–327.

Kolchin, V. F. (1966) The speed of convergence to limit distributions in the classical ball problem, *Theory Prob. Appl.*, **11**, 128–140.

Kolchin, V. F. (1967) Uniform local limit theorems in the classical ball problem for a case with varying lattices. *Theory Prob. Appl.*, **12**, 57–67.

Kolchin, V. F. (1968) A certain class of limit theorems for conditional distribution, *Litov. Mat. Sb.*, **8**, 53–63 (English translation: *Sel. Transl. Math. Stat. Prob.* (*AMS*), **11**, 1973, 185–197.)

Kolchin, V. F. (1971) A problem of the allocation of particles in cells and cycles of random permutation, *Theory Prob. Appl.*, **16**, 74–90.

Kolchin, V. F. and Chistyakov, V. P. (1973) New limit theorems in the occupancy problem, *Tr. Mosk, Inst. Elektron. Mashinostr.*, **32**, 65–72 (in Russian).

Kolchin, V. F. and Chistyakov, V. P. (1974) Combinatorial problems in probability theory, *Itogi Nauk. Tekh. Teor. Veroiatn.*, **11**, 5–45 (in Russian).

Loève, M. (1963) *Probability Theory*, 3rd ed., New York: Van Nostrand.

Moran, P. A. P. (1973) A central limit theorem for exchangeable variates with geometric application, *J. Appl. Prob.*, **10**, 837–846.

Newman, D. J. and Shepp, L. (1960) The double dixie cup problem, *Am. Math. Mon.*, **67**, 58–61.

Okamoto, M. (1952) On a non-parametric test, *Osaka Math. J.*, **4**, 1, 77–85.

Park, C. J. (1972) A note on the classical occupancy problem, *Ann. Math. Stat.*, **43**, 1698–1701.

Park, C. J. (1976) A note on the asymptotic normality of the distribution of the number of empty cells under Bose-Einstein statistics, San Diego State University (manuscript).

Pólya, G. (1931) Sur quelques points de la théorie des probabilités, *Ann. Inst. Poincaré*, **1**, 117–161.

Popova, T. Yu. (1968) Limit theorems in a model of distribution of particles of two types, *Theory Prob. Appl.*, **13**, 511–516.

Rényi, A. (1962) Three new proofs and a generalization of a theorem of Irving Weiss, *Publ. Math. Inst. Hung. Acad. Sci.*, **7**, 203–214.

Rényi, A. (1970) *Foundations of Probability*, San Francisco: Holden Day.

Ridler-Rowe, C. J. (1967) On two problems of exchangeable events, *Stud. Sci. Math. Hung.*, **2**, 415–418.

Rosén, B. (1969) Asymptotic normality in a coupon collector's problem, *Z. Wahrscheinlichkeitstheorie Verw. Geb.* **13**, 256–279.

Rosén, B. (1970) On the coupon collector's waiting time, *Ann. Math. Stat.*, **41**, 1952–1969.

Samuel-Cahn, E. (1974) Asymptotic distribution for occupancy and waiting time problems with positive probability of falling through the cells, *Ann. Prob.*, **2**, 515–521.

Sevast'yanov, B.A. (1967) Convergence of the distribution of the number of empty boxes to Gaussian and Poisson processes in the classical ball problem, *Theory Prob. Appl.*, **12**, 126–134.

Sevast'yanov, B. A. (1972) Poisson limit law for a scheme of sums of dependent random variables, *Theory Prob. Appl.*, **17**, 695–699.

Sevast'yanov, B. A. and Chistyakov, V. P. (1964) Asymptotic normality in the classical problem of balls, *Theory Prob. Appl.*, **9**, 198–211 (correction: *ibid.*, 513–514).

Sherman, B. (1950) A random variable related to the spacing of sample values, *Ann. Math. Stat.*, **21**, 339–361.

Spitzer, F. (1964) *Principles of Random Walk*, New York: Von Nostrand.

Śródka, T. (1964) Fonctions limites et degrés de convergence dans la distribution de Pólya, *Bull. Soc. Sci. Lettres, Łódź*, **15**(3), 1–9.

Sveshnikov, A. G. and Tikhonov, A. N. (1971) *The Theory of Functions of a Complex Variable*, (translated from Russian), Moscow: Mir.

Teicher, H. (1961) Identifiability of mixtures, *Ann. Math. Stat.*, **32**, 244–248.

von Mises, R. (1939) Über Aufteilungs und Besetzungswahrscheinlichkeiten, *Rev. Fac. Sci. Univ. Istanbul, N.S.* **4**, 145–163. Reprinted in *Selected Papers of R. von Mises*, Vol. 2, Providence, R.I.: American Mathematical Society, pp. 313–331.

Weiss, I. (1958) Limiting distributions in some occupancy problems, *Ann. Math. Stat.*, **29**, 878–884.

Willis, J. C. (1922) *Age and Area*, Cambridge: Cambridge University Press.

Author Index

Subject Index

Urn Models Index

Applied Probability and Statistics (*Continued*)

HAHN and SHAPIRO · Statistical Models in Engineering
HALD · Statistical Tables and Formulas
HALD · Statistical Theory with Engineering Applications
HARTIGAN · Clustering Algorithms
HILDEBRAND, LAING and ROSENTHAL · Prediction Analysis of Cross Classifications
HOEL · Elementary Statistics, *Fourth Edition*
HOLLANDER and WOLFE · Nonparametric Statistical Methods
HUANG · Regression and Econometric Methods
JAGERS · Branching Processes with Biological Applications
JOHNSON and KOTZ · Distributions in Statistics
Discrete Distributions
Continuous Univariate Distributions-1
Continuous Univariate Distributions-2
Continuous Multivariate Distributions
JOHNSON and KOTZ · Urn Models and Their Application: An Approach to Modern Discrete Probability Theory
JOHNSON and LEONE · Statistics and Experimental Design: In Engineering and the Physical Sciences, Volumes I and II, *Second Edition*
KEENEY and RAIFFA · Decisions with Multiple Objectives
LANCASTER · The Chi Squared Distribution
LANCASTER · An Introduction to Medical Statistics
LEWIS · Stochastic Point Processes
McNEIL · Interactive Data Analysis
MANN, SCHAFER and SINGPURWALLA · Methods for Statistical Analysis of Reliability and Life Data
MEYER · Data Analysis for Scientists and Engineers
OTNES and ENOCHSON · Digital Time Series Analysis
PRENTER · Splines and Variational Methods
RAO and MITRA · Generalized Inverse of Matrices and Its Applications
SARD and WEINTRAUB · A Book of Splines
SEAL · Stochastic Theory of a Risk Business
SEARLE · Linear Models
THOMAS · An Introduction to Applied Probability and Random Processes
WHITTLE · Optimization under Constraints
WONNACOTT and WONNACOTT · Econometrics
WONNACOTT and WONNACOTT · Introductory Statistics, *Third Edition*
WONNACOTT and WONNACOTT · Introductory Statistics for Business and Economics, *Second Edition*
YOUDEN · Statistical Methods for Chemists
ZELLNER · An Introduction to Bayesian Inference in Econometrics

Tracts on Probability and Statistics

BHATTACHARYA and RAO · Normal Approximation and Asymptotic Expansions
BILLINGSLEY · Convergence of Probability Measures
CRAMER and LEADBETTER · Stationary and Related Stochastic Processes
JARDINE and SIBSON · Mathematical Taxonomy
RIORDAN · Combinatorial Identities